Fixed Point Theory and Variational Principles in Metric Spaces

The fixed point theory in metric spaces came into the existence through the PhD work of Polish mathematician Stefan Banach in 1920. The outcome of the Banach contraction principle became the initial source of the theory. It evolved with time and is now important not only for nonlinear analysis but also for many other branches of mathematics. It has also been applied to sciences and engineering. Many extensions and generalizations of the Banach contraction principle are explored by mathematicians. The proposed book covers some of the main extensions and generalizations of the principle. It focuses on the basic techniques and results of topics like set-valued analysis, variational principles, and equilibrium problems. This book will be useful for researchers working in nonlinear analysis and optimization and can be a reference book for graduate and undergraduate students.

There are some excellent books available on metric fixed point theory, but the above-mentioned topics are not covered in any single resource. The book includes a brief introduction to set-valued analysis with a focus on continuity and the fixed-point theory of set-valued maps and the last part of the book focuses on the application of fixed point theory.

Qamrul Hasan Ansari is Professor in the Department of Mathematics, Aligarh Muslim University, Aligarh. He was ranked among the top 2% scientists in the world by Stanford University in 2021 as well as in 2022, and has received the AMU Outstanding Research Award in the Sciences category in 2021. He has a teaching and research experience of thirty-two years in fields like variational inequalities, complementarity problems, optimization, convex analysis, vector equilibrium problems, fixed point theory in topological vector spaces, set-valued mappings, abstract economies, game theory, topological vector spaces, nonlinear functional analysis, advanced functional analysis, variational analysis, and optimization.

D. R. Sahu is Professor in the Department of Mathematics, Banaras Hindu University, Varanasi. He was the recipient of the Young Scientist Award (DST) in 2001. He has a teaching and research experience of twenty-five years in the fields of cryptography, equilibrium problems, fixed point theory and applications, image and signal processing, iterative learning control, Newton methods, nonlinear numerical functional analysis, operator theory, optimization in linear spaces, optimization in manifolds, and variational inequality theory.

Fixed Point Theory and Variational Principles in Metric Spaces

Qamrul Hasan Ansari
D. R. Sahu

Shaftesbury Road, Cambridge CB2 8EA, United Kingdom

One Liberty Plaza, 20th Floor, New York, NY 10006, USA

477 Williamstown Road, Port Melbourne, VIC 3207, Australia

314–321, 3rd Floor, Plot 3, Splendor Forum, Jasola District Centre, New Delhi – 110025, India

103 Penang Road, #05–06/07, Visioncrest Commercial, Singapore 238467

Cambridge University Press is part of Cambridge University Press & Assessment,
a department of the University of Cambridge.

We share the University's mission to contribute to society through the pursuit of
education, learning and research at the highest international levels of excellence.

www.cambridge.org
Information on this title: www.cambridge.org/9781009351454

First published 2023

Printed in India by Avantika Printers Pvt. Ltd.

A catalogue record for this publication is available from the British Library

ISBN 978-1-009-35145-4 Hardback

To our families

Contents

Preface

The fixed point theory is an important tool not only of nonlinear analysis, but also for many other branches of modern mathematics. It has numerous applications within mathematics and has been applied in diverse fields such as medical sciences, chemistry, economics, management, engineering, game theory, and physics.

Historically the beginning of metric fixed point theory goes back two centuries, but its name was coined only in 1922 after the pioneer work of Polish mathematician Stefan Banach in his Ph.D. dissertation. Many remarkable results of fixed point theory have been obtained during nineteen sixtees and nineteen seventies such as Caristi's fixed point theorem, Nadlar's fixed point theorem, etc. A large number of research papers have already appeared in the literature on extensions and generalizations of the Banach contraction principle.

On the other hand, Ivar Ekeland established a result on the existence of an approximate minimizer of a bounded below and lower semicontinuous function in 1972. Such a result is now known as Ekeland's variational principle. It is one of the most elegant and applicable results that appeared in the area of nonlinear analysis with diverse applications in fixed point theory, optimization, optimal control theory, game theory, nonlinear equations, dynamical systems, etc. Later, it was found that several well-known results, namely, the Caristi–Kirk fixed point theorem, Takahashi's minimization theorem, the Petal theorem, and the Daneš drop theorem, from nonlinear analysis are equivalent to Ekeland's variational principle in the sense that one can be derived by using the other results.

The set-valued maps, also called multivalued maps or point-to-set maps or multifunctions, are first considered in the famous book on topology by Kuratowski. Other eminent mathematicians, namely, Painlevé, Hausdorff, and Bouligand, have also visualized the vital role of set-valued maps as one often encounters such objects in concrete and real-life problems.

During the last decade of the last century, the theory on equilibrium problems emerged as one of the popular and hot topics in nonlinear analysis, optimization, optimal control, game theory, mathematical economics, etc. The equilibrium problem is a unified model of several fundamental mathematical problems, namely, optimization problems, saddle point problems, fixed point problems, minimax inequality problems, Nash equilibrium problem, complementarity problems, variational inequality problems, etc. In 1955, Nikaido and Isoda first considered the equilibrium problem as an auxiliary problem to establish the existence results for Nash's equilibrium points in noncooperative games. In the theory of equilibrium problems, the key contribution was made by Ky Fan in 1972, whose new existence results contained the original techniques that became a basis for most further existence theorems in the setting of topological vector spaces. That is why equilibrium problem is also known as the Ky Fan type inequality.

There are some excellent books available on metric fixed point theory. However, most of them are inaccessible for the systematic study of fixed point theory, set-valued analysis, variational principles, and equilibrium problems on one platform. The main purpose of this book is to present the basic techniques and results of these topics. The idea of writing this book came into existence while the first author was teaching a course of Nonlinear Functional Analysis at the master's level from where Chapters 2, 3, and 4 were conceived.

The present book contains six chapters.

The first chapter is devoted to the basic definitions, examples, and results concerning metric spaces that are essential for rest of the book.

In Chapter 2, the Banach contraction principle and several of its generalizations, namely, Boyd–Wong fixed point theorem for ψ-contraction mappings, fixed point theorem for weakly contraction mappings, etc., are discussed. The completeness of the metric space under the condition of the Banach contraction principle is included. Caristi's fixed point theorem and some of its consequences and generalizations are also presented.

In Chapter 3, we first give an elementary treatment of set-valued maps and then provide a complete discussion on continuity of set-valued maps. The Hausdorff metric on the family of nonempty closed bounded subsets of a metric space is presented along with the continuity of set-valued maps in terms of the Hausdorff metric. The fixed point theory for set-valued maps is also discussed. In particular, Nadler's theorem, fixed point theorem for directional contraction set-valued maps, Caristi–Kirk fixed point theorem, fixed point theorem for dissipative set-valued maps, Mizoguchi–Takahashi fixed point theorem for Ψ-contraction set-valued maps, and fixed point theorem for weakly contraction set-valued maps are presented. The stationary points for set-valued maps and DMH theorem and its consequences are also given.

In Chapter 4, we present several forms of Ekeland's variational principle and their generalizations with applications to fixed point theory and optimization. The Borwein–Preiss variational principle and Takahashi's minimization principle are also discussed with their applications to fixed point theory and weak sharp minima.

Chapter 5 provides an elementary treatment of the theory of equilibrium problems and the equilibrium version of Ekeland's variational principle, also known as Ekeland's variational principle for bifunctions or extended Ekeland's variational principle. Several equivalent results of extended Ekeland's variational principle, namely, extended Takahashi's minimization theorem, Caristi–Kirk fixed point theorem, and Oettli–Théra theorem, are also presented. The concept of weak sharp solutions for equilibrium problems is discussed.

The last chapter discusses several applications of fixed point results to the system of linear equations, differential equations and delay differential equations, second order two-point boundary value problems, and various kinds of integral equations.

The book is prepared for the graduate and undergraduate students and can also be useful for the researchers working in the area of nonlinear analysis and optimization.

Aligarh, India
Varanasi, India
January 2023

Qamrul Hasan Ansari
D. R. Sahu

Acknowledgments

We wish to express our gratitude to many of our friends and collaborators, who have generously helped and assisted us during the preparation of this book. In particular, we wish to express our deepest appreciation to Prof. R. Ahmad, India, Prof. S. Al-Homidan, Saudi Arabia, Prof. M. Ashraf, India, Prof. V. Berinde, Romania, Prof. Y.J. Cho, South Korea, Prof. S. Husain, India, Prof. M.A. Khamsi, UAE, Prof. A. Petrusel, Romania, Prof. T. Som, India, Prof. B.S. Thakur, India, Prof. H.K. Xu, China, and Prof. J.C. Yao, Taiwan, for their encouragements and personal supports.

The first author would also like to express his sincere thanks to Dr. S.S. Irfan and Dr. Z. Khan, and to his Ph.D. students Dr. Monirul Islam, Dr. Feeroz Babu, Dr. Pradeep Kumar Sharma, Dr. Mohd Zeeshan, Mr. Nasir Husain, Mr. Moinuddin, Mr. Muzaffar Sarkar Raju, Mr. Waqar Ahmad, and Mr. Hardeep Singh Shaluja for their careful reading of the manuscript.

The second author would also like to express his sincere thanks to Dr. Avinash Dixit, Dr. Pankaj Gautam, and to his Ph.D. students Mr. Amit Kumar Singh, Mr. Shikher Sharma, Mr. Nitish Kumar Singh, and Mr. Abhishek Verma for their careful reading of the manuscript.

We are grateful to the researchers whose work is cited in this book.

We are very grateful to Cambridge University Press and its staff for their indefatigable cooperation, patience, and understanding. Last but not least, we wish to thank to our family members for their infinite patience, encouragement and forbearance. This work would not have seen the light of the day had it not been to their unflinching support during this project. We remain indebted to them.

Aligarh, India
Varanasi, India
January 2023

Qamrul Hasan Ansari
D. R. Sahu

Notations and Abbreviations

$\varnothing$	the empty set
$\|\|x\|\|$	the norm of the vector x in $\mathbb{R}^n$
$\langle x, y \rangle$	the scalar product of x and y
$B[a, b]$	the space of all bounded real-valued functions defined on $[a, b]$
$C[a, b]$	the space of all continuous real-valued functions defined on $[a, b]$
$P[a, b]$	the space all polynomials defined on $[a, b]$
ℓ^∞	the space of all bounded sequences of real or complex numbers
ℓ^p	the space of all sequences $\{x_n\}$ of real or complex numbers such that $\sum_{n=1}^{\infty} \|x_n\|^p < \infty$ for all $p \geq 1$
A°	the interior of the set A
int$A$	the interior of the set A
bd(A)	the boundary of the set A
$\overline{A}$	the closure of the set A
clA	the closure of the set A
cl_XA	the closure of the set A in X
A^c	the complement of the set A
$X \setminus A$	the complement of the set A in X
A'	the derived set of A
diam(A)	the diameter of the set A
c	the space of all convergent sequences of real or complex numbers
s	the space of all sequences of real or complex numbers
$S_r(x)$	the open sphere (or open ball) with center at x and radius r
$S_r[x]$	the closed sphere (or closed ball) with center at x and radius r

$\mathbb{C}$	the set of all complex numbers
$\mathbb{C}^n$	the space of ordered n-tuples complex numbers
$\mathbb{N}$	the set of all natural numbers
$\mathbb{Q}$	the set of all rational numbers
$\mathbb{Z}$	the set of all integers
$\mathbb{R}$	the set of all real numbers
$\overline{\mathbb{R}}$	the extended real line
$\mathbb{R}_+$	the set of all nonnegative real numbers
$\mathbb{R}^n$	the n-dimensional Euclidean space
2^X	the family of all subsets of X
2^X_{cl}	the family of all nonempty closed and bounded subsets of X
2^X_b	the family of all nonempty bounded subsets of X
2^X_q	the family of all nonempty compact subsets of X
sup	supremum
lim sup	limit supremum
inf	infimum
lim inf	limit infimum
dom(f)	the domain of a single-valued map f
Dom(F)	the domain of a set-valued map F
graph(f)	the graph of a single-valued map f
Graph(F)	the graph of a set-valued map F
Image(F)	the image of a set-valued map F
EVP	Ekeland's variational principle
CMP	constrained minimization problem
MP	minimization problem

Chapter 1

Basic Definitions and Concepts from Metric Spaces

In this chapter, we gather some basic definitions, concepts, and results from metric spaces which are required throughout the book. For detail study of metric spaces, we refer to [8, 46, 61, 95, 110, 150, 154].

1.1 Definitions and Examples

Definition 1.1 Let X be a nonempty set. A real-valued function $d : X \times X \to \mathbb{R}$ is said to be a *metric* on X if it satisfies the following conditions:

(M1) $d(x, y) \geq 0$ for all $x, y \in X$;
(M2) $d(x, y) = 0$ if and only if $x = y$ for all $x, y \in X$;
(M3) $d(x, y) = d(y, x)$ for all $x, y \in X$; (symmetry)
(M4) $d(x, y) \leq d(x, z) + d(z, y)$ for all $x, y, z \in X$. (triangle inequality)

The set X together with a metric d on X is called a *metric space* and it is denoted by (X, d). If there is no confusion likely to occur we, sometime, denote the metric space (X, d) by X.

Example 1.1 Let X be a nonempty set. For any $x, y \in X$, define

$$d(x, y) = \begin{cases} 0, & \text{if } x = y, \\ 1, & \text{if } x \neq y. \end{cases}$$

Then d is a metric, and it is called a *discrete metric*. The space (X, d) is called a *discrete metric space*.

The above example shows that on each nonempty set, at least one metric that is a discrete metric can be defined.

Example 1.2 Let $X = \mathbb{R}^n$, the set of ordered n-tuples of real numbers. For any $x = (x_1, x_2, \dots, x_n) \in X$ and $y = (y_1, y_2, \dots, y_n) \in X$, we define

(a) $d_1(x, y) = \sum_{i=1}^{n} |x_i - y_i|$, (called taxicab metric)

(b) $d_2(x,y) = \left(\sum_{i=1}^{n} (x_i - y_i)^2\right)^{\frac{1}{2}}$, (called usual metric)

(c) $d_p(x,y) = \left(\sum_{i=1}^{n} |x_i - y_i|^p\right)^{\frac{1}{p}}$, $p \geq 1$

(d) $d_\infty(x,y) = \max_{1 \leq i \leq n} |x_i - y_i|$. (called max metric)

Then, $d_1, d_2, d_p\ (p \geq 1), d_\infty$ are metrics on $\mathbb{R}^n$.

Example 1.3 Let ℓ^∞ be the space of all bounded sequences of real or complex numbers, that is,

$$\ell^\infty = \left\{\{x_n\} \subset \mathbb{R} \text{ or } \mathbb{C} : \sup_{1 \leq n < \infty} |x_n| < \infty\right\}.$$

Then,

$$d_\infty(x,y) = \sup_{1 \leq n < \infty} |x_n - y_n|, \quad \text{for all } x = \{x_n\},\ y = \{y_n\} \in \ell^\infty,$$

is a metric on ℓ^∞ and (ℓ^∞, d_∞) is a metric space.

Example 1.4 Let s be the space of all sequences of real or complex numbers, that is,

$$s = \{\{x_n\} \subset \mathbb{R} \text{ or } \mathbb{C}\}.$$

Then,

$$d(x,y) = \sum_{n=1}^{\infty} \frac{1}{2^n} \frac{|x_n - y_n|}{1 + |x_n - y_n|}, \quad \text{for all } x = \{x_n\},\ y = \{y_n\} \in s,$$

is a metric on s.

Example 1.5 Let $\ell^p, 1 \leq p < \infty$, denote the space of all sequences $\{x_n\}$ of real or complex numbers such that $\sum_{n=1}^{\infty} |x_n|^p < \infty$, that is,

$$\ell^p = \left\{\{x_n\} \subset \mathbb{R} \text{ or } \mathbb{C} : \sum_{n=1}^{\infty} |x_n|^p < \infty\right\}, \quad \text{for } 1 \leq p < \infty.$$

Then,

$$d(x,y) = \left(\sum_{n=1}^{\infty} |x_n - y_n|^p\right)^{\frac{1}{p}}, \quad \text{for all } x = \{x_n\},\ y = \{y_n\} \in \ell^p,$$

is a metric on ℓ^p and (ℓ^p, d) is a metric space.

Example 1.6 Let $B[a,b]$ be the space of all bounded real-valued functions defined on $[a,b]$, that is,

$$B[a,b] = \{f : [a,b] \to \mathbb{R} : |f(t)| \leq k \text{ for all } t \in [a,b] \text{ and for some constant } k \in \mathbb{R}\}.$$

Then,

$$d(f,g) = \sup_{t\in[a,b]} |f(t) - g(t)|, \quad \text{for all } f,g \in B[a,b],$$

is a metric on $B[a,b]$.

Example 1.7 Let $C[a,b]$ be the space of all continuous real-valued functions defined on $[a,b]$. For any $f,g \in C[a,b]$, we define the real-valued functions d_∞ and d_1 on $C[a,b] \times C[a,b]$ as follows:

$$d_\infty(f,g) = \sup_{t\in[a,b]} |f(t) - g(t)|$$

and

$$d_1(f,g) = \int_a^b |f(t) - g(t)|\, dt,$$

where the integral is the Riemann integral which is possible because the functions f and g are continuous on $[a,b]$. Then, d_∞ and d_1 are metrics on $C[a,b]$.

Definition 1.2 Let X be a nonempty set. A real-valued function $d : X \times X \to \mathbb{R}$ is said to be a *pseudometric* on X if it satisfies the following conditions:

(PM1) $d(x,y) \geq 0$ for all $x,y \in X$;
(PM2) $d(x,y) = 0$ if $x = y$ for all $x,y \in X$;
(PM3) $d(x,y) = d(y,x)$ for all $x,y \in X$; (symmetry)
(PM4) $d(x,y) \leq d(x,z) + d(z,y)$ for all $x,y,z \in X$. (triangle inequality)

The set X together with a pseudometric d on X is called a *pseudometric space*.

Example 1.8 Let $X = \mathbb{R}^2$ and $d(x,y) = |x_1 - y_1|$ for all $x = (x_1,x_2)$, $y = (y_1,y_2) \in X$. Then, d is not a metric on X; however, it is a pseudometric on X. Indeed, for $x = (0,0)$, $y = (0,1) \in X$, we have $d(x,y) = 0$ but $x \neq y$. Therefore, it is not a metric on X. It can be easily checked that d satisfies the conditions (PM1) – (PM4).

Definition 1.3 Let X be a nonempty set. A real-valued function $d : X \times X \to \mathbb{R}$ is said to be a *quasimetric* on X if it satisfies the following conditions:

(QM1) $d(x,y) \geq 0$ for all $x,y \in X$;
(QM2) $d(x,y) = 0$ if and only if $x = y$ for all $x,y \in X$;
(QM3) $d(x,y) \leq d(x,z) + d(z,y)$ for all $x,y,z \in X$. (triangle inequality)

The set X together with a quasimetric d on X is called a *quasimetric space*.

Example 1.9 The real-valued functions $d_1, d_2 : \mathbb{R} \times \mathbb{R} \to \mathbb{R}$ defined by

$$d_1(x,y) = \begin{cases} y - x, & \text{if } y \geq x, \\ \alpha(x - y), & \text{if } y < x, \end{cases}$$

for $\alpha > 0$, and

$$d_2(x,y) = \begin{cases} e^y - e^x, & \text{if } y \geq x, \\ e^{-y} - x^{-x}, & \text{if } y < x, \end{cases}$$

are quasimetrics on $\mathbb{R}$.

Definition 1.4 Let (X, d) be a metric space and let A and B be nonempty subsets of X. The *distance between the sets A and B* is given by

$$d(A, B) = \inf\{d(x, y) : x \in A, \ y \in B\}.$$

Since $d(x, y) = d(y, x)$, we have $d(A, B) = d(B, A)$.

If A is a singleton set $\{x\}$, then

$$d(\{x\}, B) = \inf\{d(x, y) : y \in B\}.$$

It is called the *distance of a point $x \in X$ from the set B*, and we write $d(x, B)$ in place of $d(\{x\}, B)$.

Remark 1.1 **(a)** The equation $d(x, B) = 0$ does not imply that x belongs to B.

(b) If $d(A, B) = 0$, then it is not necessary that A and B have common points.

Example 1.10 Let $A = \{x \in \mathbb{R} : x > 0\}$ and $B = \{x \in \mathbb{R} : x < 0\}$ be subsets of $\mathbb{R}$ with the usual metric. Then $d(A, B) = 0$, but A and B have no common point. If $x = 0$, then $d(x, B) = 0$; but $x \notin B$.

Definition 1.5 Let (X, d) be a metric space and A be a nonempty subset of X. The *diameter of A*, denoted by $\operatorname{diam}(A)$, is given by

$$\operatorname{diam}(A) = \sup\{d(x, y) : x, y \in A\}.$$

The set A is called *bounded* if there exists a constant k such that $\operatorname{diam}(A) \leq k < \infty$. In other words, A is bounded if its diameter is finite, otherwise it is called *unbounded*.

In particular, the metric space (X, d) is bounded if the set X is bounded.

1.2 Open Sets and Closed Sets

Definition 1.6 Let (X, d) be a metric space. Given a point $x_0 \in X$ and a real number $r > 0$, the sets

$$S_r(x_0) = \{y \in X : d(x_0, y) < r\}$$

and

$$S_r[x_0] = \{y \in X : d(x_0, y) \leq r\}$$

are called *open sphere* (or *open ball*) and *closed sphere* (or *closed ball*), respectively, with center at x_0 and radius r.

Remark 1.2 **(a)** The open and closed spheres are always nonempty, since $x_0 \in S_r(x_0) \subseteq S_r[x_0]$.

(b) Every open (respectively, closed) sphere in $\mathbb{R}$ with the usual metric is an open (respectively, closed) interval. But the converse is not true; for example, $(-\infty, \infty)$ is an open interval in $\mathbb{R}$ but not an open sphere.

Definition 1.7 Let A be a nonempty subset of a metric space X.

(a) A point $x \in A$ is said to be an *interior point* of A if x is the center of some open sphere contained in A. In other words, $x \in A$ is an interior point of A if there exists $r > 0$ such that $S_r(x) \subseteq A$.

(b) The set of all interior points of A is called *interior* of A and it is denoted by A°, that is,

$$A^\circ = \{x \in A : S_r(x) \subseteq A \text{ for some } r > 0\}.$$

(c) The set A is said to be *open* if each of its points is the center of some open sphere contained entirely in A; that is, A is an open set if for each $x \in A$, there exists $r > 0$ such that $S_r(x) \subseteq A$.

(d) Let $x \in X$. The set A is said to be a *neighborhood* of x if there exists an open sphere centered at x and contained in A, that is, if $S_r(x) \subseteq A$ for some $r > 0$. In case A is an open set, it is called an *open neighborhood* of x.

Remark 1.3 In a metric space, we have the following:

(a) An open sphere $S_r(x)$ with center at x and radius r is a neighborhood of x.

(b) The interior of A is the neighborhood of each of its points.

(c) Every open set is the neighborhood of each of its points.

(d) The set A is open if and only if each of its points is an interior point, that is, $A = A^\circ$.

(e) Arbitrary union of open sets is open.

(f) Finite intersection of open sets is open.

(g) Arbitrary intersection of open sets need not be open.

Theorem 1.1 *Let A and B be two subsets of a metric space X. Then,*

(a) $A \subseteq B$ *implies* $A^\circ \subseteq B^\circ$;

(b) $(A \cap B)^\circ = A^\circ \cap B^\circ$;

(c) $(A \cup B)^\circ \supseteq A^\circ \cup B^\circ$.

Definition 1.8 Let A be a subset of a metric space X. A point $x \in X$ is said to be a *limit point* (*accumulation point* or *cluster point*) of A if each open sphere centered at x contains at least one point of A other than x.

In other words, $x \in X$ is a limit point of A if

$$\left(S_r(x) - \{x\}\right) \cap A \neq \varnothing, \quad \text{for all } r > 0.$$

The set of all limit points of A is called *derived set* and it is denoted by A'.

Definition 1.9 A point $x \in X$ is said to be an *isolated point* of A if there exists an open sphere centered at x which contains no point of A other than x itself, that is, if $S_r(x) \cap A = \{x\}$ for some $r > 0$.

Remark 1.4 If a point $x \in X$ is not a limit point of A, then it is an isolated point. Hence every point of a metric space X is either a limit point or an isolated point of X.

Example 1.11 Consider the metric space $X = \left\{0, 1, \frac{1}{2}, \frac{1}{3}, \frac{1}{4}, \cdots\right\}$ with the usual metric given by the absolute value. Then, 0 is the only limit point of X while all other points are the isolated points of X.

Definition 1.10 Let A be a subset of a metric space X. The *closure* of A, denoted by $\overline{A}$ or cl$A$, is the union of A and the set of all limit points of A, that is, $\overline{A} = A \cup A'$.

In other words, $x \in \overline{A}$ if every open sphere $S_r(x)$ centered at x and radius $r > 0$ contains a point of A, that is, $x \in \overline{A}$ if and only if $S_r(x) \cap A \neq \varnothing$ for every $r > 0$.

Remark 1.5 Let A and B be subsets of a metric space X. Then,

(a) $\overline{\emptyset} = \emptyset$;
(b) $\overline{X} = X$;
(c) $\overline{(\overline{A})} = \overline{A}$;
(d) $A \subseteq B$ implies $\overline{A} \subseteq \overline{B}$;
(e) $\overline{A \cup B} = \overline{A} \cup \overline{B}$;
(f) $\overline{A} = \left(\overline{A}\right)'$;
(g) $\overline{A \cap B} \subseteq \overline{A} \cap \overline{B}$, but $\overline{A \cap B} \not\supseteq \overline{A} \cap \overline{B}$.

Theorem 1.2 *Let (X, d) be a metric space and A be a subset of X. Then, $x \in \overline{A}$ if and only if $d(x, A) = 0$.*

Definition 1.11 Let A be a subset of a metric space X. The set A is said to be *closed* if it contains all its limit points, that is, $A' \subseteq A$.

Remark 1.6 **(a)** Let A be a subset of a metric space X. Then clearly A is closed if and only if $\overline{A} = A$.
(b) Let A be a subset of a metric space X. Then A is closed if and only if the complement of A is an open set.
(c) In a metric space, every finite set, empty set, and whole space are closed sets.
(d) Arbitrary intersection of closed sets is closed.
(e) Finite union of closed sets is closed. However, arbitrary union of closed sets need not be closed.

Definition 1.12 Let A be a subset of a metric space X. A point $x \in X$ is called a *boundary point* of A if it is neither an interior point of A nor of $X \setminus A$, that is, $x \notin A^\circ$ and $x \notin (X \setminus A)^\circ$.

In other words, $x \in X$ is a *boundary point* of A if every open sphere centered at x intersects both A and $X \setminus A$.

The set of all boundary points of A is called the *boundary of* A and it is denoted by $\text{bd}(A)$.

Remark 1.7 It is clear that $\text{bd}(A) = \overline{A} \cap \overline{(X \setminus A)} = \overline{A} \cap \overline{A^c}$.

1.3 Complete Metric Spaces

Definition 1.13 Let (X, d) be a metric space. A sequence $\{x_n\}$ of points of X is said to be *convergent* if there is a point $x \in X$ such that for each $\varepsilon > 0$, there exists a positive integer N such that

$$d(x_n, x) < \varepsilon, \quad \text{for all } n > N.$$

The point $x \in X$ is called a *limit point* of the sequence $\{x_n\}$.

More preciously, a sequence $\{x_n\}$ in a metric space X converges to a point $x \in X$ if the sequence $\{d(x_n, x)\}$ of real numbers converges to 0.

Since $d(x_n, x) < \varepsilon$ is equivalent to $x_n \in S_\varepsilon(x)$, the definition of convergent sequence can be restated as follows:

A sequence $\{x_n\}$ in a metric space X *converges to a point* $x \in X$ if and only if for each $\varepsilon > 0$, there exists a positive integer N such that

$$x_n \in S_\varepsilon(x), \quad \text{for all } n > N.$$

For a convergent sequence $\{x_n\}$ to x, we use the following symbols:

$$x_n \to x \quad \text{or} \quad \lim_{n\to\infty} x_n = x$$

and we express it by saying that x_n approaches x or that x_n converges to x.

Definition 1.14 A sequence $\{x_n\}$ in a metric space X is said to be *bounded* if the range set of the sequence is bounded.

Remark 1.8 In a metric space, every convergent sequence is bounded.

Definition 1.15 Let (X, d) be a metric space. A sequence $\{x_n\}$ in X is said to be a *Cauchy sequence* if for each $\varepsilon > 0$, there exists a positive integer N such that

$$d(x_n, x_m) < \varepsilon, \quad \text{for all } n, m > N.$$

Theorem 1.3 *Every convergent sequence in a metric space is a Cauchy sequence.*

Exercise 1.1 Let (X, d) be a metric space and $\{x_n\}$ be a sequence in X such that $d(x_n, x_{n+1}) < \frac{1}{2^n}$ for all n. Prove that $\{x_n\}$ is a Cauchy sequence.

Proof Let $\varepsilon > 0$ and choose a positive integer N such that $\frac{1}{2^{N-1}} < \varepsilon$. Then for all $n > m > N$, we have

$$\begin{aligned} d(x_m, x_n) &\leq d(x_m, x_{m+1}) + d(x_{m+1}, x_{m+2}) + \cdots + d(x_{n-1}, x_n) \\ &< \frac{1}{2^m} + \frac{1}{2^{m+1}} + \cdots + \frac{1}{2^{n-1}} \\ &< \sum_{k=m}^{\infty} \frac{1}{2^k} = \frac{1}{2^{m-1}} < \frac{1}{2^{N-1}} < \varepsilon. \end{aligned}$$

■

Definition 1.16 A metric space (X, d) is said to be *complete* if every Cauchy sequence in X converges to a point in X.

Example 1.12 The space $\mathbb{R}^n$ with respect to all the metrics given in Example 1.2 is complete. The space $C[0, 1]$ with respect to the metric d_1 given in Example 1.7 is not complete.

Remark 1.9 A metric space (X, d) is complete if and only if every Cauchy sequence in X has a convergent subsequence.

Exercise 1.2 Let (X, d_X) and (Y, d_Y) be metric spaces. Define

$$d_{X\times Y}((x, y), (u, v)) = d_X(x, u) + d_Y(y, v), \quad \text{for all } (x, y), (u, v) \in X \times Y.$$

Prove that $d_{X\times Y}$ is a metric on $X \times Y$. Further, if (X, d_X) and (Y, d_Y) are complete, then prove that $(X \times Y, d_{X\times Y})$ is also complete.

Theorem 1.4 (Cantor's Intersection Theorem) *Let (X, d) be a complete metric space and $\{A_n\}$ be a decreasing sequence (that is, $A_{n+1} \subseteq A_n$) of nonempty closed subsets of X such that* $\text{diam}(A_n) \to 0$ *as $n \to \infty$. Then, the intersection $\bigcap_{n=1}^{\infty} A_n$ contains exactly one point.*

The converse of the above theorem is the following:

Theorem 1.5 *Let (X, d) be a metric space. If any decreasing sequence $\{A_n\}$ of nonempty closed sets in X with* $\text{diam}(A_n) \to 0$ *as $n \to \infty$ has exactly one point in its intersection, then (X, d) is complete.*

Definition 1.17 A nonempty subset A of a metric space X is said to be *dense* (or *everywhere dense*) in X if $\overline{A} = X$, that is, if every point of X is either a point or a limit point of A.

In other words, a set A is dense in X if for any given point $x \in X$, there exists a sequence of points of A that converges to x.

It can be easily seen that a subset A of X is dense if and only if A^c has empty interior.

Before giving the examples of dense sets, we provide the criteria for being dense.

Theorem 1.6 *Let A be a nonempty subset of a metric space X. The following statements are equivalent:*

(a) *For every $x \in X$, $d(x, A) = 0$.*

(b) $\overline{A} = X$.

(c) *A has nonempty intersection with every nonempty open subset of X.*

Example 1.13 **(a)** The set of all rational numbers $\mathbb{Q}$ is dense in the usual metric space $\mathbb{R}$ since $\overline{\mathbb{Q}} = \mathbb{R}$.

(b) Since $\overline{\mathbb{R} \setminus \mathbb{Q}} = \mathbb{R}$, the set of all irrational numbers $\mathbb{R} \setminus \mathbb{Q}$ is dense in the usual metric space $\mathbb{R}$.

(c) The set $A = \{a + ib \in \mathbb{C} : a, b \in \mathbb{Q}\}$ is dense in $\mathbb{C}$ since $\overline{A} = \mathbb{C}$.

(d) The set $\mathbb{Q}^n = \underbrace{\mathbb{Q} \times \mathbb{Q} \times \cdots \times \mathbb{Q}}_{n\text{-times}}$ is dense in $\mathbb{R}^n$ with the usual metric.

(e) The set

$$A = \{x = (a_1, a_2, \dots, a_n, 0, 0, \dots) : a_i \in \mathbb{Q} \text{ for all } 1 \leq i \leq n \text{ and } n \in \mathbb{N}\}$$

is dense in the space ℓ^p, $1 \leq p < \infty$, with the following metric:

$$d_p(x, y) = \left(\sum_{i=1}^{\infty} |x_i - y_i|^p \right)^{1/p},$$

where $x = \{x_1, x_2, \dots\}$ and $y = \{y_1, y_2, \dots\}$ in ℓ^p.

(f) The set $P[a, b]$ of all polynomials defined on $[a, b]$ with rational coefficients is dense in $C[a, b]$.

(g) Let (X, d) be a discrete metric space. Since every subset of X is closed, the only dense subset of X is itself.

Definition 1.18 A metric space X is said to be *separable* if there exists a countable dense set in X.

A metric space which is not separable is called *inseparable*.

Example 1.14 (a) The usual metric space $\mathbb{R}$ is separable since the set of all rational numbers $\mathbb{Q}$ is dense in $\mathbb{R}$.

(b) The usual metric space $\mathbb{C}$ is separable since the set $A = \{a + ib \in \mathbb{C} : a, b \in \mathbb{Q}\}$ is dense in $\mathbb{C}$.

(c) The Euclidean space $\mathbb{R}^n$ is separable since the set $\mathbb{Q}^n = \underbrace{\mathbb{Q} \times \mathbb{Q} \times \cdots \times \mathbb{Q}}_{n\text{-times}}$ is countable and dense in $\mathbb{R}^n$.

(d) The space $\ell^p, 1 \le p < \infty$, is separable as the set

$$A = \{x = (a_1, a_2, \dots, a_n, 0, 0, \dots) : a_i \in \mathbb{Q}, 1 \le i \le n \text{ and for all } n \in \mathbb{N}\}$$

is countable and dense in the space ℓ^p.

(e) The space $C[a, b]$ is separable since the set $P[a, b]$ of all polynomials defined on $[a, b]$ with rational coefficients is countable and dense in $C[a, b]$.

(f) A discrete metric space X is separable if and only if the set X is countable.

Example 1.15 The space ℓ^∞ of all bounded sequences of real or complex numbers with the metric

$$d_\infty(x, y) = \sup_{1 \le n < \infty} |x_n - y_n|,$$

where $x = \{x_n\}$ and $y = \{y_n\}$ in ℓ^∞, is not separable.

Definition 1.19 Two metrics d_1 and d_2 on the same underlying set X are said to be *equivalent* if for every sequence $\{x_n\}$ in X and $x \in X$,

$$\lim_{n \to \infty} d_1(x_n, x) = 0 \quad \text{if and only if} \quad \lim_{n \to \infty} d_2(x_n, x) = 0,$$

that is, a sequence converges to x with respect to the metric d_1 if and only if it converges to x with respect to the metric d_2.

The metric spaces (X, d_1) and (X, d_2) are said to be *equivalent* if the metrics d_1 and d_2 are equivalent.

Remark 1.10 If two metrics are equivalent, then the families of open sets are same in (X, d_1) and (X, d_2).

The following result provides a sufficient condition for two metrics on a set to be equivalent.

Theorem 1.7 *Two metrics d_1 and d_2 on a nonempty set X are equivalent if there exist constants $k_1, k_2 > 0$ such that*

$$k_1 d_2(x, y) \le d_1(x, y) \le k_2 d_2(x, y), \quad \textit{for all } x, y \in X. \tag{1.1}$$

1.4 Compact Spaces

Definition 1.20 Let X be a metric space and Λ be any index set.

(a) A collection $\mathscr{F} = \{G_\alpha\}_{\alpha \in \Lambda}$ of subsets of X is called a *cover* of X if $\bigcup_{\alpha \in \Lambda} G_\alpha = X$, that is, every element of X belongs to at least one member of $\mathscr{F}$. If each member of $\mathscr{F}$ is an open set in X, then it is called an *open cover* of X.

(b) A subcollection $\mathscr{C}$ of a cover $\mathscr{F}$ of X is called a *subcover* if $\mathscr{C}$ is itself a cover of X. $\mathscr{C}$ is called a *finite subcover* if it consists only a finite number of members.
In other words, if there exist $G_{\alpha_1}, G_{\alpha_2}, \dots, G_{\alpha_n} \in \mathscr{F}$ such that $\bigcup_{k=1}^{n} G_{\alpha_k} = X$, then the subcollection $\mathscr{C} = \left\{G_{\alpha_1}, G_{\alpha_2}, \dots, G_{\alpha_n}\right\}$ is called a finite subcover of X.
In this case, $\mathscr{F}$ is said to be *reducible to a finite cover* or contains a *finite subcover.*

Definition 1.21 Let X be a metric space and Y be a subset of X. A collection $\mathscr{F} = \{G_\alpha\}_{\alpha \in \Lambda}$ of subsets of X is said to be a *cover* of Y if $Y \subseteq \bigcup_{\alpha \in \Lambda} G_\alpha$.

Definition 1.22 A metric space X is said to be *compact* if every open cover of X has a finite subcover.

A nonempty subset Y of a metric space (X, d) is compact if it is a compact metric space with the metric induced on it by d.

Theorem 1.8 *Every closed subset of a compact metric space is compact.*

Definition 1.23 A collection $\mathcal{C} = \left\{C_1, C_2, \dots\right\}$ of subsets of a metric space X is said to have the *finite intersection property* if every finite subcollection of $\mathcal{C}$ has nonempty intersection, that is, for every finite collection $\left\{C_1, C_2, \dots, C_n\right\}$ of $\mathcal{C}$, we have $\bigcap_{i=1}^{n} C_i \neq \emptyset$.

Theorem 1.9 *A metric space X is compact if and only if every collection of closed sets in X having finite intersection property has nonempty intersection.*

Definition 1.24 A metric space X is said to have the *Bolzano–Weierstrass property* if every infinite subset of X has a limit point.

Definition 1.25 A metric space X is said to be *sequentially compact* if every sequence in X has a convergent subsequence.

A subset A of a metric space X is said to be *sequentially compact* if every sequence in A contains a subsequence which converges to a point in A.

It is well known that

$$\text{compactness} \Leftrightarrow \text{Bolzano–Weierstrass property} \Leftrightarrow \text{sequentially compactness}$$

Definition 1.26 Let (X, d) be a metric space and $\varepsilon > 0$ be given. A subset A of X is called an ε-*net* if A is finite and $X = \bigcup_{x \in A} S_\varepsilon(x)$, that is, if A is finite and its points are scattered through X in such a way that each point of X is distant by less than ε from at least one point of A.

In other words, a finite subset $A = \left\{x_1, x_2, \dots, x_n\right\}$ of X is an ε-net for X if for every point $y \in X$, there exists an $x_{i_0} \in A$ such that $d\left(y, x_{i_0}\right) < \varepsilon$.

Example 1.16 Let $X = \{(x, y) \in \mathbb{R} \times \mathbb{R} : x^2 + y^2 < 4\}$, that is, X is the open sphere centered at the origin and radius 2. If $\varepsilon = \frac{3}{2}$, then the set

$$A = \left\{(1,-1), (1,0), (1,1), (0,-1), (0,0), (0,1), (-1,-1), (-1,0), (-1,1)\right\}$$

is an ε-net for X.

On the other hand, if $\varepsilon = 1/2$, then A is not an ε-net for X. For example, the point $y = \left(\frac{1}{2}, \frac{1}{2}\right)$ belongs to X but the distance between y and any point in A is greater than $\frac{1}{2}$.

Definition 1.27 A metric space (X, d) is said to be *totally bounded* if it has an ε-net for each $\varepsilon > 0$.

Remark 1.11 Every totally bounded metric space X is bounded but the converse is not true in general.

Since X is totally bounded, it has an ε-net $A = \{x_1, x_2, \dots, x_n\}$ for each $\varepsilon > 0$. Then, $X = \bigcup_{i=1}^{n} S_\varepsilon(x_i)$. Since finite union of bounded sets is bounded, it follows that X is bounded.

Example 1.17 Under the usual metric $d(x, y) = |x - y|$, the real line $\mathbb{R}$ is neither bounded nor totally bounded. Under the metric $d^*(x, y) = \min\{|x - y|, 1\}$, the real line $\mathbb{R}$ is bounded but not totally bounded.

Theorem 1.10 *Every totally bounded and complete metric space is compact.*

Theorem 1.11 *Every totally bounded metric space is separable.*

Remark 1.12 A discrete metric space is compact if and only if it is finite.

1.5 Continuous Functions

Definition 1.28 Let X be a nonempty set. A function $f : X \to \mathbb{R}$ is said to be

(a) *bounded above* if there exists a real number k such that $f(x) \leq k$ for all $x \in X$;

(b) *bounded below* if there exists a real number k such that $k \leq f(x)$ for all $x \in X$;

(c) *bounded* if it is both bounded above as well as bounded below.

Definition 1.29 Let (X, d_X) and (Y, d_Y) be metric spaces. A function $f : X \to Y$ is said to be *continuous at a point* $x_0 \in X$ if for every $\varepsilon > 0$, there exists a $\delta > 0$ such that for all $x \in X$,

$$d_X(x, x_0) < \delta \quad \text{implies} \quad d_Y(f(x), f(x_0)) < \varepsilon,$$

that is,

$$x \in S_\delta(x_0) \quad \text{implies} \quad f(x) \in S_\varepsilon(f(x_0)),$$

(see Figure 1.1). In other words, f is continuous at a point $x_0 \in X$ if for every $\varepsilon > 0$, there exists a $\delta > 0$ such that

$$f(S_\delta(x_0)) \subseteq S_\varepsilon(f(x_0)).$$

The function f is said to be *continuous on* X if it is continuous at every point of X.

Theorem 1.12 *Let X and Y be metric spaces and $f : X \to Y$ be a function. The following statements are equivalent:*

(a) *f is continuous on X.*

(b) *For every sequence $\{x_n\}$ in X such that $x_n \to x \in X$ implies $f(x_n) \to f(x)$.*

(c) *$f^{-1}(B)$ is open in X wherever B is open in Y.*

(d) *$f^{-1}(D)$ is closed in X wherever D is closed in Y.*

(e) *$f(\overline{A}) \subseteq \overline{f(A)}$ for every subset A of X.*

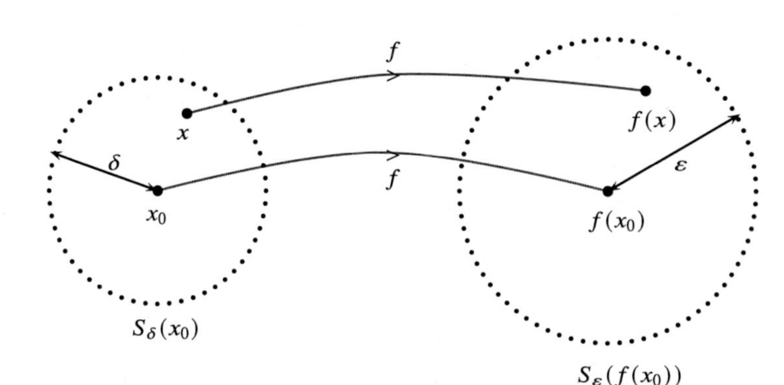

Figure 1.1 A continuous function

Theorem 1.13 *Let X and Y be metric spaces and $f : X \to Y$ be a continuous function. If A is a compact subset of X, then $f(A)$ is compact in Y.*

Exercise 1.3 Prove that a continuous real-valued function defined on a compact set is bounded and it assumes maximum and minimum values.

Proof Let $f : X \to \mathbb{R}$ be continuous and A be a compact subset of a metric space X. By Theorem 1.13, we see that $f(A)$ is a compact subset of $\mathbb{R}$. By Heine-Borel Theorem "A subset of $\mathbb{R}$ is closed and bounded if and only if it is compact", $f(A)$ is closed and bounded. Thus, $\sup f(A)$ and $\inf f(A)$ exist and belong to $f(A)$. Therefore, there exist $\hat{x}, \tilde{x} \in A$ such that for all $y \in A$, $\inf f(A) = f(\hat{x}) \leq f(y) \leq f(\tilde{x}) = \sup f(A)$. ■

Exercise 1.4 Let (X, d) be a metric space and A be a nonempty compact subset of X. Prove that for every $x_0 \in X$, there exists a $y_0 \in A$ such that

$$d(x_0, y_0) = d(x_0, A) = \inf_{y \in A} d(x_0, y).$$

Proof Consider the real-valued function $f : A \to \mathbb{R}_+$ defined by $f(x) = d(x, x_0)$ for all $x \in A$. Now $|f(x) - f(y)| = |d(x, x_0) - d(y, x_0)| \leq d(x, y)$, so f is continuous on A. But A is compact, so f has a minimum on A by Exercise 1.3. That is, there exists a $y_0 \in A$ such that $f(y_0) = d(x_0, y_0) = \inf_{y \in A} d(x_0, y) = d(x_0, A)$. ■

Definition 1.30 Let (X, d) and (Y, ρ) be metric spaces. A function $f : X \to Y$ is said to be *uniformly continuous* if for each $\varepsilon > 0$, there exists a $\delta > 0$ (depends only on ε) such that for every $x, y \in X$,

$$d(x, y) < \delta \quad \text{implies} \quad \rho(f(x), f(y)) < \varepsilon.$$

Remark 1.13 Every uniform continuous function is continuous but the converse need not be true in general.

Example 1.18 **(a)** Let X be a discrete metric space and Y be any metric space. Then, any function $f : X \to Y$ is uniformly continuous.

(b) Let $X = (0, 1)$ be a metric space with the metric induced by the usual metric on $\mathbb{R}$ and $Y = \mathbb{R}$ with the usual metric. The function $f : X \to Y$ defined by $f(x) = \frac{1}{x}$, for all $x \in X$, is not uniformly continuous.

(c) No polynomial function of degree greater than 1 is uniformly continuous on the usual metric space $\mathbb{R}$. Note that any polynomial function is continuous.

(d) The logarithmic function is not uniformly continuous on the usual metric space $X = (0, \infty)$.

Exercise 1.5 Let (X, d) be a metric space. Prove that the function $y \mapsto d(x, y)$ is uniformly continuous.

Theorem 1.14 *Let (X, d) and (Y, ρ) be metric spaces and $f : X \to Y$ be a continuous function. If X is compact, then f is uniformly continuous.*

Theorem 1.15 *Let (X, d) and (Y, ρ) be metric spaces and $f : X \to Y$ be an uniformly continuous function. If $\{x_n\}$ is a Cauchy sequence in X, then $\{f(x_n)\}$ is also a Cauchy sequence in Y.*

The following example shows that a continuous function may not map a Cauchy sequence into a Cauchy sequence.

Example 1.19 Let $X = (0, \infty)$ with the induced usual metric on $\mathbb{R}$ and $Y = \mathbb{R}$ with the usual metric. The function $f : X \to Y$ defined by $f(x) = \frac{1}{x}$, for all $x \in X$, is continuous on X. Clearly, $\left\{x_n : x_n = \frac{1}{n}\right\}_{n\in\mathbb{N}}$ is a Cauchy sequence in X. But $\left\{f\left(\frac{1}{n}\right)\right\}_{n\in\mathbb{N}} = \{n\}_{n=1}^{\infty}$ is not a Cauchy sequence in Y. Indeed, the absolute difference of any two distinct points is at least as large as 1.

Exercise 1.6 Show that the function $f(x) = e^x$ defined on the usual metric space $\mathbb{R}$ is not uniformly continuous.

Exercise 1.7 Let (X, d) be a metric space and A be a nonempty subset of X. Prove that the function $f : X \to \mathbb{R}$ defined by

$$f(x) = d(x, A), \quad \text{for all } x \in X,$$

is uniformly continuous.

In view of Theorem 1.12 (b), a function $f : X \to Y$ from a metric space X to a metric space Y is continuous at a point $x \in X$ if and only if for every sequence $\{x_n\}$ that converges to $x \in X$, we have $\lim_{n\to\infty} f(x_n) = f(x)$.

Definition 1.31 Let X be a metric space. A function $f : X \to \mathbb{R}$ is said to be

(a) *lower semicontinuous* at a point $x \in X$ if $f(x) \leq \liminf_{n\to\infty} f(x_n)$ whenever $x_n \to x$ as $n \to \infty$, equivalently,

$$f(x) \leq \liminf_{y\to x} f(y);$$

(b) *upper semicontinuous at a point* $x \in X$ if $f(x) \geq \limsup_{n\to\infty} f(x_n)$ whenever $x_n \to x$ as $n \to \infty$, equivalently,

$$f(x) \geq \limsup_{y\to x} f(y);$$

(c) *upper semicontinuous* (respectively, *lower semicontinuous*) *on* X if it is upper semicontinuous (respectively, lower semicontinuous) at each point of X.

Example 1.20 Let $f : \mathbb{R} \to \mathbb{R}$ be defined by

$$f(x) = \begin{cases} -1, & \text{if } x < 0, \\ 1, & \text{if } x \geq 0. \end{cases}$$

Then, f is upper semicontinuous at $x = 0$ but not lower semicontinuous at $x = 0$ (see Figure 1.2).

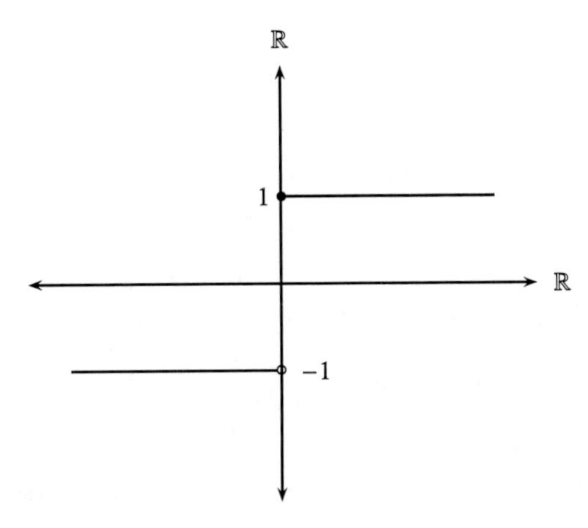

Figure 1.2 An upper semicontinuous function

Example 1.21 Let $f : \mathbb{R} \to \mathbb{R}$ be defined by

$$f(x) = \begin{cases} -1, & \text{if } x \leq 0, \\ 1, & \text{if } x > 0. \end{cases}$$

Then, f is lower semicontinuous at $x = 0$ but not upper semicontinuous at $x = 0$ (see Figure 1.3).

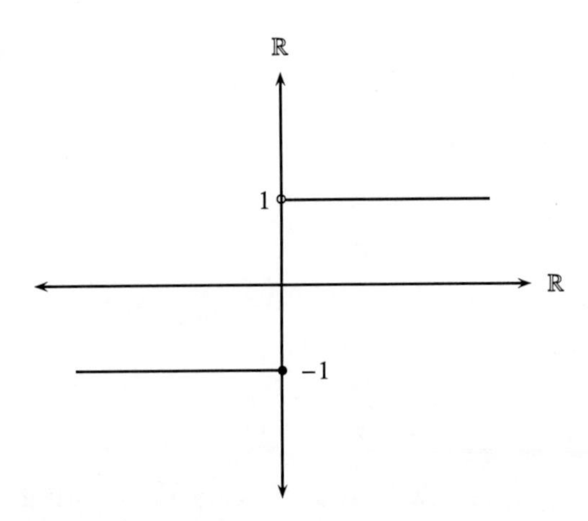

Figure 1.3 A lower semicontinuous function

Remark 1.14 (a) A function $f : X \to \mathbb{R}$ is lower (respectively, upper) semicontinuous on X if and only if the *lower level set* $\{x \in X : f(x) \le \alpha\}$ (respectively, the *upper level set* $\{x \in X : f(x) \ge \alpha\}$) is closed in X for all $\alpha \in \mathbb{R}$. Equivalently, f is lower (respectively, upper) semicontinuous on X if and only if the set $\{x \in X : f(x) > \alpha\}$ (respectively, $\{x \in X : f(x) < \alpha\}$) is open in X for all $\alpha \in \mathbb{R}$.

(b) A function $f : X \to \mathbb{R}$ is lower (respectively, upper) semicontinuous on X if and only if the *epigraph* $\{(x, \alpha) \in X \times \mathbb{R} : f(x) \le \alpha\}$ (respectively, the *hypograph* $\{(x, \alpha) \in X \times \mathbb{R} : f(x) \ge \alpha\}$) of f is closed in X.

In view of the above remark, Penot and Théra [142] gave the following definition of lower semicontinuous functions.

Definition 1.32 Let X be a metric space. A function $f : X \to \mathbb{R}$ is said to be

(a) *lower semicontinuous* on X if for each $x \in X$ and each $\alpha \in \mathbb{R}$ such that $f(x) > \alpha$, there exists $\delta > 0$ such that $f(y) > \alpha$ for all $y \in S_\delta(x)$;

(b) *upper semicontinuous* on X if for each $x \in X$ and each $\alpha \in \mathbb{R}$ such that $f(x) < \alpha$, there exists $\delta > 0$ such that $f(y) < \alpha$ for all $y \in S_\delta(x)$.

The following theorem shows that the conditions for lower semicontinuity on a metric space X given in Definition 1.31 and Definition 1.32 are equivalent.

Theorem 1.16 *Let (X, d) be a metric space, $x \in X$ and $f : X \to \mathbb{R}$ be a function. Then the following statements are equivalent:*

(a) *f is lower semicontinuous at x.*

(b) *For every $\varepsilon > 0$, there is a $\delta > 0$ such that $f(x) - f(y) < \varepsilon$ whenever $d(x, y) < \delta$.*

Proof (a) $\Rightarrow$ (b) Suppose that f is lower semicontinuous at x. Set $\lambda := \liminf_{y \to x} f(y)$, and let $\varepsilon > 0$. Then there is a $\delta > 0$ such that

$$|\lambda - \inf f(S_\delta(x))| < \varepsilon.$$

By the definition of the lower semicontinuity of f at x, we have

$$f(x) \le \lambda.$$

Thus, for every $y \in S_\delta(x)$, we have

$$f(x) - f(y) \le \lambda - f(y) \le \lambda - \inf_{z \in S_\delta(x)} f(z) < \varepsilon.$$

(b) $\Rightarrow$ (a) Let $\varepsilon > 0$ and choose a positive δ_0 such that $f(x) - f(y) < \frac{\varepsilon}{2}$ for all $y \in S_\delta(x)$. Assume that $\bar{x} \in S_{\delta_0}(x)$ such that

$$f(\bar{x}) < \inf_{z \in S_{\delta_0}(x)} f(z) + \frac{\varepsilon}{2}.$$

Hence,

$$\alpha := f(x) - \inf_{z \in S_\delta(x)} f(z) < f(x) - f(\bar{x}) + \frac{\varepsilon}{2}, \quad \text{for all positive } \delta \le \delta_0. \tag{1.2}$$

Note that the number α in (1.2) is positive. It follows that

$$\left| f(x) - \inf_{z \in S_\delta(x)} f(z) \right| < \varepsilon, \quad \text{for all positive } \delta \leq \delta_0.$$

Therefore, $f(x) \leq \liminf_{y \to x} f(y)$, that is, f is lower semicontinuous at x. ■

Similarly, we can have the following result for upper semicontinuous functions.

Theorem 1.17 *Let (X, d) be a metric space, $x \in X$ and $f : X \to \mathbb{R}$ be a function. Then the following statements are equivalent:*

(a) *f is upper semicontinuous at x.*
(b) *For every $\varepsilon > 0$, there is a $\delta > 0$ such that $f(y) - f(x) < \varepsilon$ whenever $d(x, y) < \delta$.*

The following theorem establishes that every lower semicontinuous function attains its minimum in any compact set.

Theorem 1.18 *Let K be a nonempty compact subset of a metric space (X, d) and $f : X \to \mathbb{R}$ be a lower semicontinuous function. Then, f attains its minimum on K.*

Proof Set $\lambda := \inf_{z \in K} f(z)$. If $\lambda = -\infty$, then there exists a sequence $\{x_n\}$ in K such that $\lim_{n\to\infty} f(x_n) = -\infty$. Without loss of generality, we may assume that $\{x_n\}$ converges to $x \in K$ by the compactness of K. By the lower semicontinuity of f, we have

$$f(x) \leq \lim_{n\to\infty} f(x_n) = -\infty,$$

which is not possible. Thus, $\lambda \in \mathbb{R}$. Now take a sequence $\{x_n\}$ in K such that $\lim_{n\to\infty} f(x_n) = \lambda$. Assume that $\{x_n\}$ converges to $x_0 \in K$ by the compactness of K. Again, by the lower semicontinuity of f, we have

$$f(x_0) \leq \lim_{n\to\infty} f(x_n) = \lambda = \inf_{z \in K} f(z).$$

This means that x_0 is a minimizer of f on K. ■

Exercise 1.8 Let K be a nonempty compact subset of a metric space (X, d) and $f : X \to \mathbb{R}$ be an upper semicontinuous function. Prove that f attains its maximum on K.

Chen et. al. [58] introduced the following concept of lower semicontinuity from above.

Definition 1.33 Let X be a metric space. A function $f : X \to \mathbb{R}$ is said to be

(a) *lower semicontinuous from above* at a point $x \in X$ if for any sequence $\{x_n\}$ in X converging to x and satisfying $f(x_{n+1}) \leq f(x_n)$ for all $n \in \mathbb{N}$, we have $f(x) \leq \lim_{n\to\infty} f(x_n)$;

(b) *upper semicontinuous from below* at a point $x \in X$ if for any sequence $\{x_n\}$ in X converging to x and satisfying $f(x_{n+1}) \geq f(x_n)$ for all $n \in \mathbb{N}$, we have $f(x) \geq \lim_{n\to\infty} f(x_n)$;

(c) *lower semicontinuous from above* (respectively, *upper semicontinuous from below*) on X if it is lower semicontinuous from above (respectively, upper semicontinuous from below) at every point of X.

Obviously, lower (respectively, upper) semicontinuity implies lower semicontinuity from above (respectively, upper semicontinuity from below), but the converse implications do not hold.

Example 1.22 Let $f : \mathbb{R} \to \mathbb{R}$ be a function defined by

$$f(x) = \begin{cases} x + \frac{1}{2}, & \text{if } x < 0, \\ x^2 + 1, & \text{if } x \geq 0. \end{cases}$$

Then, f is lower semicontinuous from above at $x = 0$, but not lower semicontinuous at this point.

Example 1.23 Let $f : \mathbb{R} \to \mathbb{R}$ be a function defined by

$$f(x) = \begin{cases} x - 1, & \text{if } x < 0, \\ 0, & \text{if } x = 0, \\ x + 1, & \text{if } x > 0. \end{cases}$$

Then, f is lower semicontinuous from above as well as upper semicontinuous from below at $x = 0$, but it is neither lower nor upper semicontinuous at this point.

By Theorem 1.18, we have that every bounded below and lower semicontinuous real-valued function has a minimum on a compact set. However, Chen et al. [58] showed that the Weierstrass's theorem still holds for bounded below and lower semicontinuous from above functions.

Theorem 1.19 *Let K be a nonempty compact subset of a metric space X and $f : K \to \mathbb{R}$ be bounded below and lower semicontinuous from above. Then, there exists $\bar{x} \in K$ such that $f(\bar{x}) = \inf_{y \in K} f(y)$.*

Proof Since K is compact and f is bounded below, there exists a sequence $\{x_n\}$ in K such that $x_n \to \bar{x} \in K, f(x_1) \geq f(x_2) \geq \cdots \geq f(x_n) \geq \cdots$ and $f(x_n) \to \inf_{y \in K} f(y)$.

By the lower semicontinuity from above, we have

$$f(\bar{x}) \leq \lim_{n \to \infty} f(x_n) = \inf_{y \in K} f(y).$$

Hence, $f(\bar{x}) = \inf_{y \in K} f(y)$. ■

Chen et al. [58] also showed that Ekeland's variational principle and Caristi's fixed point theorem hold for lower semicontinuity from above functions.

Definition 1.34 A function $\varphi : [0, \infty) \to [0, \infty)$ is said to be *right upper semicontinuous,* also called *upper semicontinuous from the right,* if $\varphi(t) \geq \limsup_{r \to t^+} \varphi(r)$ for all $t \geq 0$.

Example 1.24 Define a function $\varphi : [0, \infty) \to [0, \infty)$ by

$$\varphi(t) = \begin{cases} \sqrt{t}, & \text{if } t \in [0, 1), \\ \sqrt{t} + 1, & \text{if } t \in [1, \infty). \end{cases}$$

We see that $\varphi(1) = 2$ and the function ψ is discontinuous at $t = 1$. Note that $\lim_{t \to 1^-} \varphi(t) = 1$ and $\limsup_{t \to 1^+} \varphi(t) = 2$. Thus, the function φ is right upper semicontinuous.

Lemma 1.1 [89] *Let $\varphi : [0, \infty) \to [0, \infty)$ be a right upper semicontinuous function such that $\varphi(t) < t$ for all $t > 0$. Then, $\lim_{n\to\infty} \varphi^n(t) = 0$.*

Proof Since for each $t > 0$, $\varphi(t) < t$, we have $\varphi^2(t) = \varphi(\varphi(t)) < \varphi(t) < t$. By induction, we obtain a nonincreasing sequence $\{\varphi^n(t)\}$. So we can assume that $\{\varphi^n(t)\}$ decreases to a nonnegative number c. If $c > 0$, then

$$c > \varphi(c) \geq \limsup_{n\to\infty} \varphi(\varphi^n(t)) = \lim_{n\to\infty} \varphi^{n+1}(t) = c,$$

which is a contraction. Hence $c = 0$, and $\lim_{n\to\infty} \varphi^n(t) = 0$. ■

Lemma 1.2 [89] *Let $\varphi : [0, \infty) \to [0, \infty)$ be a right upper semicontinuous function such that $\varphi(t) < 1$ for all $t > 0$. Then the function $\Phi : [0, \infty) \to [0, \infty)$, defined by $\Phi(t) = \varphi(t)t$, is right upper semicontinuous and $\Phi(t) < t$ for all $t > 0$.*

Proof Since for each $t > 0$, $\varphi(t) < 1$, we have $\Phi(t) = \varphi(t)t < t$, and

$$\begin{aligned}\Phi(t) = \varphi(t)t &\geq \left(\limsup_{r\to t^+} \varphi(r)\right) t \\ &= \limsup_{r\to t^+} (\varphi(r)r) \\ &= \limsup_{r\to t^+} \Phi(r).\end{aligned}$$

Hence, $\Phi(t)$ is right upper semicontinuous. ■

Lemma 1.3 [89] *Let $\varphi : [0, \infty) \to [0, 1)$ be such that $\limsup_{r\to t^+} \varphi(r) < 1$ for all $t > 0$, and let $\Psi(t) = \max\left\{\varphi(t), \limsup_{r\to t^+} \varphi(r)\right\}$ for all $t > 0$. Then, the function $\Psi : [0, \infty) \to [0, 1)$ is right upper semicontinuous and $\Psi(t) \geq \varphi(t)$ for all $t \geq 0$.*

Proof Since for each $t > 0$, $\varphi(t) < 1$, and $\limsup_{r\to t^+} \varphi(r) < 1$ for all $t > 0$, we have $\Psi(t) = \max\left\{\varphi(t), \limsup_{r\to t^+} \varphi(r)\right\} < 1$, and Ψ is a function from $[0, \infty)$ to $[0, 1)$.

Now we prove that Ψ is right upper semicontinuous. Let $\alpha = \limsup_{r\to t^+} \Psi(r)$. Then by the definition of upper limit, there exists a nonincreasing sequence $\{t_n\}$ with limit t such that $\lim_{n\to\infty} \Psi(t_n) = \alpha$. Denote $\alpha_n = \Psi(t_n)$, then $\lim_{n\to\infty} \alpha_n = \alpha$. For each $\varepsilon > 0$, $n = 1, 2, ...$, if $\Psi(t_n) = \varphi(t_n)$, take $t'_n = t_n$, and if $\Psi(t_n) = \limsup_{r\to t^+} \varphi(r)$, by the definition of upper limit, we can choose t'_n, $t_{n-1} > t'_n \geq t_n$ such that

$$\varphi(t'_n) \geq \limsup_{r\to t'_n} \varphi(r) - \varepsilon = \alpha_n - \varepsilon.$$

In both the cases, we have $\{t'_n : t_{n-1} > t'_n \geq t_n\}$ such that $\varphi(t'_n) \geq \alpha_n - \varepsilon$ and $t'_n \to t^+$. Since

$$\limsup_{r \to t^+} \varphi(r) \geq \lim_{n\to\infty} \varphi(t'_n) \geq \lim_{n\to\infty} (\alpha_n - \varepsilon) = \alpha - \varepsilon,$$

we have

$$\Psi(t) = \max\left\{\varphi(t), \limsup_{r\to t^+} \varphi(r)\right\} \geq \alpha - \varepsilon, \quad \text{for all } \varepsilon > 0.$$

Hence, $\Psi(t) \geq \alpha = \limsup_{r\to t^+} \Psi(r)$. ∎

Chapter 2

Fixed Point Theory in Metric Spaces

The origin of metric fixed point theory goes back to the remarkable work of Polish mathematician Stefan Banach in his Ph.D. dissertation in 1920. The scholarly outcome of the dissertation is now known as the Banach contraction principle. The beauty of the Banach contraction principle is that it requires only completeness and contraction condition on the underlying metric space and mapping, respectively. With these conditions it provides the following assertions:

- The existence and uniqueness of a fixed point.
- The method to compute the approximate fixed points.
- The error estimates for approximate fixed points.

A large number of research papers have already appeared in the literature on extensions and generalizations of the Banach contraction principle. One of the most important generalizations is due to Boyd and Wong [43] for ψ-contraction mappings. Another important generalization of the Banach contraction principle is given by Rhoades [148] for weakly contraction mappings introduced by Alber and Guerre-Delabriere [2]. However, a remarkable generalization of the Banach contraction principle is the Caristi's fixed point theorem [52].

In this chapter, the Banach contraction principle and some of its extensions, namely, Boyd–Wong fixed point theorem for ψ-contraction mappings, a fixed point theorem for weakly contraction mappings, and Caristi's fixed point theorem, are presented. The converse of the Banach contraction principle that provides the characterization of completeness of the metric space is also discussed.

2.1 Fixed Points

Definition 2.1 Let X be a nonempty set and $T : X \to X$ be a mapping.

- A point $\bar{x} \in X$ is said be a *fixed point* of T if $T(\bar{x}) = \bar{x}$.
- The problem of finding a point $\bar{x} \in X$ such that $T(\bar{x}) = \bar{x}$ is called a fixed point problem.
- We denote by $\text{Fix}(T) = \{x \in X : T(x) = x\}$ the set of all fixed points of T.

Example 2.1 (a) Let $T : \mathbb{R} \to \mathbb{R}$ be a mapping defined by $T(x) = x + a$ for some fixed number $a \neq 0$. Then, T has no fixed point.

(b) Let $T : \mathbb{R} \to \mathbb{R}$ be a mapping defined by $T(x) = \frac{1}{2}x$. Then, $x = 0$ is the only fixed point of T.

(c) Let $T : \mathbb{R} \to \mathbb{R}$ be a mapping defined by $T(x) = x^2$. Then, $x = 0$ and $x = 1$ are two fixed points of T.

(d) Let $T : \mathbb{R} \to \mathbb{R}$ be a mapping defined by $T(x) = x$. Then, T has infinitely many fixed points. In fact, every point of $\mathbb{R}$ is a fixed of T.

Example 2.1 shows that a mapping may not have any fixed point, may have a unique fixed point, may have more than one fixed point or may even have infinitely many fixed points.

For each $m \in \mathbb{N}$, let

$$\text{Fix}(T^m) = \{x \in X : T^m(x) = \underbrace{T \circ T \circ \cdots \circ T}_{m\text{-times}}(x) = x\}.$$

Proposition 2.1 *Let X be a nonempty set and $T : X \to X$ be a mapping. If $x \in X$ is a unique fixed point of $T^m = \underbrace{T \circ T \circ \cdots \circ T}_{m\text{-times}}$ for any $m \in \mathbb{N}$, then it is the unique fixed point of T. In other words, if* $\text{Fix}(T^m) = \{x\}$ *for any $m \in \mathbb{N}$, then* $\text{Fix}(T) = \{x\}$.

Proof Let x be a fixed point of T^m. Then, we have

$$T(x) = T(T^m(x)) = T^m(T(x)).$$

Hence, $T(x)$ is a fixed point of T^m. By hypothesis, x is the unique fixed point of T^m; we can therefore conclude that $T(x) = x$, that is, x is a fixed point of T.

We now show that x is the unique fixed point of T. Let y be another fixed point of T. Then,

$$y = T(y) = T^2(y) = \cdots = T^m(y).$$

Thus, y is a fixed point of T^m. Since x is the unique fixed of T^m, we get $x = y$. ■

The following examples show that the linear problems and nonlinear problems can be modelled as a fixed point problem.

Example 2.2 Consider the problem of solving the equation $x^3 - x - 1 = 0$. Then the equation $x^3 - x - 1 = 0$ can be re-written as $x^3 - 1 = x$ or $(1 + x)^{1/3} = x$. Consider the mappings $T_1, T_2 : \mathbb{R} \to \mathbb{R}$ defined by

$$T_1(x) = x^3 - 1 \quad \text{and} \quad T_2(x) = (1 + x)^{1/3}, \quad \text{for all } x \in \mathbb{R}.$$

Then solving the equation $x^3 - x - 1 = 0$ is equivalent to finding a fixed point of T_1 or T_2.

Example 2.3 Consider the problem of finding a solution of the following system of n linear equations with n unknowns:

$$\left.\begin{aligned} a_{11}x_1 + a_{12}x_2 + \cdots + a_{1n}x_n &= b_1 \\ a_{21}x_1 + a_{22}x_2 + \cdots + a_{2n}x_n &= b_2 \\ \cdots\cdots\cdots\cdots\cdots\cdots\cdots\cdots& \\ a_{n1}x_1 + a_{n2}x_2 + \cdots + a_{nn}x_n &= b_n. \end{aligned}\right\} \tag{2.1}$$

This system can be written as

$$\left.\begin{aligned}
x_1 &= (1-a_{11})x_1 - a_{12}x_2 - \cdots - a_{1n}x_n + b_1\\
x_2 &= -a_{21}x_1 + (1-a_{22})x_2 - a_{23}x_3 - \cdots - a_{2n}x_n + b_2\\
x_3 &= -a_{31}x_1 - a_{32}x_2 + (1-a_{33})x_3 - \cdots - a_{3n}x_n + b_3\\
&\cdots\cdots\cdots\cdots\cdots\cdots\cdots\cdots\cdots\cdots\cdots\cdots\\
x_n &= -a_{n1}x_1 - a_{n2}x_2 - a_{n3}x_3 - \cdots + (1-a_{nn})x_n + b_n.
\end{aligned}\right\} \tag{2.2}$$

By letting $\alpha_{ij} = -a_{ij} + \delta_{ij}$, where

$$\delta_{ij} = \begin{cases} 1, & \text{for } i = j,\\ 0, & \text{for } i \neq j,\end{cases}$$

the system (2.2) is equivalent to the following system:

$$x_i = \sum_{j=1}^{n} \alpha_{ij}x_j + b_i, \quad \text{for } i = 1, 2, 3, \dots, n. \tag{2.3}$$

Let $x = (x_1, x_2, \dots, x_n) \in \mathbb{R}^n$,

$$\mathbf{A} = \begin{pmatrix} a_{11} & a_{12} & \cdots & a_{1n}\\ a_{21} & a_{22} & \cdots & a_{2n}\\ \cdots & \cdots & \cdots & \cdots\\ \cdots & \cdots & \cdots & \cdots\\ \cdots & \cdots & \cdots & \cdots\\ a_{n1} & a_{n2} & \cdots & a_{nn}\end{pmatrix},$$

and $b = (b_1, b_2, \dots, b_n) \in \mathbb{R}^n$. Then, the vectors $x = (x_1, x_2, \dots, x_n) \in \mathbb{R}^n$ and $b = (b_1, b_2, \dots, b_n) \in \mathbb{R}^n$ can be represented in the form of matrices

$$\mathbf{x} = \begin{pmatrix} x_1\\ x_2\\ \vdots\\ x_n\end{pmatrix} \quad \text{and} \quad \mathbf{b} = \begin{pmatrix} b_1\\ b_2\\ \vdots\\ b_n\end{pmatrix},$$

respectively. Therefore, the system (2.3) is equivalently written as

$$\mathbf{x} = (\mathbf{I} - \mathbf{A})\mathbf{x} + \mathbf{b}, \tag{2.4}$$

where $\mathbf{I}$ denotes the identity matrix. Let $T : \mathbb{R}^n \to \mathbb{R}^n$ be a matrix transformation defined by

$$T(\mathbf{x}) = (\mathbf{I} - \mathbf{A})\mathbf{x} + \mathbf{b}. \tag{2.5}$$

Then, finding the solution of the system (2.4) is equivalent to finding a fixed point of the matrix transformation T defined by (2.5).

Example 2.4 Let X be a linear space, $b \in X$ and $\mathcal{F} : X \to X$ be an operator. Consider the following problem:

$$\text{Find } x \in X \text{ such that } \mathcal{F}(x) = b. \tag{2.6}$$

Then, finding the solutions of the operator equation (2.6) is equivalent to finding the fixed points of the mapping $T : X \to X$ defined by

$$T(x) = x - \mu(\mathcal{F}(x) - b), \quad \text{for all } x \in X,$$

where μ is a nonzero constant.

Example 2.5 Let $X = C[0, 1]$ be a linear space of real-valued continuous functions defined on the closed interval $[0, 1]$ and $T : X \to X$ be a mapping defined by

$$T(x(s)) = x(0) + \int_0^s x(t)dt, \quad \text{for } s \in [0, 1].$$

Clearly, finding the solutions of the integral equation

$$x(s) = x(0) + \int_0^s x(t)dt, \quad \text{for } s \in [0, 1],$$

is equivalent to finding a fixed point of the mapping T. In fact, $x(t) = ae^t$ is a fixed point of T for an arbitrary constant $a \in \mathbb{R}$.

2.2 Banach Contraction Principle

Definition 2.2 Let (X, d) be a metric space. A mapping $T : X \to X$ is said to be *Lipschitz continuous* if there exists a constant $\alpha > 0$ such that

$$d(T(x), T(y)) \leq \alpha d(x, y), \quad \text{for all } x, y \in X. \tag{2.7}$$

- If $\alpha \in (0, 1)$, then T is said to be *contraction.*
- If $\alpha = 1$, then T is said to be *nonexpansive.*
- If $d(T(x), T(y)) < d(x, y)$ for all $x \neq y$, then T is said to be *contractive.*

The number $\alpha > 0$ that satisfies (2.7) is called *Lipschitz constant* of T. If $\alpha \in (0, 1)$ and satisfies (2.7), then it called *contractivity constant* or *contraction constant.*

Clearly, a contraction maps points closer together. In particular, for any $x \in X$ and every $r > 0$, all points y in the sphere $S_r(x)$ are mapped into a sphere $S_s(f(x))$ with $s < r$. It is illustrated in Figure 2.1.

Remark 2.1 **(a)** It is clear that every Lipschitz continuous mapping is continuous but the converse need not be true; see Example 2.6.

(b) It is clear from the definition that every contraction mapping is contractive but the converse may not be true. For example, consider the metric space $X = [0, \infty)$ with the usual metric and the mapping $T : X \to X$ is given by $T(x) = \frac{1}{1+x^2}$. Then, T is contractive but not a contraction.

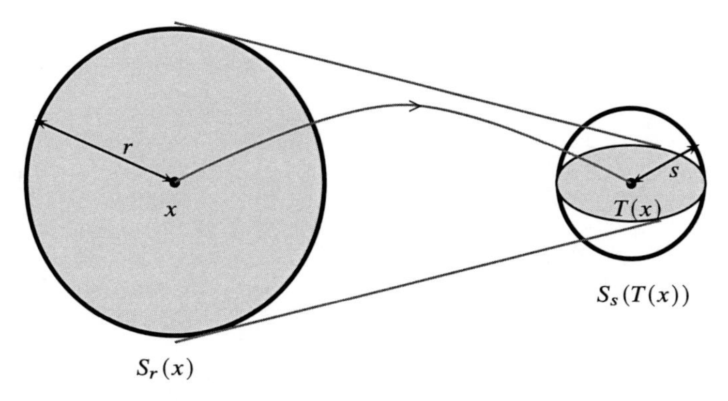

Figure 2.1 A contraction mapping

(c) Every contraction mapping is nonexpansive but the converse need not be true; see Example 2.8.

(d) A continuous mapping need not be a contraction. For example, consider the metric space $X = [0,1]$ with the usual metric and define a mapping $T : X \to X$ by $T(x) = x^2$ for all $x \in X$. Then, T is Lipschitz continuous with Lipschitz constant $\alpha = 2$, but not a contraction. Indeed,

$$|T(x) - T(y)| = |x^2 - y^2| \leq 2|x - y|, \quad \text{for all } x, y \in X.$$

Of course, T is continuous, and 0 and 1 are the fixed points of T.

(e) Every constant mapping T defined on a metric space (X, d) into itself is a contraction.

The following example shows that a continuous mapping may not be Lipschitz continuous.

Example 2.6 Let $X = \left[-\frac{1}{\pi}, \frac{1}{\pi}\right]$ be a metric space with the usual metric and $T : X \to X$ be a mapping defined by

$$T(x) = \begin{cases} \frac{x}{2} \sin \frac{1}{x}, & \text{if } x \neq 0, \\ 0, & \text{if } x = 0, \end{cases} \tag{2.8}$$

(see Figure 2.2). Then, T is continuous on X with $\mathrm{Fix}(T) = \{0\}$, but not Lipschitz continuous.

Indeed, assume the contrary that the mapping T defined by (2.8) is Lipschitz continuous with constant α. Then, for sequences $\{x_n\}$ and $\{y_n\}$ in X, we have

$$|T(x_n) - T(y_n)| \leq \alpha |x_n - y_n|, \quad \text{for all } n \in \mathbb{N}. \tag{2.9}$$

Consider two sequences $\{x_n\} = \left\{\frac{1}{2n\pi}\right\}$ and $\{y_n\} = \left\{\frac{1}{2n\pi + \frac{\pi}{2}}\right\}$. Then,

$$|x_n - y_n| = \left| \frac{1}{2n\pi} - \frac{1}{2n\pi + \frac{\pi}{2}} \right| = \frac{\pi}{2(2n\pi)(2n\pi + \frac{\pi}{2})} = \frac{1}{2n(4n+1)\pi},$$

and

$$|T(x_n) - T(y_n)| = \left| \frac{\sin(2n\pi)}{4n\pi} - \frac{\sin(2n\pi + \frac{\pi}{2})}{(4n+1)\pi} \right| = \left| 0 - \frac{1}{(4n+1)\pi} \right| = \frac{1}{(4n+1)\pi}.$$

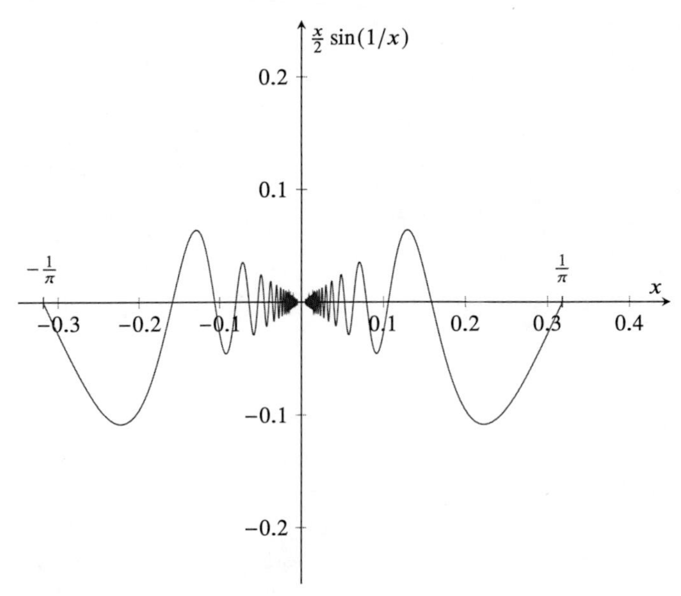

Figure 2.2 The graph of the function $T(x) = \frac{x}{2}\sin(1/x)$

Thus, from (2.9), we have

$$\frac{1}{(4n+1)\pi} \leq \frac{\alpha}{2n(4n+1)\pi}, \quad \text{for all } n \in \mathbb{N},$$

that is,

$$2 \leq \frac{\alpha}{n}, \quad \text{for all } n \in \mathbb{N}.$$

Letting limit as $n \to \infty$, we get a contradiction. Hence, T is not Lipschitz continuous with constant α. Of course, 0 is the only fixed point of T.

Example 2.7 Let $X = [0, 1]$ be a metric space with the usual metric. Define a mapping $T : X \to X$ by

$$T(x) = \frac{x^2}{4}, \quad \text{for all } x \in X. \tag{2.10}$$

Then for all $x, y \in X$, we have

$$d(T(x), T(y)) = \left|\frac{x^2}{4} - \frac{y^2}{4}\right| = \frac{x+y}{4}|x-y| \leq \frac{1}{2}|x-y| = \frac{1}{2}d(x,y).$$

Hence, the mapping T defined by (2.10) is a contraction on (X, d) with contraction constant $\alpha = \frac{1}{2}$, and 0 is the only fixed point of T.

The following example shows that a nonexpansive mapping may not be a contraction.

Example 2.8 Let $X = [0, 1]$ be a metric space with the usual metric and $T : X \to X$ be a mapping defined by

$$T(x) = \frac{x^2}{2}, \quad \text{for all } x \in X. \tag{2.11}$$

Then for all $x, y \in X$, we have

$$d(T(x), T(y)) = \left| \frac{x^2}{2} - \frac{y^2}{2} \right| = \frac{x + y}{2} |x - y| \leq d(x, y).$$

Thus, T is a nonexpansive mapping, but not a contraction. Also, 0 is the only fixed point of T.

The following example shows that a nonexpansive mapping may not have a fixed point.

Example 2.9 Let $X = \mathbb{R}$ be a metric space with the usual metric and $T : X \to X$ be a mapping defined by

$$T(x) = 1 + x, \quad \text{for all } x \in X.$$

Then, T is a nonexapnsive mapping but has no fixed point.

The following example shows that a contractive mapping may not be a contraction and also it may not have a fixed point.

Example 2.10 Consider the metric space $X = [1, \infty)$ with the usual metric and the mapping $T : X \to X$ is given by $T(x) = x + \frac{1}{x}$. Then, T is a contractive mapping but not a contraction. In fact,

$$|T(x) - T(y)| = |x - y| \left(1 - \frac{1}{xy}\right) < |x - y|, \quad \text{for all } x, y \in X \text{ with } x \neq y.$$

So, there would be no $\alpha \in (0, 1)$ such that

$$|T(x) - T(y)| \leq \alpha |x - y|, \quad \text{for all } x, y \in X.$$

Hence, T is not a contraction. However, T is contractive. Also, T has no fixed point.

Exercise 2.1 Let $X = \mathbb{R}$ be a metric space with the usual metric and $T : X \to X$ be defined by

$$T(x) = \frac{\pi}{2} + x - \tan^{-1} x, \quad \text{for all } x \in X.$$

Prove that T is a contractive mapping, but not a contraction.
Hint:

$$T'(x) = 1 - \frac{1}{1 + x^2} = \frac{x^2}{1 + x^2} < 1, \quad \text{for all } x \in X.$$

By mean value theorem, we have

$$d(T(x), T(y)) < d(x, y), \quad \text{for all } x, y \in X \text{ with } x \neq y.$$

Proposition 2.2 *Let (X, d) be a metric space and $T : X \to X$ be a contraction mapping with constant $\alpha \in (0, 1)$. If T has a fixed point, then the fixed point is unique.*

Proof Suppose that T has two distinct fixed points, say, x and y, that is, $T(x) = x$ and $T(y) = y$ with $x \neq y$. Since T is a contraction mapping and $\alpha \in (0, 1)$, we have

$$d(x, y) = d(T(x), T(y)) \leq \alpha d(x, y),$$

a contradiction. Therefore, the fixed point of T is unique. ∎

Theorem 2.1 (Banach Contraction Principle) *Let (X, d) be a complete metric space and $T : X \to X$ be a contraction mapping with constant $\alpha \in (0, 1)$. Then, T has a unique fixed point.*

Proof We construct a sequence $\{x_n\}$ by the following iterative method.

Choose an arbitrary point $x_0 \in X$. Then $x_0 \neq T(x_0)$, otherwise x_0 would be a fixed point of T and there is nothing to prove. So, we define

$$x_1 = T(x_0),\ x_2 = T(x_1),\ x_3 = T(x_2),\ \dots,\ x_n = T(x_{n-1}), \quad \text{for all } n \in \mathbb{N}.$$

We claim that this sequence $\{x_n\}$ of points of X is a Cauchy sequence.

Since T is a contraction mapping with contraction constant $\alpha \in (0, 1)$, for all $p = 1, 2, \dots$, we have

$$\begin{aligned} d(x_{p+1}, x_p) &= d(T(x_p), T(x_{p-1})) \leq \alpha d(x_p, x_{p-1}) \\ &= \alpha d(T(x_{p-1}), T(x_{p-2})) \leq \alpha^2 d(x_{p-1}, x_{p-2}) \\ &\ \ \vdots \\ &= \alpha^{p-1} d(T(x_1), T(x_0)) \leq \alpha^p d(x_1, x_0). \end{aligned}$$

Let m and n be any positive integers with $n < m$. Then by the triangle inequality, we have

$$\begin{aligned} d(x_m, x_n) &\leq d(x_m, x_{m-1}) + d(x_{m-1}, x_{m-2}) + \cdots + d(x_{n+1}, x_n) \\ &\leq \left(\alpha^{m-1} + \alpha^{m-2} + \cdots + \alpha^n\right) d(x_1, x_0) \\ &\leq \alpha^n \left(\alpha^{m-n-1} + \alpha^{m-n-2} + \cdots + 1\right) d(x_1, x_0) \\ &\leq \frac{\alpha^n}{1 - \alpha} d(x_1, x_0). \end{aligned}$$

Since $\lim_{n\to\infty} \alpha^n = 0$ and $d(x_1, x_0)$ is fixed, the right hand side of the above inequality approaches 0 as n tends to ∞. It follows that $\{x_n\}$ is a Cauchy sequence in X. Since X is complete, there exists $\bar{x} \in X$ such that $x_n \to \bar{x}$. We show that this limit point $\bar{x}$ is a fixed point of T. Since every contraction mapping is continuous, we have

$$\bar{x} = \lim_{n\to\infty} x_{n+1} = \lim_{n\to\infty} T(x_n) = T\left(\lim_{n\to\infty} x_n\right) = T(\bar{x}),$$

and thus, $\bar{x}$ is a fixed point of T. By Proposition 2.2, $\bar{x}$ is a unique fixed point of T. ∎

Remark 2.2 (a) If X is not complete in Theorem 2.1, then T may not have a fixed point. For example, consider the metric space $X = (0, 1)$ with the usual metric and the mapping $T : X \to X$ is defined by $T(x) = \frac{x}{2}$. Then, X is not a complete metric space and T does not have any fixed point.

(b) If T is not a contraction in Theorem 2.1, then it may not have a fixed point. For example, consider the Example 2.10, where $X = [1, \infty)$ is a complete metric space with the usual metric and the mapping $T : X \to X$ defined by $T(x) = x + \frac{1}{x}$ is a contractive mapping but not a contraction. Of course, T does not have any fixed point.

Remark 2.3 If $T : X \to X$ is a contraction mapping with constant α, then for each $m \in \mathbb{N}$, $T^m : X \to X$ is also a contraction with constant α^m.

Indeed, for any $m \in \mathbb{N}$ and all $x, y \in X$, we have

$$d\left(T^m(x), T^m(y)\right) \le \alpha d\left(T^{m-1}(x), T^{m-1}(y)\right) \le \cdots \le \alpha^m d(x, y).$$

The following example shows that if T^m is a contraction, then T may not be a contraction mapping.

Example 2.11 Let $X = \mathbb{R}$ be a metric space with the usual metric and $T : X \to X$ be a mapping defined as

$$T(x) = \begin{cases} 1, & \text{if } x \text{ is rational,} \\ 0, & \text{if } x \text{ is irrational.} \end{cases}$$

Then, T is not continuous, and hence not a contraction mapping. Now,

$$T^2(x) = T(T(x)) = \begin{cases} T(1) = 1, & \text{if } x \text{ is rational,} \\ T(0) = 1, & \text{if } x \text{ is irrational.} \end{cases}$$

Then, T^2 is a contraction mapping but both T^2 and T have the same fixed point, which is 1.

In the above example, X is a complete metric space and $T : X \to X$ is not a contraction mapping but $T^2 = T \circ T$ is a contraction; even then T has a fixed point. It motivates us to present the following result.

Theorem 2.2 *Let (X, d) be a complete metric space and $T : X \to X$ be a mapping such that for some positive integer m, $T^m = \underbrace{T \circ T \circ \cdots \circ T}_{m\text{-times}}$ is a contraction mapping. Then, T has a unique fixed point.*

Proof By Theorem 2.1, T^m has a unique fixed point $x \in X$, that is, $T^m(x) = x$. Then by Proposition 2.1, $x \in X$ is a unique fixed point of T. ■

Exercise 2.2 Let (X, d) be a complete metric space and $x_0 \in X$. For $r > 0$, let $S_r[x_0] = \{x \in X : d(x, x_0) \le r\}$ be a closed sphere, and let $T : S_r[x_0] \to X$ be a contraction mapping on $S_r[x_0]$ with constant $\alpha \in (0, 1)$. Assume that

$$d(T(x_0), x_0) \le (1 - \alpha)r. \tag{2.12}$$

Then, prove that the following statements hold:

(a) T has a unique fixed point $\bar{x} \in S_r[x_0]$.

(b) For x_0, the sequence $\{x_n\}$ in $S_r[x_0]$ defined by

$$x_{n+1} = T(x_n), \quad \text{for all } n = 0, 1, 2, \ldots \tag{2.13}$$

converges to $\bar{x}$.

Proof (a) We first show that T is mapping from $S_r[x_0]$ into itself. For this, let $x \in S_r[x_0]$. Then,

$$\begin{aligned} d(T(x), x_0) &\leq d(T(x), T(x_0)) + d(T(x_0), x_0) \\ &\leq \alpha d(x, x_0) + (1 - \alpha)r \\ &\leq \alpha r + (1 - \alpha)r = r. \end{aligned}$$

Thus, $T(x) \in S_r[x]$, and therefore T is a contraction mapping with constant α from $S_r[x_0]$ into itself. Then by Banach Contraction Principle 2.1, T has a unique fixed point $\bar{x} \in S_r[x_0]$.
(b) Since $T : S_r[x_0] \to S_r[x_0]$ is a contraction mapping, it follows that the sequence $\{x_n\}$ in $S_r[x_0]$ defined by (2.13) converges to $\bar{x}$. ∎

Remark 2.4 A contractive mapping on a complete metric space into itself may not have a fixed point. For example, consider the metric space $X = [0, \infty)$ with the usual metric and the mapping $T : X \to X$ is given by $T(x) = x + \frac{1}{1+x}$. Then, X is complete and T is contractive but T has no fixed point in X. Note that T is not a contraction mapping.

It is natural to ask whether the Banach Contraction Principle 2.1 could be modified if the contraction condition (2.7) holds for sufficiently close points only. To give the answer of this question, we present the following definitions and results.

Definition 2.3 Let (X, d) be a metric space and $\varepsilon > 0$ be given. A finite sequence $\{x_0, x_1, \ldots, x_n\}$ of elements of X is called an *ε-chain* joining x_0 and x_n if

$$d(x_{i-1}, x_i) < \varepsilon, \quad \text{for all } i = 1, 2, \ldots, n.$$

The metric space (X, d) is said to be *ε-chainable* if for every pair (x, y) of points of X, there exists an ε-chain joining x and y.

Definition 2.4 Let (X, d) be a metric space. A mapping $T : X \to X$ is said to be *locally contraction* if for every $z \in X$, there exist $\varepsilon > 0$ and $\alpha \in [0, 1)$, which may depend on z, such that

$$d(T(x), T(y)) \leq \alpha d(x, y), \quad \text{for all } x, y \in S_\varepsilon(z) := \{u \in X : d(z, u) < \varepsilon\}. \tag{2.14}$$

Definition 2.5 Let (X, d) be a metric space. A mapping $T : X \to X$ is said to be *(ε, α)-uniformly locally contraction* if it is locally contraction and both $\varepsilon > 0$ and $\alpha \in [0, 1)$ do not depend on z, that is, if there exist $\varepsilon > 0$ and $\alpha \in [0, 1)$ such that

$$d(T(x), T(y)) \leq \alpha d(x, y), \quad \text{for all } x, y \in X \text{ with } d(x, y) < \varepsilon.$$

Theorem 2.3 [72] *Let (X, d) be a complete ε-chainable metric space and $T : X \to X$ be a (ε, α)-uniformly locally contraction mapping. Then, T has a unique fixed point $\bar{x} \in X$ and $\bar{x} = \lim_{n\to\infty} T^n(x_0)$, where x_0 is an arbitrary element in X.*

Proof Let x be an arbitrary point in X. Consider the ε-chain $x = x_0, x_1, \dots, x_n = T(x)$. Then,

$$d(x_{i-1}, x_i) < \varepsilon, \quad \text{for all } i = 1, 2, \dots, n.$$

By the triangle inequality, we have

$$d(x, T(x)) \leq \sum_{i=1}^{n} d(x_{i-1}, x_i) < n\varepsilon.$$

For each pair of consecutive points of the ε-chain, the (ε, α)-uniformly contraction of T implies that

$$d(T(x_{i-1}), T(x_i)) \leq \alpha d(x_{i-1}, x_i) < \alpha\varepsilon.$$

Inductively, we have

$$d\left(T^m(x_{i-1}), T^m(x_i)\right) \leq \alpha d\left(T^{m-1}(x_{i-1}), T^{m-1}(x_i)\right) < \alpha^m \varepsilon, \quad \text{for all } m \in \mathbb{N}. \tag{2.15}$$

Therefore, we obtain

$$d\left(T^m(x), T^{m+1}(x)\right) \leq \sum_{i=1}^{n} d\left(T^m(x_{i-1}), T^m(x_i)\right) < \alpha^m n\varepsilon. \tag{2.16}$$

It follows that the sequence of iterates $\{T^i(x)\}$ is a Cauchy sequence. Indeed, if j and k $(j < k)$ are positive integers, then

$$\begin{aligned} d\left(T^j(x), T^k(x)\right) &\leq \sum_{i=j}^{k-1} d\left(T^i(x), T^{i+1}(x)\right) \\ &< \left(\alpha^j + \alpha^{j+1} + \cdots + \alpha^{k-1}\right) n\varepsilon \\ &< \frac{\alpha^j}{1-\alpha} n\varepsilon \to 0 \text{ as } j \to \infty. \end{aligned}$$

By the completeness of X, we have $\lim_{i\to\infty} T^i(x) = \bar{x} \in X$. Since T is continuous, we have

$$T(\bar{x}) = T\left(\lim_{i\to\infty} T^i(x)\right) = \lim_{i\to\infty} T^{i+1}(x) = \lim_{i\to\infty} T^i(x) = \bar{x}.$$

To prove the uniqueness of the fixed point $\bar{x}$, we assume the contrary that there exists another fixed point $\tilde{x} \neq \bar{x}$ of T. Then, $d(\bar{x}, \tilde{x}) > 0$. Let $\bar{x} = x_0, x_1, \dots, x_k = \tilde{x}$ be an ε-chain. Then, from (2.15), we have

$$\begin{aligned} d(\bar{x}, \tilde{x}) &= d\left(T^m(\bar{x}), T^m(\tilde{x})\right) \\ &\leq \sum_{i=1}^{k} d\left(T^m(x_{i-1}), T^m(x_i)\right) \\ &< \alpha^m k\varepsilon \to 0 \text{ as } m \to \infty, \end{aligned}$$

a contradiction. Therefore, $\bar{x} = \tilde{x}$. ∎

The following theorem guarantees the existence of a fixed point for a contractive mapping defined on a compact metric space.

Theorem 2.4 *Let (X, d) be a compact metric space and $T : X \to X$ be a contractive mapping. Then, T has a unique fixed point $\bar{x}$. Moreover, for any $x \in X$, $\lim_{n\to\infty} T^n(x) = \bar{x}$, that is, the successive iterates $T(x), T^2(x), \dots, T^n(x), \dots$ converge to the unique fixed point $\bar{x}$ of T.*

Proof Define a mapping $\psi : X \to [0, \infty)$ by

$$\psi(x) = d(x, T(x)), \quad \text{for all } x \in X.$$

Then ψ is continuous. Indeed, by contractiveness of T, we have

$$\begin{aligned} |\psi(x) - \psi(y)| &= |d(x, T(x)) - d(y, T(y))| \\ &\le |d(x, T(x)) - d(T(x), y)| + |d(T(x), y) - d(y, T(y))| \\ &\le d(x, y) + d(T(x), T(y)) \\ &< 2d(x, y). \end{aligned}$$

Let $\varepsilon > 0$ be given. Then,

$$|\psi(x) - \psi(y)| < 2d(x, y) < \varepsilon \quad \text{whenever} \quad d(x, y) < \delta = \frac{\varepsilon}{2}.$$

Therefore, ψ is continuous. Clearly, ψ is bounded below. Since X is compact and ψ is continuous, then by Theorem A.6, there exists a minimizer $\bar{x} \in X$ of ψ, that is, there exists $\bar{x} \in X$ such that $\psi(\bar{x}) \le \psi(y)$ for all $y \in X$. We show that $\bar{x}$ is a fixed point of T.

Suppose the contrary that $\bar{x}$ is not a fixed point of T. Then, $T(\bar{x}) \neq \bar{x}$. By contractiveness of T, we have

$$\psi\left(T(\bar{x})\right) = d\left(T(\bar{x}), T^2(\bar{x})\right) < d(\bar{x}, T(\bar{x})) = \psi\left(\bar{x}\right),$$

which contradicts that $\bar{x}$ is a minimizer of ψ. Hence, $\bar{x}$ is a fixed point of T. The uniqueness follows on the lines of the proof of Proposition 2.2.

Let $x \in X$. If $T^n(x) \neq \bar{x}$; then

$$d\left(T^{n+1}(x), \bar{x}\right) = d\left(T^{n+1}(x), T(\bar{x})\right) < d(T^n(x), \bar{x}), \quad \text{for all } n \in \mathbb{N}.$$

Therefore, $\left\{d\left(T^{n+1}(x), \bar{x}\right)\right\}$ is a strictly decreasing sequence of nonnegative real numbers and so converges to its infimum. Suppose that $a = \lim_{n\to\infty} d\left(T^{n+1}(x), \bar{x}\right)$. Since $\{T^n(x)\}$ is a sequence of points of a compact metric space, there exists a subsequence $\{T^{n_k}(x)\}$ which converges to some point, say, $y \in X$. Now,

$$d(y, \bar{x}) = \lim_{k\to\infty} d\left(T^{n_k}(x), \bar{x}\right) = \lim_{k\to\infty} d\left(T^{n_k+1}(x), \bar{x}\right) = a.$$

If $a \neq 0$, then $y \neq \bar{x}$, and hence,

$$\begin{aligned} a = d(y, \bar{x}) &> d(T(y), T(\bar{x})) = d(T(y), \bar{x}) \\ &= \lim_{k\to\infty} d\left(T(T^{n_k}(x)), \bar{x}\right) = \lim_{k\to\infty} d\left(T^{n_k+1}(x), \bar{x}\right) = a, \end{aligned}$$

a contradiction. Thus, $a = 0$, and therefore, $\lim_{n\to\infty} T^n(x) = \bar{x}$. ■

Definition 2.6 Let (X, d) be a metric space. A mapping $T : X \to X$ is said to be ε-*contractive* if there exists $\varepsilon > 0$ such that

$$d(T(x), T(y)) < d(x, y), \quad \text{for all } x \neq y \in X \text{ with } d(x, y) < \varepsilon.$$

Theorem 2.5 [73] *Let (X, d) be a metric space and $T : X \to X$ be an ε-contractive mapping. Let $x_0 \in X$ be a point such that the sequence $\{T^n(x_0)\}$ has a subsequence that converges to a point $\bar{x}$ of X. Then, $\bar{x}$ is a periodic point of T, that is, there exists a positive integer k such that $T^k(\bar{x}) = \bar{x}$.*

Proof Let $\{n_i\}$ be a strictly increasing sequence in $\mathbb{N}$ and $\{T^{n_i}(x_0)\}$ be a subsequence of the sequence $\{T^n(x_0)\}$ that converges to $\bar{x}$. Define

$$x_i = T^{n_i}(x_0), \quad \text{for all } i \in \mathbb{N}.$$

Then for given $\varepsilon > 0$, there exists $N \in \mathbb{N}$ such that

$$d(x_i, \bar{x}) = d\left(T^{n_i}(x_0), \bar{x}\right) < \frac{\varepsilon}{4}, \quad \text{for all } i \geq N.$$

Choose any $i \geq N$ and let $k = n_{i+1} - n_i$. Then,

$$\begin{aligned} d\left(x_{i+1}, T^k(\bar{x})\right) &= d\left(T^{n_{i+1}}(x_0), T^k(\bar{x})\right) = d\left(T^{k+n_i}(x_0), T^k(\bar{x})\right) \\ &= d\left(T^k(x_i), T^k(\bar{x})\right) < d(x_i, \bar{x}) < \frac{\varepsilon}{4}. \end{aligned}$$

Hence, by the triangle inequality, we have

$$d\left(\bar{x}, T^k(\bar{x})\right) \leq d(\bar{x}, x_{i+1}) + d\left(x_{i+1}, T^k(\bar{x})\right) < \frac{\varepsilon}{4} + \frac{\varepsilon}{4} = \frac{\varepsilon}{2}.$$

Assume the contrary that $\tilde{x} = T^k(\bar{x}) \neq \bar{x}$. Then, $0 < d(\bar{x}, \tilde{x}) = d(\bar{x}, T^k(\bar{x}) < \frac{\varepsilon}{2}$. Since T is ε-contractive, we have

$$d(T(\bar{x}), T(\tilde{x})) < d(\bar{x}, \tilde{x}), \quad \text{or} \quad \frac{d(T(\bar{x}), T(\tilde{x}))}{d(\bar{x}, \tilde{x})} < 1.$$

Set $Y := X \times X \setminus \Delta$, where $\Delta = \{(x, y) \in X \times X : x = y\}$ is the diagonal. Define $r : Y \to \mathbb{R}$ by

$$r(x, y) = \frac{d(T(x), T(y))}{d(x, y)}, \quad \text{for all } (x, y) \in Y.$$

Then clearly, r is continuous on Y, and hence, r is continuous at $(\bar{x}, \tilde{x})$ with $r(\bar{x}, \tilde{x}) < 1$. So there exist $\delta > 0$ and $\alpha \in (0, 1)$ such that

$$d(x, \bar{x}) < \delta, \; d(y, \tilde{x}) < \delta \quad \Rightarrow \quad d(T(x), T(y)) < \alpha d(x, y).$$

Since $\lim_{j \to \infty} T^k(x_j) = T^k(\bar{x}) = \tilde{x}$, there exists $N' \geq N$ such that

$$d(x_j, \bar{x}) < \delta \quad \text{and} \quad d\left(T^k(x_j), \tilde{x}\right) < \delta, \quad \text{for all } j \geq N'.$$

Hence,

$$d\left(T(x_j), T\left(T^k(x_j)\right)\right) < \alpha d\left(x_j, T^k(x_j)\right). \tag{2.17}$$

By the triangular inequality, we have

$$d\left(x_j, T^k(x_j)\right) \le d(x_j, \bar{x}) + d\left(\bar{x}, T^k(x_j)\right) + d\left(T^k(x_j), T^k(x_j)\right)$$
$$< \frac{\varepsilon}{4} + \frac{\varepsilon}{2} + \frac{\varepsilon}{4} = \varepsilon, \quad \text{for all } j \ge N' > N. \tag{2.18}$$

From (2.17) and (2.18), we obtain

$$d\left(T(x_j), T(T^k(x_j))\right) < \alpha d\left(x_j, T^k(x_j)\right) < \alpha\varepsilon < \varepsilon, \quad \text{for all } j \ge N'.$$

Since T is ε-contractive, we have

$$d\left(T^m(x_j), T^m(T^k(x_j))\right) < \alpha d\left(x_j, T^k(x_j)\right), \quad \text{for all } j \ge N', \tag{2.19}$$

where $m \in \mathbb{N}$. Set $m = n_{j+1} - n_j$ in (2.19), we get

$$d\left(x_{j+1}, T^k(x_{j+1})\right) = d\left(T^{n_{j+1}-n_j}(x_j), T^{n_{j+1}-n_j}(T^k(x_j))\right)$$
$$< \alpha d\left(x_j, T^k(x_j)\right) < \alpha\varepsilon, \quad \text{for all } j \ge N'.$$

Hence, for $j, s \ge N'$ with $s \ge j$, we have

$$d\left(x_s, T^k(x_s)\right) < \alpha d\left(x_{s-1}, T^k(x_{s-1})\right) < \cdots < \alpha^{s-j} d\left(x_j, T^k(x_j)\right) < \alpha^{s-j}\varepsilon.$$

Thus,

$$d(\bar{x}, \tilde{x}) \le d(\bar{x}, x_s) + d\left(x_s, T^k(x_s)\right) + d\left(T^k(x_s), \tilde{x}\right) \to 0 \text{ as } s \to \infty,$$

a contradiction. Therefore, $\bar{x} = T^k(\bar{x})$. ■

Exercise 2.3 Let (X, d) be a complete metric space and $T : X \to X$ be a mapping such that

$$d\left(T(x), T(y)\right) \ge \alpha d(x, y), \quad \text{for some } \alpha > 1 \text{ and for all } x, y \in X.$$

Assume that $T(X) = X$. Prove that T is one-to-one. Moreover, prove that T has a unique fixed point $\bar{x} \in X$ with $\lim_{n\to\infty} T^{-n}(x) = \bar{x}$ for some $x \in X$.

2.2.1 Characterization of Completeness in Terms of the Banach Contraction Principle

In 1959, Bessage [31] established the converse of the Banach contraction principle that provides the characterization of completeness of the metric space. There are at least three different proofs of such a result. The first one, given by Bessage [31], uses a special form of the Axiom of Choice. The second one, from the book [71], is a special case of that given by Wong [175], and it uses Zorn's lemma. In fact, Wong [175] extended Bessage's theorem to a finite family of commuting mappings. The third one, due to Janos [103], is based on combinatorial techniques with a use of Ramsey's theorem.

Jachymski [99] gave the simple proof of Bessage's theorem by modifying the proof presented in the book [71]. Such proof is presented here.

Proposition 2.3 [99] *Let X be a nonempty set, $T : X \to X$ be a mapping and $\alpha \in (0, 1)$. Then, the following statements are equivalent:*

(a) *There exists a complete metric d on X such that*

$$d(T(x), T(y)) \leq \alpha d(x, y), \quad \text{for all } x, y \in X. \tag{2.20}$$

(b) *There exists a function $\varphi : X \to [0, \infty)$ such that $\varphi^{-1}(0)$ is a singleton set and the inequality*

$$\varphi(T(x)) \leq \alpha\varphi(x), \quad \text{for all } x \in X \tag{2.21}$$

holds.

Proof (a) ⇒ (b). Suppose that there exists a complete metric d on X such that (2.20) holds. Then by Banach Contraction Principle 2.1, T has a unique fixed point $\bar{x} \in X$. Define $\varphi : X \to [0, \infty)$ by

$$\varphi(x) = d(x, \bar{x}), \quad \text{for all } x \in X.$$

Then,

$$\varphi(T(x)) = d(T(x), \bar{x}) = d(T(x), T(\bar{x})) \leq \alpha d(x, \bar{x}) = \alpha\varphi(x), \quad \text{for all } x \in X,$$

that is, the inequality (2.21) holds. It is easy to see that $\varphi^{-1}(0)$ is a singleton set.

(b) ⇒ (a). Suppose that there exists a function $\varphi : X \to [0, \infty)$ such that $\varphi^{-1}(0)$ is a singleton set and the inequality (2.21) holds. Define $d : X \times X \to [0, \infty)$ by

$$d(x, y) = \begin{cases} \varphi(x) + \varphi(y), & \text{if } x \neq y, \\ 0, & \text{if } x = y. \end{cases}$$

It is easy to see that d is a metric on X. For $x, y \in X$, from (2.21), we have

$$d(T(x), T(y)) \leq \alpha d(x, y).$$

We now claim that (X, d) is complete. Let $\{x_n\}$ be a Cauchy sequence in X. Without loss of generality, we may assume that the set $\{x_n : n \in \mathbb{N}\}$ is infinite, otherwise $\{x_n\}$ would contain a constant subsequence and so it will converge. Therefore, there is a subsequence $\{x_{n_i}\}$ of distinct elements of $\{x_n\}$ such that

$$d(x_{n_i}, x_{n_j}) = \varphi(x_{n_i}) + \varphi(x_{n_j}), \quad \text{for } i \neq j.$$

Hence, $\varphi(x_{n_i}) \to 0$. By the hypothesis, $\varphi^{-1}(0) = \{z\}$ for some $z \in X$. Since

$$d(x_{n_i}, z) = \varphi(x_{n_i}) + \varphi(z) = \varphi(x_{n_i}) \to 0,$$

we conclude that $\{x_n\}$ converges to z. Therefore, (X, d) is complete. ■

Theorem 2.6 [31] *Let X be a nonempty set, $\alpha \in (0, 1)$ and $T : X \to X$ be a mapping such that T^m has a unique fixed point $z \in X$ for every $m \geq 1$. Then, there exists a complete metric d on X such that*

$$d(T(x), T(y)) \leq \alpha d(x, y), \quad \text{for all } x, y \in X. \tag{2.22}$$

Proof Let Z be the subset of X consisting of all elements $x \in X$ such that $T^k(x) = z$ for some $k \in \mathbb{N}$. Define an equivalence relation on the set $X \setminus Z$ as follows:

$$x \sim y \quad \Leftrightarrow \quad T^n(x) = T^m(y), \quad \text{for some } m, n \in \mathbb{N}.$$

For $x \in X$, let $[x]$ denote an equivalence class, that is,

$$[x] = \{y \in X : y \sim x\}.$$

Note that if $T^n(x) = T^m(y)$ and $T^{n'}(x) = T^{m'}(y)$, then

$$T^{n+m'}(x) = T^{m+n'}(x).$$

Since $x \notin Z$, we have $n+m' = m+n'$, that is, $n-m = n'-m'$. At this point, by means of the axiom of choice, we select an element from each equivalence class. We now define the distance of z from a generic $x \in X$ by

$$d(x, z) = \begin{cases} 0, & \text{if } x = z, \\ \alpha^{-j}, & \text{if } x \in Z \text{ with } x \neq z, \\ \alpha^{n-m}, & \text{if } x \notin Z, \end{cases}$$

where $j = \min\{p \geq 1 : T^p(x) = z\}$ and $m, n \in \mathbb{N}$ with $T^n(\bar{x}) = T^m(x)$, where $\bar{x}$ is the selected representative of the equivalence class $[x]$. Clearly, d is well defined. Now, for any $x, y \in X$, we set

$$d(x, y) = \begin{cases} d(x, z) + d(y, z), & \text{if } x \neq y, \\ 0, & \text{if } x = y. \end{cases}$$

It can be easily seen that d is a metric. We also observe that the only Cauchy sequences that do not converge to z are ultimately constant. We now claim that T is a contraction. Let $x \in X$ with $x \neq z$. Then we have the following cases:

CASE I: If $x \in Z$ and $T(x) = z$, then

$$0 = d(T(x), z) < \alpha d(x, z).$$

CASE II: If $x \in Z$ and $T(x) \neq z$, then since there is a smallest $j \geq 2$ such that $z = T^j(x) = T^{j-1}(T(x))$, it follows that

$$d(T(x), z) = \alpha^{-j+1} = \alpha d(x, z).$$

CASE III $x \notin Z$. Suppose that $\bar{x}$ is the representative of $[x]$. Note there exist $m, n \in \mathbb{N}$ such that

$$z = T^n(\bar{x}) = T^m(x) = T^{m-1}T(x).$$

It follows that $d(T(x), z) = \alpha^{n-m+1} = \alpha d(x, z)$.

Thus, we see that

$$d(T(x), z) \leq \alpha d(x, z), \quad \text{for all } x \in X.$$

For $x, y \in X$ with $x \neq y$, we have

$$\begin{aligned} d(T(x), T(y)) &\leq d(T(x), z) + d(T(y), z) \\ &\leq \alpha[d(x, z) + d(y, z)] = \alpha d(x, y). \end{aligned}$$

Therefore, T is a contraction. ■

The following interesting result is established by Janos [102].

Theorem 2.7 *Let (X, d) be a compact metric space and $T : X \to X$ be a continuous mapping such that $\bigcap_{n=1}^{\infty} T^n(X) = \{x_0\}$. Then for each $\alpha \in (0, 1)$, there exists a metric d_α such that (X, d_α) is a compact space and T is a contraction mapping with respect to d_α.*

Proof We divide the proof into several claims.

CLAIM 1. There exists a new metric d^* which is equivalent to d and satisfies the following property:

$$d^*(T(x), T(y)) \leq d^*(x, y), \quad \text{for all } x, y \in X. \tag{2.23}$$

Define $d^* : X \times X \to [0, \infty)$ by

$$d^*(x, y) = \sup_{n=0}^{\infty} d(T^n(x), T^n(y)), \quad \text{for all } x, y \in X.$$

Then, obviously d^* is a metric on X and satisfies (2.23).

To prove that d^* is equivalent to d, it is sufficient to show that any Cauchy sequence with respect to d^* is a Cauchy sequence with respect to d and vice versa. Since $d^*(x, y) \geq d(x, y)$, it follows that any Cauchy sequence with respect to d^* is a Cauchy sequence with respect to d. To prove the converse, let $\{x_n\}$ be a Cauchy sequence with respect to d. Since X is compact, we can assume that $x_n \xrightarrow{d} x$. Now we show that $x_n \xrightarrow{d^*} x$. Assume the contrary that it is not true. Then, there exists $\varepsilon > 0$ such that $d^*(x_{n_k}, x) \geq \varepsilon$. Without loss of generality, we can denote this subsequence again by $\{x_n\}$. We define a sequence $\{k_n\}$ of integers such that

$$d^*(x_n, x) = d\left(T^{k_n}(x_n), T^{k_n}(x)\right)$$

and if $\{k_n\}$ is bounded, then there exists some k which is repeated infinitely often and thus, using the continuity of T with respect to d, we obtain

$$d\left(T^k(x_n), T^k(x)\right) \geq \varepsilon,$$

which is a contradiction.

If $\{k_n\}$ is unbounded, then this contradicts the property

$$\bigcap_{n=1}^{\infty} T^n(X) = \{x_0\}.$$

This proves the claim.

CLAIM 2. For each $\alpha \in (0, 1)$, there exists a metric d_α which is equivalent to d and satisfies the following property:

$$d_\alpha(T(x), T(y)) \leq \alpha d_\alpha(x, y), \quad \text{for all } x, y \in X. \tag{2.24}$$

Consider the sequence of sets

$$M_0 = X,\ M_1 = T(X),\ \dots,\ M_n = T^n(X), \dots$$

and define the functions

$$n(x) = \max\{n : x \in M_n\}$$

and with this function

$$n(x, y) = \min\{n(x), n(y)\}.$$

Then, obviously, we have

$$n(T(x), T(y)) \geq n(x, y) + 1.$$

We define a mapping $d_\alpha : X \times X \to [0, \infty)$ by

$$d_\alpha(x, y) = \alpha^{n(x,y)} d^*(x, y), \quad \text{for all } x, y \in X.$$

Then, it is easy to see that this mapping is not a metric on X, but satisfies the following property:

$$d_\alpha(T(x), T(y)) < \alpha d_\alpha(x, y), \quad \text{for all } x, y \in X.$$

We modify this mapping d_α to obtain a metric in the following way: For any $x, y \in X$, consider a chain

$$C_{x,y} = \{x = x_0, x_1, \dots, x_n = y\},$$

and put

$$D_\alpha(C_{x,y}) = \sum_{i=0}^{n} d_\alpha(x_{i-1}, x_i).$$

Now we define the mapping $d_\alpha^* : X \times X \to [0, \infty)$ by

$$d_\alpha^*(x, y) = \inf\{D_\alpha(C_{x,y})\}, \quad \text{for all } x, y \in X,$$

where the inf is taken over all possible chains. Then, we show that d_α^* is the required metric.

Clearly, for all $x, y, z \in X$, we have

$$d_\alpha^*(x, x) = 0 \quad \text{and} \quad d_\alpha^*(x, y) = d_\alpha^*(y, x),$$

and the triangle inequality follows from the fact that the union of chains $C_{x,y}$ with chains $C_{y,x}$ is a subset of the set of chains $X_{x,z}$

To show that $d_\alpha^*(x, y) > 0$ for all $x \neq y$, we may assume, without loss of generality, that $n(x) < n(y)$. Now if $y = x_0$, then,

$$d_\alpha^*(x, x_0) \geq \alpha^{n(y)} \min\left\{d^*(x, y)\, d^*(y, M_{n(y)+1})\right\},$$

which is positive. If $y \neq x_0$, then every chain $C_{x,y}$ lies entirely in the set $M_{n(y)+1}$ or it does not; hence

$$d^*(x, y) \geq \alpha^{n(y)} \min\left\{d^*(x, y)\, d^*(y, M_{n(y)+1})\right\}$$

and the assertion that d_α^* is a metric is proved.

Now we prove that these metrics are equivalent and we remark that T is a contraction mapping with respect to d^*_α.

It is clear from the definition of d^*_α that $d^*_\alpha \le d_\alpha \le d^*$. To prove that d^*_α and d^* are equivalent, we show that every Cauchy sequence $\{x_n\}$ with respect to d^*_α is a Cauchy sequence with respect to d^*.

Since X is a compact space with respect to d^*, we can choose a subsequence, again denoted by $\{x_n\}$, that converges to a point y. We note that $x_n \xrightarrow{d^*_\alpha} x$. Since $x_n \xrightarrow{d^*_\alpha} y$, we obtain a contradiction and hence the assertion is proved.

Since $D_\alpha(T(C_{x,y})) \le \alpha D_\alpha(C_{x,y})$, we have, by taking inf, that T is a contraction mapping with respect to d^*_α. ■

Theorem 2.8 *Let (X, d) be a compact metric space and $T : X \to X$ be a continuous mapping such that $\{x_0\}$ is the only nonempty invariant set in X. Then, there exists an equivalent metric d^* on X such that (X, d^*) is a compact space and T is a contraction mapping with respect to d^*.*

Proof Since the sequence $\{T^n(X)\}$ of compact sets is decreasing and X is compact, X has nonempty intersection. Let $M = \bigcap_{n=1}^{\infty} T^n(X)$. By Theorem 2.7, it suffices to prove that $T(M) = M$.

It is obvious that $T(M) \subseteq M$. For the converse, let $x \in M$ be arbitrary. Consider the sequence $\{G_n\}$ of the sets

$$G_n = T^{-1}(x) \cap T^n(X)$$

which are nonempty and compact. Note that $\emptyset \neq \bigcap_{n=1}^{\infty} G_n = T^{-1}(x) \cap M$ gives us that $x \in T(M)$. ■

2.2.2 Criteria for Contraction Mappings

The following result provides the criteria to examine whether a real-valued differentiable function is a contraction or not.

Theorem 2.9 *Let $X = [a, b]$ be a metric space with the usual metric and $T : [a, b] \to [a, b]$ be a differentiable function on (a, b). Then, T is a contraction mapping on X if and only if there exists a real number $\alpha \in (0, 1)$ such that $|T'(x)| \le \alpha$ for all $x \in (a, b)$.*

Proof Let T be a contraction mapping. Then, there exists $\alpha \in (0, 1)$ such that

$$|T(x) - T(y)| \le \alpha |x - y|, \quad \text{for all } x, y \in [a, b].$$

Then for any x and $x + \Delta x$ in $[a, b]$, we have

$$\big|T(x + \Delta x) - T(x)\big| \le \alpha\, |x + \Delta x - x| = \alpha |\Delta x|.$$

Hence, for $\Delta x \neq 0$, we have

$$\left|\frac{T(x + \Delta x) - T(x)}{\Delta x}\right| \le \alpha,$$

and therefore,

$$|T'(x)| = \lim_{\Delta x \to 0} \left|\frac{T(x + \Delta x) - T(x)}{\Delta x}\right| \le \alpha.$$

Conversely, assume that there exists a real number $\alpha \in (0,1)$ such that $|T'(x)| \leq \alpha$ for all $x \in (a,b)$. By mean value theorem, for any $x \neq y$ in $[a,b]$, there exists $c \in (a,b)$ such that

$$T'(c) = \frac{T(x) - T(y)}{x - y}.$$

Since $|T'(c)| \leq \alpha$, we have

$$\left| \frac{T(x) - T(y)}{x - y} \right| = |T'(c)| \leq \alpha,$$

and hence,

$$|T(x) - T(y)| \leq \alpha |x - y|.$$

Therefore, T is a contraction mapping on $[a,b]$. ■

Remark 2.5 The conclusion of Theorem 2.9 does not necessarily hold if the domain of T is the entire real line.

Example 2.12 (Fibonacci's Rabbits) Leonardo Pisano, better known as Fibonacci, considered the problem of finding the number of pairs of rabbits that can be grown from one pair in one year. He figured out that each pair breeds a pair every month, but a newborn pair only breeds in the second month after birth. Let b_n denote the number of rabbit pairs at time n. Let $b_0 = 1$ and in the first month they breed one pair, so $b_1 = 2$. At time $n = 2$, again one pair is breed (from the one that was around at time $n = 1$, the other one does not yet have the required age to breed). Hence, $b_2 = b_1 + b_0$. Subsequently, $b_{n+1} = b_n + b_{n-1}$. Expecting the growth to be exponential we would like to see how fast these numbers grow, by calculating $a_n = b_{n+1}/b_n$. Namely, if $b_n \to cd^n$ as $n \to \infty$ for some c, d, then $b_{n+1}/b_n \to d$. We have

$$a_{n+1} = \frac{b_{n+2}}{b_{n+1}} = \frac{1}{a_n} + 1.$$

Thus $\{a_n\}_{n \in \mathbb{N}}$ is the orbit of $a_0 = 1$ of the map $T(x) = \frac{1}{x} + 1$. Then $T'(x) = -\frac{1}{x^2}$, and so T is not a contraction on $(0, \infty)$. But we note that $a_1 = 2$ and consider the map T on the closed interval $[3/2, 2]$. Then $T(3/2) = 5/3 > 3/2$ and $T(2) = 3/2$, and hence $T([3/2, 2]) \subset [3/2, 2]$. Furthermore, for $x \in [3/2, 2]$, we have $|T'(x)| = 1/x^2 \leq 4/9 < 1$, and so T is a contraction mapping. Therefore, by Banach Contraction Principle 2.1, there exists a unique fixed point of T, and so $\lim\limits_{n \to \infty} a_n$ exists. The solution is a fixed point of T, which yields $x^2 - x - 1 = 0$. The only positive root of this equation is $x = (1 + \sqrt{5})/2$.

Exercise 2.4 Let $X = [1, \infty)$ be a metric space with the usual metric and $T : X \to X$ be a mapping defined by $T(x) = \frac{10}{11}\left(x + \frac{1}{x}\right)$ for all $x \in X$. Prove that T is a contraction mapping with contraction constant $\alpha = \frac{10}{11}$.

Exercise 2.5 Let $X = [0, 1]$ be a metric space with the usual metric and $T : X \to X$ be a mapping defined by

$$T(x) = \left(\frac{9}{10} + \frac{x}{5} - \frac{x^2}{4}\right)^3, \quad \text{for all } x \in X. \tag{2.25}$$

Show that the mapping T is a contraction on X.

Exercise 2.6 Let $X = [0, 1]$ be a metric space with the usual metric and $T : X \to X$ be a mapping defined by $T(x) = \frac{1}{7}\left(x^3 + x^2 + 1\right)$ for all $x \in X$. Prove that T is a contraction mapping with contraction constant $\alpha = \frac{5}{7}$.
Hint: For any $x, y \in \mathbb{R}$ with $x \neq y$, either use $\left|x^3 - y^3\right| = |x - y| \left|x^2 + xy + y^2\right|$ or Theorem 2.9.

Exercise 2.7 Let $X = \{x \in \mathbb{Q} : x \geq 1\}$ be a metric space with the usual metric and $T : X \to X$ be a mapping defined by $T(x) = \frac{x}{2} + \frac{1}{x}$ for all $x \in X$. Prove that T is a contraction mapping with contraction constant $\alpha = \frac{1}{2}$.

Exercise 2.8 Let $X = \mathbb{R}$ be a metric space with the usual metric and $T : X \to X$ be a mapping defined by $T(x) = \cos x$ for all $x \in X$. Show that the mapping T is not a contraction but contractive. Also show that the mapping $T(x) = \frac{99}{100} \cos x$ for all $x \in X$ is a contraction mapping on X.
Hint:

$$\begin{aligned}
|\cos x - \cos y| &= \left|2 \sin\left(\frac{x-y}{2}\right) \sin\left(\frac{x+y}{2}\right)\right| \\
&= \left|2 \sin\left(\frac{x-y}{2}\right)\right| \left|\sin\left(\frac{x+y}{2}\right)\right| \\
&\leq 2\left|\sin\left(\frac{x-y}{2}\right)\right| \\
&< 2\left|\frac{x-y}{2}\right| \quad (\text{since } |\sin w| < |w| \text{ for all } w \neq 0) \\
&= |x - y|.
\end{aligned}$$

Exercise 2.9 Let $X = \mathbb{R}$ be a metric space with the usual metric and $T : X \to X$ be a mapping defined by $T(x) = \sin x$ for all $x \in X$. Show that the mapping T is contractive.

Exercise 2.10 Let $X = \mathbb{R}^3$ be a metric space with the usual metric and $T : X \to X$ be a mapping defined by $T(x) = T(x_1, x_2, x_3) = \left(\frac{1}{2} \cos x_2 + 1, \frac{2}{3} \sin x_3, \frac{3}{4} x_1\right)$ for all $x = (x_1, x_2, x_3) \in X$. Show that the mapping T is a contraction on X and deduce that the simultaneous equations

$$x_1 = \frac{1}{2}\cos x_2 + 1,$$
$$x_2 = \frac{2}{3}\sin x_3,$$
$$x_3 = \frac{3}{4}x_1$$

have a unique solution.

2.3 Some Extension of the Banach Contraction Principle

2.3.1 ψ-contraction Mappings and Fixed Point Results

As a generalization of the contraction condition, Boyd and Wong [43] introduced the following definition of ψ-contraction mappings.

Definition 2.7 Let (X, d) be a metric space and $\psi : [0, \infty) \to [0, \infty)$ be an upper semicontinuous function such that $0 \leq \psi(t) < t$ for all $t > 0$. A mapping $T : X \to X$ is said to be *ψ-contraction* if

$$d(T(x), T(y)) \leq \psi(d(x, y)), \quad \text{for all } x, y \in X. \tag{2.26}$$

Remark 2.6 (a) $\psi(t) = \frac{t}{1+t}$ for all $t \geq 0$ satisfies the condition in Definition 2.7.

(b) If $\psi(t) = \alpha t$ for all $t \geq 0$, where $\alpha \in (0, 1)$, then every ψ-contraction mapping is contraction with constant α.

(c) Every ψ-contraction mapping is nonexpansive.

Theorem 2.10 (Boyd–Wong Fixed Point Theorem) *Let (X, d) be a complete metric space and $T : X \to X$ be a ψ-contraction mapping, where $\psi : [0, \infty) \to [0, \infty)$ is an upper semicontinuous function such that $0 \leq \psi(t) < t$ for all $t > 0$. Then, T has a unique fixed point $\bar{x} \in X$. Moreover, $\{T^n(x)\}$ converges to $\bar{x}$ for all $x \in X$.*

Proof For any fixed $x \in X$, define a sequence $\{x_n\}$ in X by $x_n = T^n(x)$ for all $n \in \mathbb{N}$. Set $a_n := d(x_n, x_{n+1}) = d\left(T^n(x), T^{n+1}(x)\right)$ for all $n \in \mathbb{N}$. We divide the proof into three steps:

STEP 1. We prove $\lim_{n\to\infty} a_n = 0$.

For all $n > 1$, we have

$$\begin{aligned} a_n = d(x_n, x_{n+1}) &= d(T(x_{n-1}), T(x_n)) \\ &\leq \psi\left(d(x_{n-1}, x_n)\right) = \psi(a_{n-1}) < a_{n-1}. \end{aligned}$$

Hence, $\{a_n\}$ is monotonically decreasing and bounded below, so it is convergent, that is, $\lim_{n\to\infty} a_n$ exists. Let $\lim_{n\to\infty} a_n = a \geq 0$. Assume that $a > 0$. By the upper semicontinuity of ψ, we have

$$a = \lim_{n\to\infty} a_{n+1} \leq \lim_{n\to\infty} \psi(a_n) \leq \psi(a) < a,$$

a contradiction. Hence, $a = 0$.

STEP 2. We show that $\{x_n\}$ is a Cauchy sequence.

Assume that $\{x_n\}$ is not a Cauchy sequence. Then, there exists $\varepsilon > 0$ such that for any $k \in \mathbb{N}$, there exist $m_k, n_k \in \mathbb{N}$ with $m_k > n_k \geq k$ and

$$d(x_{n_k}, x_{m_k}) \geq \varepsilon, \quad \text{for all } k = 1, 2, \dots . \tag{2.27}$$

Assume that for each k, m_k is the smallest number greater than n_k for which (2.27) holds. Let $d_k := d(x_{n_k}, x_{m_k})$. Since $\lim_{n\to\infty} d(x_n, x_{n+1}) = \lim_{n\to\infty} a_n = 0$, there exists k_0 such that

$$d(x_{m_k-1}, x_{n_k}) < \varepsilon, \quad \text{for all } k \geq k_0.$$

Therefore, for all $k \geq k_0$, we have

$$\begin{aligned} \varepsilon \leq d_k &\leq d(x_{m_k}, x_{m_k-1}) + d(x_{m_k-1}, x_{n_k}) \\ &< a_{m_k-1} + \varepsilon. \end{aligned}$$

Thus, $d_k \to \varepsilon$ from above as $k \to \infty$. Moreover,

$$\begin{aligned} d_k &\leq d(x_{m_k+1}, x_{n_k+1}) + d(x_{m_k+1}, x_{m_k}) + d(x_{n_k+1}, x_{n_k}) \\ &\leq d(T(x_{m_k}), T(x_{n_k})) + a_{m_k} + a_{n_k} \\ &\leq \psi(d_k) + a_{m_k} + a_{n_k}. \end{aligned}$$

Letting $k \to \infty$ and using the upper semicontinuity of ψ, we obtain

$$\begin{aligned} \varepsilon &= \lim_{k\to\infty} d_k = \lim_{k\to\infty} d(x_{m_k}, x_{n_k}) \\ &\leq \lim_{k\to\infty} \psi(d_k) + \lim_{k\to\infty} (a_{m_k} + a_{n_k}) \leq \psi(\varepsilon), \end{aligned}$$

a contradiction of the hypothesis since $\varepsilon > 0$. Hence, $\{x_n\}$ is a Cauchy sequence in X.

STEP 3. Existence and uniqueness of a fixed point.

Since $\{x_n\}$ is a Cauchy sequence and X is complete, $\lim_{n\to\infty} x_n = \bar{x} \in X$. By the continuity of T, we have $\bar{x} = T(\bar{x})$.

If $\bar{x}$ and $\tilde{x}$ are two distinct fixed points of T, then, by (2.26), we have

$$d(\bar{x}, \tilde{x}) = d(T(\bar{x}), T(\tilde{x})) \leq \psi(d(\bar{x}, \tilde{x})) < d(\bar{x}, \tilde{x}),$$

a contradiction. Hence, $\bar{x} = \tilde{x}$. ∎

Remark 2.7 In the original paper of Boyd and Wong [43], the upper semicontinuity from the right of ψ is considered. So, Theorem 2.10 is slightly different from the one in [43].

Remark 2.8 If we replace the condition $\psi(t) < t$ by the condition $\psi(t_0) < t_0$ for at least one value to t_0, then Theorem 2.10 may fail. In this case, T may have no fixed point or else more than one fixed point.

Example 2.13 Let $X = (-\infty, -1] \cup [1, \infty)$ be a metric space with the absolute value metric. Let

$$T_1(x) = \begin{cases} \frac{1}{2}(x+1), & \text{if } x \geq 1, \\ \frac{1}{2}(x-1), & \text{if } x \leq -1, \end{cases}$$

and

$$T_2(x) = -T_1(x), \quad \text{for all } x \in X.$$

Then, T_1 and T_2 satisfy condition (2.26) with

$$\psi(t) = \begin{cases} \frac{1}{2}t, & \text{if } t < 2, \\ \frac{1}{2}t + 1, & \text{if } t \geq 2. \end{cases}$$

The function ψ satisfies all the conditions of Theorem 2.10 except that $\psi(2) = 2$. We observe that T_1 has two fixed points -1 and 1, while T_2 has no fixed point.

The following example shows that the Boyd–Wong fixed point theorem is applicable but not the Banach contraction principle.

Example 2.14 Let $X = [0, 1]$ be a metric space with the usual metric and $T : X \to X$ be defined by

$$T(x) = x - x^2, \quad \text{for all } x \in X.$$

Then, T has a unique fixed point $\bar{x} = 0$, but T is not a contraction. However, T satisfies all the conditions of Theorem 2.10 with $\psi(t) = t - t^2$.

Matkowski [126] established the following variant of Theorem 2.10.

Theorem 2.11 *Let X be a complete metric space and $\psi : (0, \infty) \to (0, \infty)$ be monotonically nondecreasing such that $\lim_{n\to\infty} \psi^n(t) = 0$ for all $t > 0$. If $T : X \to X$ satisfies*

$$d(T(x), T(y)) \leq \psi(d(x, y)), \quad \textit{for all } x, y \in X,$$

then it has a unique fixed point $\bar{x}$ and $\lim_{n\to\infty} d(T^n(x), \bar{x}) = 0$ for all $x \in X$.

Proof For any fixed $x \in X$, let $x_n = T^n(x)$, $n = 1, 2, \ldots$. Then $x_1 = T(x) \neq x$, otherwise x would be a fixed point of T. For all $n \in \mathbb{N}$, we have

$$\begin{aligned} d\left(T^n(x), T^{n+1}(x)\right) &\leq \psi\left(d\left(T^{n-1}(x), T^n(x)\right)\right) \\ &\leq \psi^2\left(d\left(T^{n-2}(x), T^{n-1}(x)\right)\right) \\ &\;\;\vdots \\ &\leq \psi^n(d(x, T(x))) = \psi^n(d(x, x_1)). \end{aligned}$$

Therefore,

$$0 \leq \lim_{n\to\infty} d(x_n, x_{n+1}) = \lim_{n\to\infty} d\left(T^n(x), T^{n+1}(x)\right) \leq \lim_{n\to\infty} \psi^n(d(x, x_1)) = 0.$$

Thus, $\lim_{n\to\infty} d(x_n, x_{n+1}) = 0$. We show that $\{x_n\}$ is a Cauchy sequence.

Since $\psi^n(t) \to 0$ for all $t > 0$, we have $\psi(\varepsilon) < \varepsilon$ for any $\varepsilon > 0$. Since $\lim_{n\to\infty} d(x_n, x_{n+1}) = 0$, for any given $\varepsilon > 0$, we can choose n such that

$$d(x_{n+1}, x_n) \leq \varepsilon - \psi(\varepsilon).$$

Let $S_\varepsilon[x_n] = \{x \in X : d(x, x_n) \leq \varepsilon\}$. If $z \in S_\varepsilon[x_n]$, then $d(z, x_n) \leq \varepsilon$ and

$$\begin{aligned} d(T(z), x_n) &\leq d(T(z), T(x_n)) + d(T(x_n), x_n) \\ &\leq \psi(d(z, x_n)) + d(x_{n+1}, x_n) \quad \text{as } T(x_n) = x_{n+1} \\ &\leq \psi(\varepsilon) + (\varepsilon - \psi(\varepsilon)) = \varepsilon. \end{aligned}$$

Therefore, $T(z) \in S_\varepsilon[x_n]$ and so $T : S_\varepsilon[x_n] \to S_\varepsilon[x_n]$ is a mapping from $S_\varepsilon[x_n]$ into itself. It follows that $d(x_m, x_n) \leq \varepsilon$ for all $m \geq n$. Since ε was arbitrary, $\{x_n\}$ is a Cauchy sequence.

The rest of the proof follows as the proof of Theorem 2.10. ■

Remark 2.9 Some other generalizations of the contraction condition and the Banach contraction principle are studied by Altman [6], Geraghty [86], Jachymski and Jóźwik [101], Rakotch [146] and the references therein.

2.3.2 Weakly Contraction Mappings and Fixed Point Results

Let (X, d) be a metric space and $T : X \to X$ be a mapping. Then the inequality (2.7) can be written as

$$d(T(x), T(y)) \leq d(x, y) - \beta d(x, y), \quad \text{for all } x, y \in X, \tag{2.28}$$

where $\beta = 1 - \alpha$.

Inspired by this formulation, Alber and Guerre-Delabriere [2] introduced the following concept of weakly contraction mappings in the setting of Hilbert spaces and later it has been extended to metric space settings by Rhoades [148].

Definition 2.8 Let (X, d) be a metric space. A mapping $T : X \to X$ is said to be *weakly contraction* if

$$d(T(x), T(y)) \leq d(x, y) - \varphi(d(x, y)), \quad \text{for all } x, y \in X, \tag{2.29}$$

where $\varphi : [0, \infty) \to [0, \infty)$ is a continuous and nondecreasing function such that $\varphi(0) = 0$, $\varphi(t) > 0$ for $t > 0$ and $\lim_{t\to\infty} \varphi(t) = \infty$.

Example 2.15 Consider the metric space $X = [0, 1]$ with the usual metric and a mapping $T : X \to X$ defined by $T(x) = \sin x$ for all $x \in X$. Then, T is weakly contraction. Indeed, let $x, y \in X$ with $0 < y < x < 1$. Since

$$0 \leq \sin x = x - \frac{x^3}{3!} + \frac{x^5}{5!} - \frac{x^7}{7!} + \cdots,$$

we have

$$\sin x - \sin y = x - y - \left(\frac{x^3}{3!} - \frac{y^3}{3!}\right) + \left(\frac{x^5}{5!} - \frac{y^5}{5!}\right) - \left(\frac{x^7}{7!} - \frac{y^7}{7!}\right) + \cdots.$$

Observe that

$$\frac{x^{2n+1}}{(2n+1)!} - \frac{y^{2n+1}}{(2n+1)!} \geq \frac{x^{2n+3}}{(2n+3)!} - \frac{y^{2n+3}}{(2n+3)!}, \quad \text{for all } n \in \mathbb{N},$$

and therefore,

$$\begin{aligned}
\sin x - \sin y &= x - y - \left(\frac{x^3}{3!} - \frac{y^3}{3!}\right) + \left(\frac{x^5}{5!} - \frac{y^5}{5!}\right) - \left(\frac{x^7}{7!} - \frac{y^7}{7!}\right) + \cdots \\
&\leq x - y - \left(\frac{x^3}{3!} - \frac{y^3}{3!}\right) + \left(\frac{x^5}{5!} - \frac{y^5}{5!}\right) - \left(\frac{x^9}{9!} - \frac{y^9}{9!}\right) + \left(\frac{x^9}{9!} - \frac{y^9}{9!}\right) + \cdots \\
&\leq x - y - \left(\frac{x^3}{3!} - \frac{y^3}{3!}\right) + \left(\frac{x^5}{5!} - \frac{y^5}{5!}\right).
\end{aligned}$$

For $a, b \in (0, \infty)$ with $a > b$, we have

$$a^p - b^p \leq pa^{p-1}(a - b), \quad p > 1.$$

In particular,

$$\begin{aligned}
x^5 - y^5 &\leq 5x^4(x - y) \\
&\leq 5x^2(x - y) \\
&\leq 5(x^2 + xy + y^2)(x - y) = 5(x^3 - y^3).
\end{aligned}$$

Thus,

$$\frac{x^5 - y^5}{5!} \leq \frac{x^3 - y^3}{4!}.$$

Using the above fact, we obtain

$$\sin x - \sin y \leq x - y - \left(\frac{x^3}{3!} - \frac{y^3}{3!}\right) + \frac{x^3 - y^3}{4!} = x - y - \frac{1}{8}(x^3 - y^3).$$

Note that $x^3 - y^3 \geq (x - y)^3$. Then,

$$|\sin x - \sin y| \leq |x - y| - \frac{1}{8}|x - y|^3.$$

Therefore, T is weakly contractive.

Remark 2.10 **(a)** If T is defined on a nonempty bounded subset K of a metric space X, then the hypothesis $\lim\limits_{t \to \infty} \varphi(t) = \infty$ is not necessary in the definition of a weakly contraction mapping.

(b) If $\varphi(t) = \beta t$ for all $t \geq 0$, where $\beta \in (0, 1)$, then T is a contraction mapping with constant $1 - \beta$. Also, if $\varphi(t) = (1 - \alpha)t$, where $\alpha \in (0, 1)$ is contractivity constant, then a weakly contraction mapping is a contraction.

(c) Clearly, every weakly contraction mapping is nonexpansive.

(d) It is clear that weakly contraction mappings lie between contraction mappings and contractive mappings.

(e) Weakly contraction mappings are closely related to the condition of Theorem 2.10. If φ is lower semicontinuous, then $\psi(t) = t - \varphi(t)$ is upper semicontinuous and in this case condition (2.29) reduces to the condition (2.26).

Theorem 2.12 *Let (X, d) be a complete metric space and $T : X \to X$ be a weakly contraction mapping. Then, T has a unique fixed point.*

Proof Let $x_0 \in X$ be arbitrary and define $x_{n+1} := T(x_n)$ for all $n = 0, 1, 2, ...$ Then, from (2.29), we have

$$d(x_{n+1}, x_{n+2}) = d\left(T(x_n), T(x_{n+1})\right) \leq d(x_n, x_{n+1}) - \varphi\left(d(x_n, x_{n+1})\right).$$

Set $a_n := d(x_n, x_{n+1})$. Then,

$$a_{n+1} \leq a_n - \varphi(a_n) \leq a_n. \tag{2.30}$$

Therefore, $\{a_n\}$ is a nonincreasing sequence and so it has a limit. Let $\lim_{n\to\infty} a_n = a$. Assume that $a > 0$. Since φ is nondecreasing, we have $\varphi(a_n) \geq \varphi(a) > 0$. Therefore, from (2.30), we have

$$a_{n+1} \leq a_n - \varphi(a).$$

Thus, $a_{N+m} \leq a_m - N\varphi(a)$, a contradiction for N sufficiently large. Therefore, $a = 0$.

Let $\varepsilon > 0$ be fixed and choose N so that $d(x_N, x_{N+1}) \leq \min\{\varepsilon/2, \varphi(\varepsilon/2)\}$. Let $S_\varepsilon[x_N] = \{x \in X : d(x, x_N) \leq \varepsilon\}$ be a closed ball. Then we show that T is a mapping from $S_\varepsilon[x_N]$ into itself.

Let $x \in S_\varepsilon[x_N]$. Then we consider the following two cases.

Case 1. If $d(x, x_N) \leq \varepsilon/2$, then we have

$$\begin{aligned} d(T(x), x_N) &\leq d(T(x), T(x_N)) + d(T(x_N), x_N) \\ &\leq d(x, x_N) - \varphi\left(d(x, x_N)\right) + d(x_{N+1}, x_N) \\ &< \varepsilon/2 + \varepsilon/2 = \varepsilon. \end{aligned}$$

Case 2. If $\varepsilon/2 < d(x, x_N) \leq \varepsilon$, then $\varphi\left(d(x, x_N)\right) \geq \varphi(\varepsilon/2)$, and therefore

$$\begin{aligned} d(T(x), x_N) &\leq d(x, x_N) - \varphi\left(d(x, x_N)\right) + d(x_{N+1}, x_N) \\ &\leq d(x, x_N) - \varphi(\varepsilon/2) + \varphi(\varepsilon/2) \\ &= d(x, x_N) \leq \varepsilon. \end{aligned}$$

Thus, T is a mapping from $S_\varepsilon[x_N]$ into itself. Therefore, $x_n \in S_\varepsilon[x_N]$ for $n > N$. Since ε was arbitrary, $\{x_n\}$ is a Cauchy sequence. Since X is complete, $\{x_n\}$ converges to a point in X, that is, $\lim_{n\to\infty} x_n = x \in X$. By the continuity of T, x is a fixed point of T. ■

Remark 2.11 There are several other generalizations of a weakly contraction mapping in the literature; see, for example, [27–30, 113, 145] and the references therein.

2.4 Caristi's Fixed Point Theorem

We now present a different kind of principle in which the contraction condition is imposed only at the first step.

Theorem 2.13 *Let X be a complete metric space and $T : X \to X$ be a continuous mapping such that*

$$d(x, T(x)) \leq \varphi(x) - \varphi(T(x)), \quad \text{for all } x \in X, \tag{2.31}$$

and for some function $\varphi : X \to [0, \infty)$. Then, $\{T^n(x)\}$ converges to a fixed point of T for all $x \in X$.

Proof For any fixed $x \in X$, let $x_n = T^n(x)$, $n = 1, 2, ...$ By the inequality (2.31), we have

$$0 \leq \varphi(x) - \varphi(T(x)) \text{ if and only if } \varphi(T(x)) \leq \varphi(x), \quad \text{for all } x \in X,$$

and therefore,

$$\varphi(x_{n+1}) = \varphi\left(T^{n+1}(x)\right) = \varphi\left(T(T^n(x))\right) = \varphi(T(x_n)) \leq \varphi(x_n).$$

Thus, $\{\varphi(T^n(x))\} = \{\varphi(x_n)\}$ is monotonically decreasing and bounded below. Hence, $\lim_{n\to\infty} \varphi(T^n(x)) = r \geq 0$. By the triangle inequality, if $m, n \in \mathbb{N}$ and $m > n$, then

$$\begin{aligned}
d\left(T^n(x), T^m(x)\right) &\leq d\left(T^n(x), T^{n+1}(x)\right) + d\left(T^{n+1}(x), T^{n+2}(x)\right) \\
&\quad + \cdots + d\left(T^{m-1}(x), T^m(x)\right) \\
&\leq \varphi\left(T^n(x)\right) - \varphi\left(T^{n+1}(x)\right) + \varphi\left(T^{n+1}(x)\right) - \varphi\left(T^{n+2}(x)\right) \\
&\quad + \cdots + \varphi\left(T^{m-1}(x)\right) - \varphi\left(T^m(x)\right) \\
&\leq \varphi\left(T^n(x)\right) - \varphi\left(T^m(x)\right),
\end{aligned}$$

so, $\lim_{m,n\to\infty} d(T^n(x), T^m(x)) = 0$. It follows that $\{T^n(x)\} = \{x_n\}$ is a Cauchy sequence in X. Since X is complete, there exists $\bar{x} \in X$ such that $\lim_{n\to\infty} T^n(x) = \bar{x}$, and by the continuity of T, $\bar{x} = T(\bar{x})$. ■

Remark 2.12 In Theorem 2.13, we can obtain an estimate on the rate of convergence of $\{T^n(x)\}$ by referring back to the inequality

$$\sum_{i=n}^{m-1} d\left(T^i(x), T^{i+1}(x)\right) \leq \varphi\left(T^n(x)\right) - \varphi\left(T^m(x)\right).$$

This yields

$$d\left(T^n(x), T^m(x)\right) \leq \varphi\left(T^n(x)\right) - \varphi\left(T^m(x)\right) \leq \varphi\left(T^n(x)\right),$$

and if $T(\bar{x}) = \bar{x}$, upon letting $m \to \infty$, we get

$$d\left(T^n(x), \bar{x}\right) \leq \varphi\left(T^n(x)\right).$$

Remark 2.13 If $T : X \to X$ is a contraction mapping, then it is continuous and satisfies inequality (2.31).

Indeed, since T is a contraction mapping, we have

$$d\left(T(x), T^2(x)\right) \leq \alpha d(x, T(x)), \quad \text{for all } x \in X.$$

By adding $d(x, T(x))$ to both the sides of the above inequality, we obtain

$$d(x, T(x)) + d\left(T(x), T^2(x)\right) \leq d(x, T(x)) + \alpha d(x, T(x));$$

equivalently,

$$d(x, T(x)) - \alpha d(x, T(x)) \leq d(x, T(x)) - d\left(T(x), T^2(x)\right).$$

This in turn is equivalent to

$$d(x, T(x)) \leq \frac{1}{1-\alpha}\left[d(x, T(x)) - d\left(T(x), T^2(x)\right)\right].$$

Define the function $\varphi : X \to [0, \infty)$ by

$$\varphi(x) = \frac{1}{1-\alpha} d(x, T(x)), \quad \text{for all } x \in X.$$

This gives us the basic inequality

$$d(x, T(x)) \leq \varphi(x) - \varphi(T(x)), \quad \text{for all } x \in X.$$

Theorem 2.14 (Caristi's Theorem) *Let (X, d) be a complete metric space and $T : X \to X$ be a mapping (not necessarily continuous). Suppose that there exists a lower semicontinuous function $\varphi : X \to [0, \infty)$ such that*

$$d(x, T(x)) \leq \varphi(x) - \varphi(T(x)), \quad \textit{for all } x \in X. \tag{2.32}$$

Then, T has a fixed point.

Proof For any $x, y \in X$, define the relation $\preccurlyeq$ on X by

$$x \preccurlyeq y \quad \text{if and only if} \quad d(x, y) \leq \varphi(x) - \varphi(y).$$

It is easy to see that this ordering is (i) reflexive, that is, for all $x \in X$, $x \preccurlyeq x$; (ii) antisymmetric, that is, for all $x, y \in X$, $x \preccurlyeq y$ and $y \preccurlyeq x$ imply that $x = y$; (iii) transitive, that is, for all $x, y, z \in X$, $x \preccurlyeq y$ and $y \preccurlyeq z$ imply that $x \preccurlyeq z$. Then $(X, \preccurlyeq)$ is a partial ordered set. Let $x_0 \in X$ be an arbitrary but fixed element of X. Then by Zorn's Lemma B.1, we obtain a maximal totally ordered subset M of X containing x_0. Let $M = \{x_\alpha\}_{\alpha \in \Lambda}$, where Λ is totally ordered and

$$x_\alpha \preccurlyeq x_\beta \quad \text{if and only if} \quad \alpha \preccurlyeq \beta, \quad \text{for all } \alpha, \beta \in \Lambda.$$

Since $\{\varphi(x_\alpha)\}_{\alpha \in \Lambda}$ is a decreasing net in $[0, \infty)$, there exists $r \geq 0$ such that $\varphi(x_\alpha) \to r$ as α increases.

Let $\varepsilon > 0$ be given. Then, there exists $\alpha_0 \in \Lambda$ such that

$$\alpha \succcurlyeq \alpha_0 \quad \text{implies} \quad r \le \varphi(x_\alpha) \le r + \varepsilon.$$

Let $\beta \succcurlyeq \alpha \succcurlyeq \alpha_0$. Then,

$$d\left(x_\alpha, x_\beta\right) \le \varphi\left(x_\alpha\right) - \varphi\left(x_\beta\right) \le r + \varepsilon - r = \varepsilon,$$

which implies that $\{x_\alpha\}$ is a Cauchy net in X. Since X is complete, there exists $x \in X$ such that $x_\alpha \to x$ as α increases. From the lower semicontinuity of φ, we deduce that $\varphi(x) \le r$. If $\beta \succcurlyeq \alpha$, then $d(x_\alpha, x_\beta) \le \varphi(x_\alpha) - \varphi(x_\beta)$. Letting β increase, we obtain

$$d(x_\alpha, x) \le \varphi(x_\alpha) - r \le \varphi(x_\alpha) - \varphi(x),$$

which gives us $x_\alpha \preccurlyeq x$ for $\alpha \in \Lambda$. In particular, $x_0 \preccurlyeq x$. Since M is maximal, $x \in M$. Moreover, the condition (2.32) implies that

$$x_\alpha \preccurlyeq x \preccurlyeq T(x), \quad \text{for all } \alpha.$$

Again by the maximality, $T(x) \in M$. Since $x \in M$, $T(x) \preccurlyeq x$ and hence $T(x) = x$. ∎

Remark 2.14 **(a)** As in Remark 2.13, every contraction mapping $T : X \to X$ satisfies the condition (2.32) of the above theorem by choosing $\varphi(x) = \frac{1}{1-k} d(x, T(x))$.
(b) Wong [174] provided a refinement of the proof of Caristi's Fixed Point Theorem 2.14.

Weston [173] established the following theorem.

Theorem 2.15 *Let (X, d) be a complete metric space and $\varphi : X \to [0, \infty)$ be a lower semicontinuous function. Then, there exists a point $\bar{x} \in X$ such that*

$$d(\bar{x}, x) > \varphi(\bar{x}) - \varphi(x), \quad \textit{for all } x \in X \setminus \{\bar{x}\}. \tag{2.33}$$

A point $\bar{x} \in X$ which satisfies (2.33) is called a *d-point* for φ.

Proof Choose any $x_1 \in X$ and construct a sequence $\{x_n\}$ in the following way: For each $n \in \mathbb{N}$, let

$$c_n = \inf\left\{\varphi(x) : \varphi(x_n) - \varphi(x) \ge d(x_n, x) > 0\right\},$$

and let x_{n+1} be a point such that

$$\varphi(x_n) - \varphi(x_{n+1}) \ge d\left(x_n, x_{n+1}\right), \tag{2.34}$$

and

$$\varphi(x_{n+1}) < c_n + \frac{1}{n}. \tag{2.35}$$

(If x_n is a d-point for φ, then x_{n+1} must be x_n, and $c_n = \infty$). From (2.34), it follows that the sequence $\{\varphi(x_n)\}$ is nonincreasing, and that if $m \ge n$, then

$$\varphi(x_n) - \varphi(x_m) \ge d\left(x_n, x_m\right). \tag{2.36}$$

Since the sequence $\{\varphi(x_n)\}$ is bounded below, it is convergent. Hence, by (2.36) and the completeness of X, the sequence $\{x_n\}$ converges to some point, say, x_0. Now, we claim that

$$\varphi(x_n) - \varphi(x_0) \geq d(x_n, x_0), \quad \text{for all } n \in \mathbb{N}. \tag{2.37}$$

Assume the contrary that for some n,

$$\varphi(x_n) - \varphi(x_0) < d(x_n, x_0) - \varepsilon,$$

where $\varepsilon > 0$. Then by the lower semicontinuity of φ, there would be a neighborhood U of x_0 such that

$$\varphi(x_n) - \varphi(x) < d(x_n, x) - \varepsilon, \quad \text{for all } x \in U.$$

Then, m could be such that $x_m \in U$ and $d(x_m, x_0) < \varepsilon$, so that

$$\varphi(x_n) - \varphi(x_m) < d(x_n, x_0) - \varepsilon < d(x_n, x_m),$$

a contradiction to (2.36).

If x_0 is not a d-point for φ, then for some x, we have

$$\varphi(x_0) - \varphi(x) \geq d(x_0, x) > 0. \tag{2.38}$$

From (2.37) (with $n+1$ in place of n) and (2.35), we obtain

$$\varphi(x) \leq \varphi(x_{n+1}) + \varphi(x) - \varphi(x_0) < c_n + \frac{1}{n} + \varphi(x) - \varphi(x_0).$$

Hence, by (2.38), we can choose n so that $\varphi(x) < c_n$. From (2.37) and (2.38), $\varphi(x_n) > \varphi(x)$, so that $x_n \neq x$ and therefore $d(x_n, x) > 0$ and, moreover,

$$\varphi(x_n) - \varphi(x) \geq d(x_n, x) > 0.$$

It follows from the definition of c_n that $\varphi(x) \geq c_n$, and we have a contradiction. Thus x_0 is a d-point for φ. ■

Dancs, Hegedüs, and Medvegyev [67] proved the following result.

Theorem 2.16 *Caristi's Fixed Point Theorem 2.14 is equivalent to Theorem 2.15.*

Proof Theorem 2.14 implies Theorem 2.15: Suppose that the hypothesis of Theorem 2.14 holds and assume to the contrary that there is no $\bar{x} \in X$ such that (2.33) holds. Then for all $x \in X$, there would be a point $T(x) \neq x$ in X such that

$$d(x, T(x)) \leq \varphi(x) - \varphi(T(x)),$$

a contradiction to Theorem 2.14.

Theorem 2.15 implies Theorem 2.14: Suppose that the hypothesis of Theorem 2.15 holds and assume that $\bar{x} \in X$ satisfies (2.33). Then, $\bar{x}$ is a fixed point of T satisfying (2.32) since otherwise the inequality

$$d(\bar{x}, T(\bar{x})) > \varphi(\bar{x}) - \varphi(T(\bar{x}))$$

would hold, contradicting (2.32). ■

Theorem 2.15 implies the following result.

Theorem 2.17 *Let (X, d) be a complete metric space and $\varphi : X \to [0, \infty)$ be a lower semicontinuous function. If $\tilde{x} \in X$ is an arbitrary point, then there exists a point $\bar{x} \in X$ such that*

$$d(\tilde{x}, \bar{x}) \le \varphi(\tilde{x}) - \varphi(\bar{x}) \tag{2.39}$$

and

$$d(\bar{x}, x) > \varphi(\bar{x}) - \varphi(x), \quad \text{for all } x \in X \setminus \{\bar{x}\}. \tag{2.40}$$

Proof By the lower semicontinuity of φ, the set $S := \{x \in X : d(\tilde{x}, x) \le \varphi(\tilde{x}) - \varphi(x)\}$ is closed. Hence, the metric space (S, d) is complete. Applying Theorem 2.15 for the space S, we obtain a point $\bar{x} \in X$ such that $d(\tilde{x}, \bar{x}) \le \varphi(\tilde{x}) - \varphi(\bar{x})$ and $d(\bar{x}, x) > \varphi(\bar{x}) - \varphi(x)$ for all $x \in S \setminus \{\bar{x}\}$. To complete the proof, we need to prove that the last inequality holds for all $x \in X \setminus S$, as well.

If for $x \in X \setminus S$, the inequality $d(\bar{x}, x) \le \varphi(\bar{x}) - \varphi(x)$ would be true, then adding it to the inequality $d(\tilde{x}, \bar{x}) \le \varphi(\tilde{x}) - \varphi(\bar{x})$, we get $d(\tilde{x}, x) \le \varphi(\tilde{x}) - \varphi(x)$, contrary to $x \notin S$. ■

Weston [173] gave the following characterization of completeness of the metric space.

Theorem 2.18 *If (X, d) is not complete, then there is a uniformly continuous function $\varphi : X \to \mathbb{R}$ which is bounded below but has no d-point $\bar{x} \in X$, that is,*

$$d(\bar{x}, x) > \varphi(\bar{x}) - \varphi(x), \quad \text{for all } x \in X \setminus \{\bar{x}\}. \tag{2.41}$$

Proof Let $\{x_n\}$ be a Cauchy sequence in the metric space (X, d) which is not convergent. For any $x \in X$, $\{2d(x, x_n)\}$ is a Cauchy sequence in $\mathbb{R}$: let $\varphi(x)$ be its limit. Then $\varphi(x) > 0$, so the function φ is bounded below. Also, if $x_0 \in X$, then

$$|\varphi(x_0) - \varphi(x)| \le 2d(x_0, x),$$

so φ is uniformly continuous; and

$$\frac{1}{2}\{\varphi(x_0) + \varphi(x)\} \ge d(x_0, x),$$

so,

$$\varphi(x_0) - \varphi(x) \ge d(x_0, x) + \frac{1}{2}\{\varphi(x_0) - 3\varphi(x)\}.$$

Now, by the definition of φ, $\varphi(x_m) \to 0$ as $m \to \infty$; therefore $3\varphi(x) < \varphi(x_0)$ if $x = x_m$ and m is large. Thus, x_0 is not a d-point for φ. ■

Remark 2.15 **(a)** Let $X = \mathbb{R}$ be a metric space with the usual metric. Then, a function $\varphi : X \to \mathbb{R}$ which is uniformly continuous but not bounded below may or may not have a d-point.

(b) Let (X, d) be a metric space and $\varphi : X \to \mathbb{R}$ be a function. Define a relation $\preccurlyeq$ on X as follows:

$$x \preccurlyeq y \quad \Leftrightarrow \quad \varphi(y) - \varphi(x) \ge d(x, y) > 0.$$

Then, this relation is an order relation which is transitive, antisymmetric, and strictly irreflexive. Clearly, a point of X is a d-point for φ if and only if it is minimal with respect to $\preccurlyeq$.

(c) For a mapping $T : X \to X$, we can choose d and φ so that the relation $\preccurlyeq$ has the property that if $T(x) \neq x$, then $T(x) \preccurlyeq x$ (or, more general, $y \preccurlyeq x$ for some y), and then any d-point for φ is a fixed point of T.

Suzuki [164] established the following generalized forms of Caristi's fixed point theorem.

Theorem 2.19 [164, Theorem 2] *Let (X, d) be a complete metric space and $T : X \to X$ be a mapping. Let $f : X \to [0, \infty)$ be a lower semicontinuous function and $h : X \to [0, \infty)$ be a function satisfying*

$$\sup\left\{h(x) : x \in X,\ f(x) \leq \inf_{w \in X} f(w) + \eta\right\} < \infty$$

for some $\eta > 0$. Suppose that

$$d(x, T(x)) \leq h(x)(f(x) - f(T(x))), \quad \textit{for all } x \in X.$$

Then, T has a fixed point of T.

Proof Let $x \in X$. For $h(x) > 0$, from the assumption, we have $f(T(x)) \leq f(x)$. In the case of $h(x) = 0$, we have $d(x, T(x)) = 0$, and hence, $f(T(x)) = f(x)$ as $f(T(x)) \leq f(x)$ for all $x \in X$. Therefore,

$$f(T(x)) \leq f(x), \quad \text{for all } x \in X. \tag{2.42}$$

Define

$$Y := \left\{x \in X : f(x) \leq \inf_{w \in X} f(w) + \eta\right\} \text{ and } \gamma := \sup_{x \in Y} h(x) < \infty.$$

Then, Y is closed. Hence, Y is complete since X is complete and f is lower semicontinuous. From (2.42), we see that Y is nonempty and $T(Y) \subseteq Y$. Note that

$$d(x, T(x)) \leq \gamma(f(x) - f(T(x))), \quad \text{for all } x \in X.$$

Since γf is lower semicontinuous, there exists a fixed point $\bar{x} \in Y \subseteq X$ of T by Theorem 2.14. ∎

Theorem 2.20 [21, 164] *Let (X, d) be a complete metric space and $T : X \to X$ be a mapping. Let $f : X \to [0, \infty)$ be a lower semicontinuous function and $c : X \to [0, \infty)$ be an upper semicontinuous from the right satisfying the following inequality:*

$$d(x, T(x)) \leq \max\{c(f(x)), c(f(T(x)))\}(f(x) - f(T(x))), \quad \textit{for all } x \in X.$$

Then, T has a fixed point.

Proof Set $t_0 := \inf_{w \in X} f(w)$ and fix $\gamma > c(t_0)$. Since c is upper semicontinuous from the right, there exists $\eta > 0$ such that $c(t) \leq \gamma$ for $t \in [t_0, t_0 + \eta]$. Define a function $h : X \to [0, \infty)$ by

$$h(x) = \max\{c(f(x)), c(f(T(x)))\}, \quad \text{for all } x \in X.$$

Note that (2.42) is true. Hence, for $x \in X$ with $f(x) \leq t_0 + \eta$, we have $f(T(x)) \leq t_0 + \eta$, which shows that $h(x) \leq \gamma$. Thus,

$$\sup\left\{h(x) : x \in X,\ f(x) \leq \inf_{w \in X} f(w) + \eta\right\} \leq \gamma < \infty.$$

Therefore, Theorem 2.20 follows from Theorem 2.19. ∎

For further generalizations of Caristi's fixed point theorem, we refer to [21, 81, 98, 101, 110, 164] and the references therein.

Chapter 3

Set-valued Analysis

Continuity and Fixed Point Theory

The first natural instance when set-valued maps occur is the inverse f^{-1} of a single-valued map f from a set X to another set Y. We can always define f^{-1} as a set-valued map which associates with any data y the (possibly empty) set of solutions

$$f^{-1}(y) = \{x \in X : f(x) = y\}$$

to the equation $f(x) = y$. Kuratowski realized the importance of set-valued maps, also called multi-valued maps or point-to-set maps or multifunctions, and devoted considerable space in his famous book on topology. Other eminent mathematicians, namely, Painlevé, Hausdorff, and Bouligand, have also visualized the vital role of set-valued maps as one often encounters such objects in concrete and real-life problems. However, the authors of Bourbaki's volume *Topologie Génerale* emphasized the study of properties of set-valued maps as single-valued maps from a set to the power set of another set, or factorizing single-valued maps to make them bijective. This came as a stumbling block in the progress of this field. The set-valued maps started becoming popular among mathematicians working in the areas of game theory and economics when Kakutani's fixed point theorem [109], an extension of Brouwer's fixed point theorem to set-valued maps, came in existence, which was motivated by the work of Von Neumann. In 1969, Nadler [131] extended the Banach contraction principle to set-valued maps which plays an important and vital role in the theory of variational inequalities, differential inclusions, control theory, fractal geometry, etc. Since then, the theory of set-valued maps became an important subject for mathematicians working in nonlinear analysis, optimization theory, game theory, control theory, economics, fractal geometry, etc.

In view of such a wide variety of applications, most of the basic notions and results of single-valued maps have been extended to set-valued maps. These include:

- Limits and continuity
- Nonlinear functional analysis (existence and approximation of solutions to equations and inclusions; fixed point theory; etc.)
- Tangents and normals
- Differentiation of set-valued maps
- Convergence of a sequence of set-valued maps
- Measures and integration
- Differential inclusions

In this chapter, we focus our attention on the basic properties, continuity, and fixed point theory for set-valued maps. For applications of set-valued maps to fractal geometry, see [25]. For other topics mentioned above and the applications of set-valued maps, we refer to [15–18, 22, 24–26, 47, 65, 81, 88, 89, 95, 98, 112, 121, 129, 131, 136, 140, 147, 151, 152, 156, 157, 166, 168, 171, 183] and the references therein.

3.1 Basic Concepts and Definitions

Definition 3.1 Let X and Y be two nonempty sets. A *set-valued map* or *multivalued map* or *point-to-set map* or *multifunction* F from X to Y, denoted by $F : X \rightrightarrows Y$, is a map that associates with any $x \in X$ a subset $F(x)$ of Y; the set $F(x)$ is called the *image of* x under F. F is called *proper* if there exists at least one element $x \in X$ such that $F(x) \neq \emptyset$, that is, F is not the constant map $\emptyset$. In this case, the set $\text{Dom}(F) = \{x \in X : F(x) \neq \emptyset\}$ is called the *domain* of F. Actually, a set-valued map F is characterized by its *graph*, the subset of $X \times Y$ defined by

$$\text{Graph}(F) = \{(x, y) : y \in F(x)\}.$$

Indeed, if A is a nonempty subset of the product set $X \times Y$, then the graph of a set-valued map F is defined by

$$y \in F(x) \quad \text{if and only if} \quad (x, y) \in A.$$

The domain of F is the projection of Graph(F) on X, and the *image* of F is the subset of Y defined by

$$\text{Image}(F) = \bigcup_{x \in X} F(x) = \bigcup_{x \in \text{Dom}(F)} F(x),$$

is the projection of Graph(F) on Y. A set-valued map $F : X \rightrightarrows Y$ is called *strict* if $\text{Dom}(F) = X$, that is, if the image $F(x)$ is nonempty for all $x \in X$. Let K be a nonempty subset of X and $F : K \rightrightarrows Y$ be a strict set-valued map. Then the extension of F on the whole set X, denoted by $F_K : X \rightrightarrows Y$, is defined by

$$F_K(x) = \begin{cases} F(x), & \text{when } x \in K, \\ \emptyset, & \text{when } x \notin K, \end{cases}$$

whose domain $\text{Dom}(F_K)$ is K.

When $F : X \rightrightarrows Y$ is a set-valued map and $K \subseteq X$, we denote by $F_{|K}$ its restriction to K.

Example 3.1 (a) Let $F : \mathbb{R} \rightrightarrows \mathbb{R}$ be defined by

$$F(x) = \begin{cases} \{1\}, & \text{if } x < 0, \\ \{0, 1\}, & \text{if } x = 0, \\ \{0\}, & \text{if } x > 0. \end{cases}$$

Then, F is a set-valued map (see Figure 3.1).

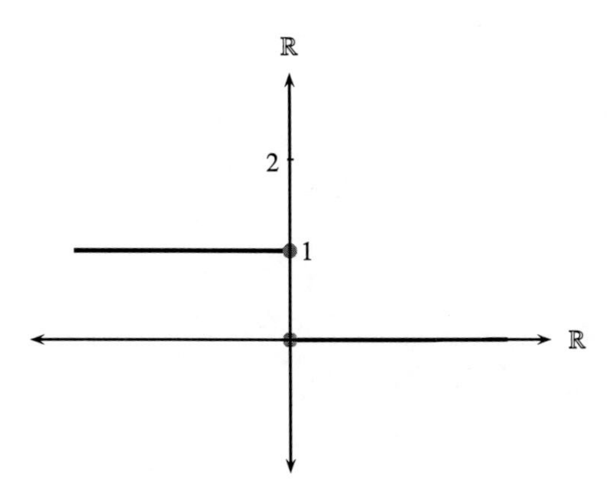

Figure 3.1 A set-valued map

(b) Let $F : \mathbb{R} \rightrightarrows \mathbb{R}$ be defined by

$$F(x) = \begin{cases} \{1\}, & \text{if } x < 0, \\ [0, 1], & \text{if } x = 0, \\ \{0\}, & \text{if } x > 0. \end{cases}$$

Then, F is a set-valued map (see Figure 3.2).

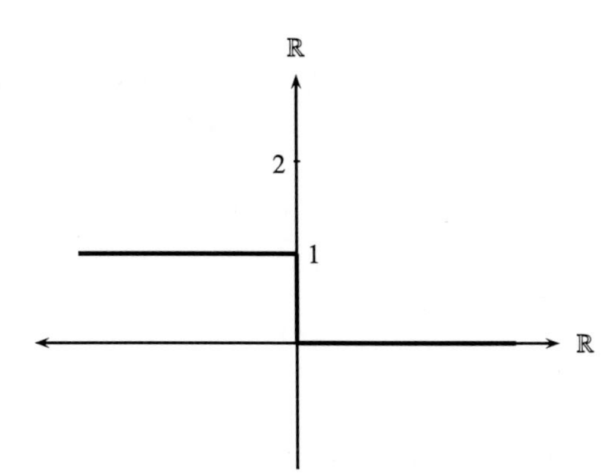

Figure 3.2 A set-valued map

(c) Let $F : [0, 1] \rightrightarrows [0, 1]$ be defined by

$$F(x) = \begin{cases} \left[0, \frac{1}{2}\right], & \text{if } x \neq \frac{1}{2}, \\ [0, 1], & \text{if } x = \frac{1}{2}. \end{cases}$$

Then, F is a set-valued map (see Figure 3.3).

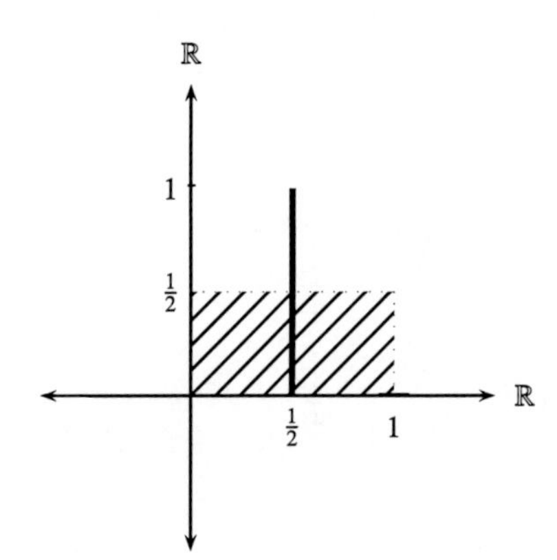

Figure 3.3 A set-valued map

(d) Let $F : [0,1] \rightrightarrows [0,1]$ be defined by

$$F(x) = [x, 1], \quad \text{for all } x \in [0,1].$$

Then, F is a set-valued map (see Figure 3.4).

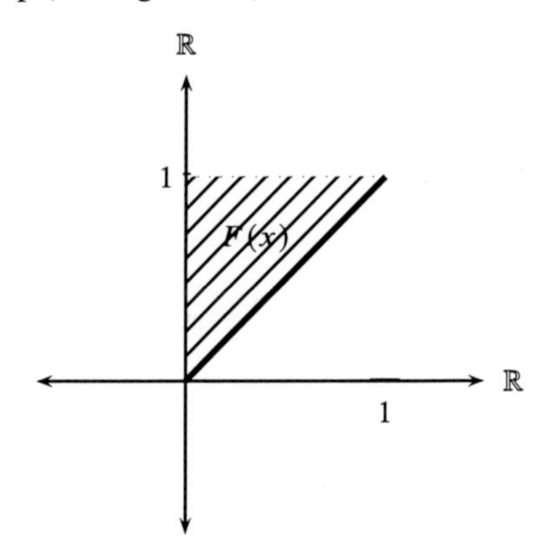

Figure 3.4 A set-valued map

(e) Let $X = [0, \infty)$, $Y = \mathbb{R}$ and $F : X \rightrightarrows Y$ be defined by

$$F(x) = [e^{-x}, 1], \quad \text{for all } x \in X.$$

Then, F is a set-valued map.

(f) Let $I^2 = \{(x,y) : 0 \leq x \leq 1 \text{ and } 0 \leq y \leq 1\}$ and $F : I^2 \rightrightarrows I^2$ be defined as follows:

$$\begin{aligned} F(x,y) &= \text{the line segment in } I^2 \text{ from the point } (x,0) \text{ to the point } (0,y), \\ &\quad \text{for all } (x,y) \in I^2. \end{aligned}$$

Then, F is a set-valued map.

Example 3.2 (Inverse Function) If $f : X \to Y$ is a single-valued map, then its inverse f^{-1} can be considered as a set-valued map $F : Y \rightrightarrows X$ defined by

$$F(y) = f^{-1}(y), \quad \text{for all } y \in \text{Image}(f).$$

This set-valued map is strict whenever f is surjective, and single-valued whenever f is injective. This map plays an important role when we study the equation $f(x) = y$ and the behavior of the set of solutions of $f^{-1}(y)$ as y ranges over Y.

Example 3.3 (Metric Projection) Let (X, d) be a metric space and K be a nonempty compact subset of X. Then by Exercise 1.4, for every $x \in X$, there exists a $y \in K$ such that $d(x, y) = d(x, K)$, where $d(x, K) = \inf_{z \in K} d(x, z)$ is the distance from a point x to the set K. The *metric projection* is a set-valued map $P : X \rightrightarrows K$ defined by

$$P(x) = \{y \in K : d(x, y) = d(x, K)\}, \quad \text{for all } x \in X.$$

Definition 3.2 Let $F : X \rightrightarrows Y$ be a set-valued map. For a nonempty subset A of X, we write

$$F(A) = \bigcup_{x \in A} F(x).$$

The set $F(A)$ is called the *image* of A under the set-valued map F.

By convention, $F(\emptyset) = \emptyset$.

In the rest of this section, we assume that X, Y and Z are nonempty sets.

Definition 3.3 Let $F : X \rightrightarrows Y$ be a set-valued map.

(a) If $F(X) = Y$, then F is called *quasi-surjective*.
(b) If $F(x_1) \cap F(x_2) \neq \emptyset$ implies $x_1 = x_2$ for any $x_1, x_2 \in X$, then F is called *hyperinjective*.

Example 3.4 Let $F : \mathbb{R} \rightrightarrows \mathbb{R}$ be a set-valued map defined by

$$F(x) = \begin{cases} \left\{\frac{1}{x}\right\}, & \text{if } x \neq 0, \\ \{0\}, & \text{if } x = 0. \end{cases}$$

Then, F is hyperinjective.

Remark 3.1 If $x_1 \neq x_2$ implies $F(x_1) \cap F(x_2) = \emptyset$, then F is hyperinjective.

Theorem 3.1 *Let $\{A_\alpha\}_{\alpha \in \Lambda}$ be a family of nonempty subsets of X and $F : X \rightrightarrows Y$ be a set-valued map. Then the following assertions hold.*

(a) *If $A_1 \subseteq A_2$, then $F(A_1) \subseteq F(A_2)$.*

(b) $F\left(\bigcup_{\alpha \in \Lambda} A_\alpha\right) = \bigcup_{\alpha \in \Lambda} F(A_\alpha)$.

(c) $F\left(\bigcap_{\alpha\in\Lambda} A_\alpha\right) \subset \bigcap_{\alpha\in\Lambda} F(A_\alpha)$.

(d) $F(X \setminus A_1) \supseteq F(X) \setminus F(A_1)$. *Further, if F is hyperinjective, then* $F(X \setminus A_1) = F(X) \setminus F(A_1)$.

Proof (a) If $y \in F(A_1)$, then $y \in F(x)$ for some $x \in A_1$. Since $A_1 \subseteq A_2$, we have $y \in F(x)$ for some $x \in A_2$, hence, $y \in F(A_2)$.

(b) If $y \in F\left(\bigcup_{\alpha\in\Lambda} A_\alpha\right)$, then $y \in F(x)$ for some $x \in \bigcup_{\alpha\in\Lambda} A_\alpha$, hence, for at least one index α_0, we have $y \in F(A_{\alpha_0})$. Therefore, $y \in \bigcup_{\alpha\in\Lambda} F(A_\alpha)$.
By reversing the argument, we can deduce the desired formula.

(c) If $y \in F\left(\bigcap_{\alpha\in\Lambda} A_\alpha\right)$, then $y \in F(x)$ for some x such that $x \in A_\alpha$ for all $\alpha \in \Lambda$, and so $y \in F(A_\alpha)$ for all $\alpha \in \Lambda$, hence, $y \in \bigcap_{\alpha\in\Lambda} F(A_\alpha)$ and so the result is proved.

(d) Let $y \in F(X)\backslash F(A_1)$. Then $y \in \bigcup_{x\in X} F(x)\backslash \bigcup_{x\in A_1} F(x)$. Therefore, $y \in F(x)$ for some $x \in X$ but $x \notin A_1$. This implies that $y \in F(x)$ for some $x \in X\backslash A_1$, and hence, $y \in \bigcup_{x\in X\backslash A_1} F(x)$. Therefore, $y \in F(X\backslash A_1)$, and so $F(X)\backslash F(A_1) \subseteq F(X\backslash A_1)$.

If F is hyperinjective, then $y \in F(X\backslash A_1)$ implies that there exists $x \in X\backslash A_1$ such that $y \in F(x)$. Hence $y \notin F(A_1)$ and so $y \in F(X)\backslash F(A_1)$. Thus, $F(X\backslash A_1) = F(X)\backslash F(A_1)$. ■

Definition 3.4 Let $F_1, F_2 : X \rightrightarrows Y$ be set-valued maps.

(a) The *union* of F_1 and F_2 is a set-valued map $F_1 \cup F_2 : X \rightrightarrows Y$ defined by

$$(F_1 \cup F_2)(x) = F_1(x) \cup F_2(x), \quad \text{for all } x \in X.$$

(b) The *intersection* of F_1 and F_2 is a set-valued map $F_1 \cap F_2 : X \rightrightarrows Y$ defined by

$$(F_1 \cap F_2)(x) = F_1(x) \cap F_2(x), \quad \text{for all } x \in X.$$

(c) The *Cartesian product* of F_1 and F_2 is a set-valued map $F_1 \times F_2 : X \rightrightarrows Y \times Y$ defined by

$$(F_1 \times F_2)(x) = F_1(x) \times F_2(x), \quad \text{for all } x \in X.$$

(d) If F_1 is a set-valued map from X to Y and F_2 is another set-valued map from Y to Z, then the *composition product* of F_2 by F_1 is a set-valued map $F_2 \circ F_1 : X \rightrightarrows Z$ defined by

$$(F_2 \circ F_1)(x) = F_2(F_1(x)), \quad \text{for all } x \in X.$$

Theorem 3.2 *Let* $F_1, F_2 : X \rightrightarrows Y$ *be set-valued maps and A is a nonempty subset of X. Then,*

(a) $(F_1 \cup F_2)(A) = F_1(A) \cup F_2(A)$;
(b) $(F_1 \cap F_2)(A) \subseteq F_1(A) \cap F_2(A)$;
(c) $(F_1 \times F_2)(A) \subseteq F_1(A) \times F_2(A)$;
(d) $(F_2 \circ F_1)(A) = F_2(F_1(A))$.

Theorem 3.3 *If one of the set-valued maps $F_1, F_2 : X \rightrightarrows Y$ is hyperinjective, then the map $F_1 \cap F_2$ and $F_1 \times F_2$ are hyperinjective.*

Proof If x and y are two distinct elements of X, then

$$\left(F_1(x) \cap F_2(x)\right) \cap \left(F_1(y) \cap F_2(y)\right) = \left(F_1(x) \cap F_1(y)\right) \cap \left(F_2(x) \cap F_2(y)\right) = \emptyset.$$

The proof for the Cartesian product $F_1 \times F_2$ is similar. ∎

Definition 3.5 Let $F : X \rightrightarrows Y$ be a set-valued map. The *inverse* $F^{-1} : Y \rightrightarrows X$ of F is defined by

$$F^{-1}(y) = \{x \in X : y \in F(x)\}, \quad \text{for all } y \in Y.$$

Further, let B be a subset of Y. The *upper inverse image* $F^{-1}(B)$ and *lower inverse image* $F_{+}^{-1}(B)$ of B under F are defined by

$$F^{-1}(B) = \{x \in X : F(x) \cap B \neq \emptyset\}$$

and

$$F_{+}^{-1}(B) = \{x \in X : F(x) \subseteq B\},$$

respectively. We also write $F^{-1}(\emptyset) = \emptyset$ and $F_{+}^{-1}(\emptyset) = \emptyset$. It is clear that $(F^{-1})^{-1} = F$ and that $y \in F(x)$ if and only if $x \in F^{-1}(y)$.

We have the following relations between domains, graphs and images of F and F^{-1}.

$$\text{Dom}(F^{-1}) = \text{Image}(F), \qquad \text{Image}(F^{-1}) = \text{Dom}(F) \text{ and} \tag{3.1}$$

$$\text{Graph}(F^{-1}) = \{(y, x) \in F(X) \times X \subseteq Y \times X : (x, y) \in \text{Graph}(F)\}, \tag{3.2}$$

(see Figures 3.5 and 3.6).

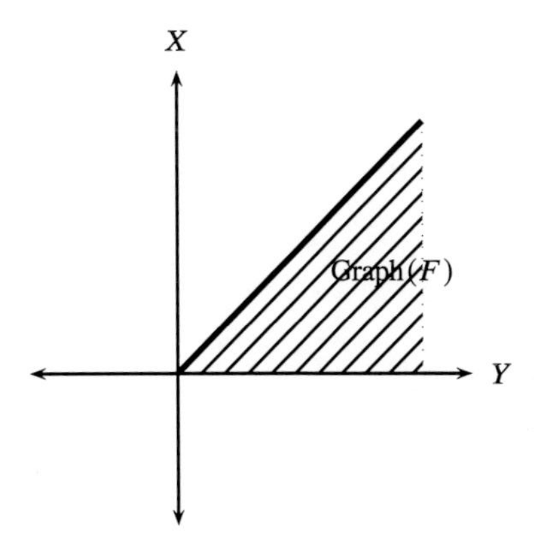

Figure 3.5 Graph of F

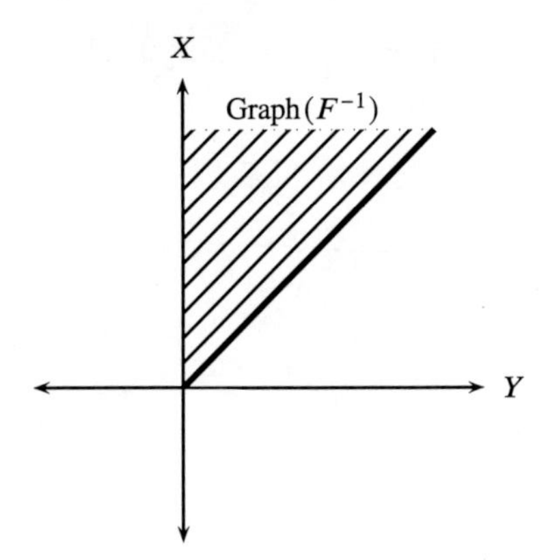

Figure 3.6 Graph of F^{-1}

Example 3.5 (a) Let $X = Y = \mathbb{R}_+ = [0, \infty)$ and $F : X \rightrightarrows Y$ be a set-valued map defined by

$$F(x) = [0, x], \quad \text{for all } x \in X.$$

Since $F(\mathbb{R}_+) = \mathbb{R}_+$, the map F^{-1} is defined on all the nonnegative half-axis. For any $y \in \mathbb{R}_+$,

$$F^{-1}(y) = \{x \in X : y \in [0, x]\} = [y, +\infty).$$

(b) Let $F : [0, 1] \rightrightarrows [0, 2]$ be a set-valued map defined by

$$F(x) = [0, x + 1], \quad \text{for all } x \in [0, 1].$$

Then, $F\left([0, 1]\right) = \bigcup_{x \in [0,1]} F(x) = [0, 2]$, and hence, F is quasisurjective.

$$F^{-1}\left([0, 3/2]\right) = \left\{x \in [0, 1] : F(x) \cap [0, 3/2] \neq \emptyset\right\} = [0, 1].$$

$$F^{-1}\left(\frac{1}{2}\right) = \left\{x \in [0, 1] : 1/2 \in F(x)\right\} = [0, 1].$$

Remark 3.2 If $F_1 : X \rightrightarrows Y$ and $F_2 : Y \rightrightarrows Z$ are set-valued maps, then $\left(F_2 \circ F_1\right)^{-1} = F_1^{-1} \circ F_2^{-1}$.

Theorem 3.4 (a) *If $f : X \to Y$ is single-valued, then $f^{-1} : Y \rightrightarrows X$ is hyperinjective.*
(b) *If the set-valued map $F : X \rightrightarrows Y$ is hyperinjective, then F^{-1} is single-valued.*

Proof (a) If f is single-valued, then

$$x \neq y \quad \text{implies} \quad f^{-1}(x) \cap f^{-1}(y) = \emptyset,$$

and hence, f^{-1} is hyperinjective.
(b) If F is hyperinjective, then the set $F^{-1}(y) = \{x \in X : y \in F(x)\}$ has only one element, and therefore, F^{-1} is single-valued. ∎

Theorem 3.5 *Let $\{B_\alpha\}_{\alpha\in\Lambda}$ be a family of nonempty subsets of Y, $A \subseteq X$ and $B \subseteq Y$. Let $F : X \rightrightarrows Y$ be a set-valued map.*

(a) *If $B_1 \subseteq B_2$, then $F^{-1}(B_1) \subseteq F^{-1}(B_2)$;*
(b) $A \subset F_+^{-1}(F(A))$;
(c) $B \supset F(F_+^{-1}(B))$;
(d) $X \setminus F_+^{-1}(B) = F_+^{-1}(Y \setminus B)$;
(e) $F_+^{-1}\left(\bigcup_{\alpha\in\Lambda} B_\alpha\right) \supset \bigcup_{\alpha\in\Lambda} F_+^{-1}(B_\alpha)$;
(f) $F_+^{-1}\left(\bigcap_{\alpha\in\Lambda} B_\alpha\right) = \bigcap_{\alpha\in\Lambda} F_+^{-1}(B_\alpha)$;
(g) $F^{-1}(F(A)) \supset A$;
(h) $F(F^{-1}(B)) \supset B \cap F(X)$;
(i) $X \setminus F^{-1}(B) = F_+^{-1}(Y \setminus B)$;
(j) $F^{-1}\left(\bigcap_{\alpha\in\Lambda} B_\alpha\right) \subset \bigcap_{\alpha\in\Lambda} F^{-1}(B_\alpha)$;
(k) $F^{-1}\left(\bigcup_{\alpha\in\Lambda} B_\alpha\right) = \bigcup_{\alpha\in\Lambda} F^{-1}(B_\alpha)$.

Theorem 3.6 *Let $F_1, F_2 : X \rightrightarrows Y$ be set-valued maps such that $(F_1 \cap F_2)(x) \neq \emptyset$ for all $x \in X$ and let $B \subseteq Y$. Then,*

(a) $(F_1 \cup F_2)^{-1}(B) = F_1^{-1}(B) \cup F_2^{-1}(B)$;
(b) $(F_1 \cap F_2)^{-1}(B) \subset F_1^{-1}(B) \cap F_2^{-1}(B)$;
(c) $(F_1 \cup F_2)_+^{-1}(B) = F_{1+}^{-1}(B) \cap F_{2+}^{-1}(B)$;
(d) $(F_1 \cap F_2)_+^{-1}(B) \supset F_{1+}^{-1}(B) \cup F_{2+}^{-1}(B)$;

Theorem 3.7 *Let $F_1 : X \rightrightarrows Y$ and $F_2 : Y \rightrightarrows Z$ be set-valued maps. Then for any $B \subseteq Z$, we have*

(a) $(F_2 \circ F_1)_+^{-1}(B) = F_{1+}^{-1}\left(F_{2+}^{-1}(B)\right)$;
(b) $(F_2 \circ F_1)^{-1}(B) = F_1^{-1}\left(F_2^{-1}(B)\right)$.

Theorem 3.8 *Let $F_1 : X \rightrightarrows Y$ and $F_2 : X \rightrightarrows Z$ be set-valued maps. Then for any $B \subseteq Y$ and $D \subseteq Z$, we have*

(a) $(F_1 \times F_2)_+^{-1}(B \times D) = F_{1+}^{-1}(B) \cap F_{2+}^{-1}(D)$;
(b) $(F_1 \times F_2)^{-1}(B \times D) = F_1^{-1}(B) \cap F_2^{-1}(D)$.

3.2 Continuity of Set-valued Maps

Let (X, d) be a metric space and K be a nonempty subset of X. The distance from a point $x \in X$ to the set K is defined by

$$d(x, K) = \inf_{y \in K} d(x, y).$$

The *open sphere of radius* $r > 0$ *around* K, denoted by $S_r(K)$, is defined by

$$S_r(K) = \{x \in X : d(x, K) < r\}.$$

Note that for any open neighborhood O of a compact set K, there exists $r > 0$ such that $S_r(K) \subseteq O$.

The *closed sphere of radius* $r > 0$ *around* K, denoted by $S_r[K]$, is defined by

$$S_r[K] = \{x \in X : d(x, K) \leq r\}.$$

The closed sphere $S_r[K]$ around K is a neighborhood of K. If K is compact, then $S_r[K]$ is closed.

Indeed, if K is compact, then it is totally bounded. Therefore, the set $S_r[K]$ can be expressed as the union of a finite number of closed sets. Thus, it is closed.

However, if K is compact, then $S_r[K]$ need not be compact. For example, consider the metric space $X = \mathbb{R}$ with the discrete metric and let K be a nonempty finite subset of X. Choose $r > 1$; then $S_r[K] = \mathbb{R}$ is closed, but not totally bounded and hence not compact.

When K is compact, each neighborhood of K contains such a closed sphere around K.

If the images of a set-valued map F are closed, bounded, compact, and so on, we say that F is *closed valued, bounded valued, compact valued*, and so on.

Recall that for single-valued maps, continuity amounts to mapping convergent sequences to convergent sequences or, equivalently, to the celebrated incantation:

$$\text{for any } \epsilon > 0, \text{ there exists } \delta > 0, \ldots$$

Unfortunately, this characterization is no longer true for set-valued maps.

There are two distinct ways to extend the concept of continuity: The first one, encompassing the above "convergence" idea, is called *lower semicontinuous*. The second, extending the "(for any $\epsilon > 0$, there exists $\delta > 0, \ldots$) definition," leads to the so-called *upper semicontinuous* of set-valued maps. The concept of two kinds of semicontinuity of a set-valued map was introduced by G. Bouligand [42] and K. Kuratowski [114].

Definition 3.6 Let X and Y be metric spaces. A set-valued map $F : X \rightrightarrows Y$ is called *upper semicontinuous* at $x_0 \in \mathrm{Dom}(F)$ if for any neighborhood U of $F(x_0)$, there exists $\delta > 0$ such that $F(x) \subseteq U$ for all $x \in S_\delta(x_0)$.

It is said to be *upper semicontinuous* on X if it is upper semicontinuous at every point of $\mathrm{Dom}(F)$.

When $F(x)$ is compact for all $x \in X$, F is upper semicontinuous at x_0 if and only if for any $\epsilon > 0$, there exists $\delta > 0$ such that $F(x) \subseteq S_\varepsilon(F(x_0))$ for all $x \in S_\delta(x_0)$.

We observe that this definition is a natural adaptation of the definition of a continuous single-valued map. Why then do we use the adjective upper semicontinuous instead of continuous? One of the reason is that the celebrated characterization of continuous maps – "a single-valued map f is continuous at x if and only if it maps sequences converging to x to sequences converging to

$f(x)$" – does not hold true any longer in the set-valued case. Indeed, the set-valued version of this characterization leads to the following definition.

Definition 3.7 Let X and Y be metric spaces. A set-valued map $F : X \rightrightarrows Y$ is called *lower semicontinuous* at $x_0 \in \mathrm{Dom}(F)$ if for any $y \in F(x_0)$ and for any sequence of elements $x_n \in \mathrm{Dom}(F)$ converging to x_0, there exists a sequence of elements $y_n \in F(x_n)$ converging to y.

It is said to be *lower semicontinuous* on X if it is lower semicontinuous at every point $x \in \mathrm{Dom}(F)$.

The above definition of lower semicontinuity could be interpreted as follows:

F is *lower semicontinuous* at x_0 if for any $y_0 \in F(x_0)$ and any neighborhood $V(y_0)$ of y_0, there exists $\delta > 0$ such that $F(x) \cap V(y_0) \neq \emptyset$ for all $x \in S_\delta(x_0)$.

Actually, as in the single-valued case, this definition is equivalent to the following definition.

For any open subset $U \subseteq Y$ such that $U \cap F(x_0) \neq \emptyset$, there exists $\delta > 0$ such that $F(x) \cap U \neq \emptyset$ for all $x \in S_\delta(x_0)$.

Whenever $F(x)$ is compact, F is lower semicontinuous at x_0 if and only if for any $\varepsilon > 0$, there exists $\delta > 0$ such that $F(x) \cap S_\varepsilon(F(x_0)) \neq \emptyset$ for all $x \in S_\delta(x_0)$.

The following examples show that the notions of upper semicontinuity and lower semicontinuity are not equivalent.

Example 3.6 (a) Let $X = \mathbb{R}$ be a metric space with the usual metric and $F : X \rightrightarrows X$ be a set-valued map defined by

$$F(x) = \begin{cases} [-1, 1], & \text{if } x = 0, \\ \{0\}, & \text{if } x \neq 0. \end{cases}$$

Then, F is upper semicontinuous at zero but not lower semicontinuous at zero (see Figure 3.7).

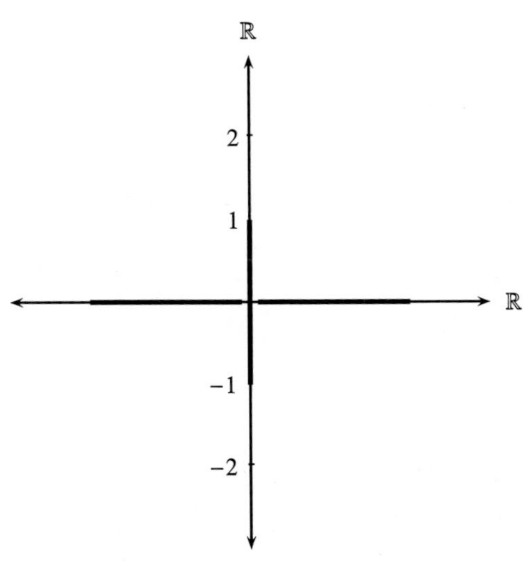

Figure 3.7 An upper semicontinuous set-valued map at zero but not lower semicontinuous at zero

(b) Let $X = [0,1]$ be a metric space with the usual metric and $F : X \rightrightarrows X$ be a set-valued map defined by

$$F(x) = \begin{cases} \frac{1}{2}x, & \text{if } 0 \leq x < \frac{1}{2}, \\ \left[\frac{1}{4}, \frac{3}{4}\right], & \text{if } x = \frac{1}{2}, \\ \frac{1}{2}(x+1), & \text{if } \frac{1}{2} < x \leq 1. \end{cases}$$

Then, F is upper semicontinuous on X but not lower semicontinuous.

(c) Let $X = \mathbb{R}$ be a metric space with the usual metric. The set-valued map $F : X \rightrightarrows X$ defined by

$$F(x) = \begin{cases} \{0\}, & \text{if } x = 0, \\ [-1,1], & \text{if } x \neq 0 \end{cases}$$

is lower semicontinuous at zero but not upper semicontinuous at zero (see Figure 3.8).

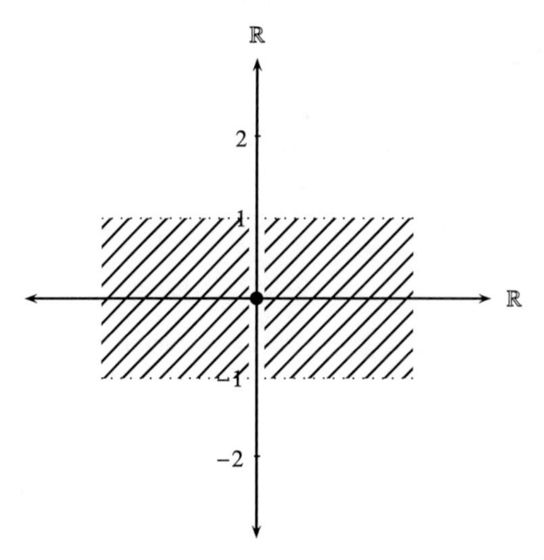

Figure 3.8 A lower semicontinuous set-valued map at zero but not upper semicontinuous at zero

(d) Let $X = [0,1]$ be a metric space with the usual metric. The set-valued map $F : X \rightrightarrows X$ defined by

$$F(x) = \begin{cases} [0,x], & \text{if } 0 \leq x < 1, \\ \{0\} & \text{if } x = 1 \end{cases}$$

is lower semicontinuous on X but not upper semicontinuous.

Definition 3.8 Let X and Y be metric spaces. A set-valued map $F : X \rightrightarrows Y$ is said to be *continuous* at $x_0 \in X$ if it is both lower semicontinuous as well as upper semicontinuous at x_0. It is said to be *continuous* on X if it is continuous at every point $x \in X$.

It is not easy for the readers to check whether a set-valued map is upper semicontinuous or lower semicontinuous by using the definition of these continuities. So, we have the following characterizations for these definitions.

Theorem 3.9 *Let X and Y be metric spaces. A set-valued map $F : X \rightrightarrows Y$ is lower semicontinuous if and only if $F^{-1}(G)$ is open for every open subset G of Y.*

Proof Suppose that F is lower semicontinuous and G is an open subset of Y. Clearly, $F^{-1}(G)$ is open if it is empty. So, we assume that $F^{-1}(G) \neq \emptyset$. If $x_0 \in F^{-1}(G)$, then $F(x_0) \cap G \neq \emptyset$. Let $y_0 \in F(x_0) \cap G$. Since G is open, it is a neighborhood of each of its points. Therefore, G is a neighborhood of y_0. So, for any $y_0 \in F(x_0)$ and any neighborhood G of y_0, there exists $\delta > 0$ such that $F(x) \cap G \neq \emptyset$ for all $x \in S_\delta(x_0)$. Hence $S_\delta(x_0) \subseteq F^{-1}(G)$. This implies that $F^{-1}(G)$ is open.

Conversely, assume that $F^{-1}(G)$ is open for each open subset G of Y such that $F(x_0) \cap G \neq \emptyset$. Then, $x_0 \in F^{-1}(G)$. Since $F^{-1}(G)$ is open, it is a neighborhood of each of its points. We have $F^{-1}(G)$ is a neighborhood of x_0. Then for any $x_0 \in F^{-1}(G)$, there exists $\delta > 0$ such that $S_\delta(x_0) \subseteq F^{-1}(G)$. Therefore, for all $x \in S_\delta(x_0)$, we have $x \in F^{-1}(G)$. This implies that $F(x) \cap G \neq \emptyset$, and hence, F is lower semicontinuous. ■

Corollary 3.1 *Let X and Y be metric spaces. A set-valued map $F : X \rightrightarrows Y$ is lower semicontinuous if and only if $F_+^{-1}(H)$ is closed for every closed subset H of Y.*

Proof It follows from Theorem 3.9 and Theorem 3.5 (i). ■

Theorem 3.10 *Let X and Y be metric spaces and $F : X \rightrightarrows Y$ be a set-valued map such that $F(x)$ is compact for all $x \in X$. Then F is upper semicontinuous if and only if for each open subset G of Y, the set*

$$F_+^{-1}(G) = \{x \in X : F(x) \subseteq G\}$$

is open.

Proof Suppose that F is upper semicontinuous. If $F_+^{-1}(G) = \emptyset$, then clearly it is open. Assume that $F_+^{-1}(G) \neq \emptyset$ and let $x_0 \in F_+^{-1}(G)$. Then $F(x_0) \subseteq G$. Since G is open, it is a neighborhood of each of its points. Therefore, G is a neighborhood of $F(x_0)$. Then by upper semicontinuity of F, there exists $\delta > 0$ such that $F(x) \subseteq G$ for all $x \in S_\delta(x_0)$. This implies that $S_\delta(x_0) \subseteq F_+^{-1}(G)$, and so $F_+^{-1}(G)$ is an open neighborhood of each of its points, and hence, $F_+^{-1}(G)$ is open.

Conversely, assume that for each open set G in Y, the set $F_+^{-1}(G)$ is open and that $F(x)$ is compact for each $x \in X$. Let $x_0 \in X$ and let G be an open set containing $F(x_0)$. Since $F(x_0)$ is compact, G is a neighborhood of $F(x_0)$. Then $F_+^{-1}(G)$ is a neighborhood of x_0, and therefore there exists $\delta > 0$ such that $S_\delta(x_0) \subseteq F_+^{-1}(G)$. Thus for all $x \in S_\delta(x_0)$, $x \in F_+^{-1}(G)$. This implies that $F(x) \subseteq G$ for all $x \in S_\delta(x_0)$, and hence, F is upper semicontinuous. ■

Corollary 3.2 *Let X and Y be metric spaces and $F : X \rightrightarrows Y$ be a set-valued map such that $F(x)$ is compact for all $x \in X$. Then, F is upper semicontinuous if and only if for each closed subset H of Y the set $F^{-1}(H)$ is closed.*

Proof It follows from Theorem 3.5 (i) and Theorem 3.10. ■

Theorem 3.11 *Let X and Y be metric spaces and $F : X \rightrightarrows Y$ be an upper semicontinuous set-valued map such that for all $x \in X$, $F(x)$ is compact. Then, the image $F(K)$ of a compact subset K of X is compact.*

Proof Let $\{G_\alpha\}_{\alpha\in\Lambda}$ be an open cover of $F(K)$. Since, for every $x \in K$, the set $F(x)$ is compact, $F(x)$ can be covered by a finite number of G_α's. Let G_x denote the union of the sets in such a finite family. Then $\{F_+^{-1}(G_x) : x \in K\}$ is an open covering of K, and so it contains a finite covering $F_+^{-1}\left(G_{x_1}\right), F_+^{-1}\left(G_{x_2}\right), \dots, F_+^{-1}\left(G_{x_n}\right)$. The sets $G_{x_1}, G_{x_2}, \dots, G_{x_n}$ cover $F(K)$ and so $F(K)$ can be covered by a finite number of G_α's. ■

Now, we mention some properties of lower semicontinuous and upper semicontinuous set-valued maps which can be easily proved (proofs are given in [26] and [88]).

1. If F_1 is a lower semicontinuous (upper semicontinuous) set-valued map from a metric space X to another metric space Y and F_2 is another lower semicontinuous (upper semicontinuous) set-valued map from Y to a metric space Z, then the composition product $F = F_2 \circ F_1$ is also lower semicontinuous (upper semicontinuous) set-valued map from X to Z.
2. The arbitrary union $F = \bigcup_{\alpha\in\Lambda} F_\alpha$ of a family of lower semicontinuous set-valued maps F_α from a metric space X to another metric space Y is also lower semicontinuous.
   But the only finite union $F = \bigcup_{i=1}^{n} F_i$ of upper semicontinuous set-valued maps F_i from a metric space X to another metric space Y is upper semicontinuous.
3. The arbitrary intersection $F = \bigcap_{\alpha\in\Lambda} F_\alpha$ of a family of upper semicontinuous set-valued maps F_α from a metric space X to another metric space Y is upper semicontinuous. However, the finite intersection of lower semicontinuous set-valued maps need not be lower semicontinuous (see Example 3.7).
4. The Cartesian product $F = \prod_{i=1}^{n} F_i$ of finite number of lower semicontinuous (upper semicontinuous) set-valued maps F_i from a metric space X to another metric space Y_i is lower semicontinuous (upper semicontinuous) set-valued map from X to $Y = \prod_{i=1}^{n} Y_i$.

Example 3.7 Let $X = [0, \pi]$ and $Y = \mathbb{R}^2$ be the metric spaces with the usual metrics. Consider two set-valued maps $F_1, F_2 : [0, \pi] \rightrightarrows \mathbb{R}^2$ defined by

$$F_1(t) = \left\{(x, y) \in \mathbb{R}^2 : y \geq 0 \text{ and } x^2 + y^2 \leq 1\right\}, \quad \text{for all } t \in [0, \pi],$$

and

$$F_2(t) = \left\{(x, y) \in \mathbb{R}^2 : x = \lambda \cos t,\ y = \lambda \sin t,\ \lambda \in [-1, 1]\right\}, \quad \text{for all } t \in [0, \pi].$$

Then, F_1 is a constant set-valued map, and hence, continuous, F_2 is lower semicontinuous but $F_1 \cap F_2$ is no longer lower semicontinuous (to see it consider $t = 0$ or $t = \pi$).

Definition 3.9 Let X and Y be metric spaces. A set-valued map $F : X \rightrightarrows Y$ is said to be *closed* if whenever $x_0 \in X$, $y_0 \in Y$, $y_0 \notin F(x_0)$, there exist two neighborhoods $U(x_0)$ of x_0 and $V(y_0)$ of y_0 such that $F(x) \cap V(y_0) = \emptyset$ for all $x \in U(x_0)$.

Example 3.8 **(a)** If $f : X \times Y \to \mathbb{R}$ is a continuous single-valued function from a metric space X to another metric space Y, then the set-valued map $F : X \rightrightarrows Y$ defined by $F(x) = \{y \in Y : f(x, y) \leq 0\}$ is closed.

(b) If f is a continuous function from a metric space (X, d) to $\mathbb{R}$, then the set-valued map F from a metric space X to itself defined by

$$F(x) = \{y \in X : d(x, y) - f(x) \leq 0\}$$

is closed.

We state some useful characterizations of closed set-valued maps.

Remark 3.3 (a) The set-valued map F is closed if and only if its graph Graph(F) is a closed set.

(b) The set-valued map F is closed if and only if for any sequences $\{x_n\}$ and $\{y_n\}$ such that $x_n \to x_0$, $y_n \to y_0$ and $y_n \in F(x_n)$ for all n, we have $y_0 \in F(x_0)$.

The following example shows that the closedness of the set-valued map F defined on a subset K of X does not always imply that the set K is closed.

Example 3.9 Let $X = \mathbb{R}$ be a metric space with the usual metric and $K = (0, \infty)$. Consider the set-valued map $F : K \rightrightarrows K$ defined by

$$F(x) = \left(\frac{1}{x}, +\infty\right), \quad \text{for all } x > 0.$$

Then the graph of F is the set

$$\text{Graph}(F) = \{(x, y) \in X \times X : xy \geq 1\}.$$

The set-valued map F is closed as the Graph(F) is closed, while the set K is not closed.

Remark 3.4 It follows from (3.2) that the closedness of F and F^{-1} are equivalent.

Theorem 3.12 *Let X and Y be metric spaces and $F : X \rightrightarrows Y$ be a closed set-valued map. If K is a compact subset of X, then the set $F(K)$ is closed.*

Proof Let $y_n \in F(K)$ for all $n = 1, 2, \ldots$, and $y_n \to y$. Suppose that x_n is an element of $F^{-1}(y_n) \cap K$. Without loss of generality, we may assume that $\{x_n\}$ converges to some point x. It follows from the definition of closedness of F that $y \in F(x)$, and hence, $y \in F(K)$. ■

Corollary 3.3 *Let X and Y be metric spaces and $F : X \rightrightarrows Y$ be a closed set-valued map. Then, the set $F(x)$ is closed for all $x \in X$.*

The converse of the above corollary is not valid as shown by the following example.

Example 3.10 Let $X = \mathbb{R}_+$ be a metric space with the usual metric and $F : X \rightrightarrows X$ be a set-valued map defined by

$$F(x) = \begin{cases} [0, 1], & \text{when } x \neq 0, \\ \{0\}, & \text{when } x = 0. \end{cases}$$

Then, the set $F(x)$ is closed for all $x \in X$. However, the set-valued map F is not closed.

Remark 3.5 If F is a closed set-valued map, then the image of a closed (respectively, compact) set need not be closed (respectively, compact) (even if the set $F(x)$ is compact for all $x \in X$).

Example 3.11 Let $X = [1, +\infty)$ and $Y = \mathbb{R}_+$ be metric spaces with the usual metric and $F : X \rightrightarrows Y$ be a set-valued map defined as $F(x) = \left[\frac{1}{x}, 1\right]$ for all $x \in X$. Then, F is closed but the image $F(X)$ of the closed set X is not closed.

Example 3.12 Let $X = \mathbb{R}_+$ be a metric space with the usual metric, and $F : X \rightrightarrows \mathbb{R}$ be a set-valued map defined as

$$F(x) = \begin{cases} \{0\}, & \text{when } x = 0, \\ \left[-\frac{1}{x}, \frac{1}{x} + 1\right], & \text{when } x \neq 0. \end{cases}$$

Then, the map F is closed, but the image of the compact set $[0, 1]$ is not compact. Note that the set $F(x)$ is compact for all $x \in X$.

Exercise 3.1 For each $i = 1, 2, \dots, m$, let $F_i : X \rightrightarrows Y$ be a closed set-valued map from a metric space X to another metric space Y. Prove that $\bigcap_{i=1}^{m} F_i$ is also a closed set-valued map.

The composition of two closed set-valued maps need not be closed.

Example 3.13 Let $X = Y = Z = \mathbb{R}_+$ be metric spaces with the usual metric. Consider the set-valued maps $F_1 : X \rightrightarrows Y$ and $F_2 : Y \rightrightarrows Z$ defined by

$$F_1(x) = \begin{cases} \left\{\frac{1}{x}\right\}, & \text{if } x \neq 0, \\ \{0\}, & \text{if } x = 0, \end{cases}$$

and

$$F_2(x) = \begin{cases} \left\{\frac{1}{x}\right\}, & \text{if } x \neq 0, \\ \{1\}, & \text{if } x = 0, \end{cases}$$

respectively. Then, F_1 and F_2 are closed set-valued maps. However, the set-valued map $F_2 \circ F_1$ given by

$$\left(F_2 \circ F_1\right)(x) = \begin{cases} \{x\}, & \text{if } x \neq 0, \\ \{1\}, & \text{if } x = 0, \end{cases}$$

is not closed. Also, F_1 is not upper semicontinuous.

Theorem 3.13 *Let X and Y be metric spaces. If $F : X \rightrightarrows Y$ is an upper semicontinuous set-valued map with compact values, then it is closed.*

Proof Let $x_0 \in X$ be an arbitrary point and assume that $y_0 \notin F(x_0)$. Since $F(x_0)$ is compact, there exist an open set G in Y containing $F(x_0)$ and a neighborhood $V(y_0)$ of y_0 such that $G \cap V(y_0) = \emptyset$. Since F is upper semicontinuous, there exists a neighborhood $U(x_0)$ of x_0 such that $F(x) \subseteq G$ for all $x \in U(x_0)$. Then $F(x) \cap V(y_0) = \emptyset$ for all $x \in U(x_0)$. Therefore, F is closed. ∎

Remark 3.6 In general, if $f : X \to Y$ is a single-valued continuous map from X onto Y, then the inverse map $f^{-1} : Y \rightrightarrows X$ is a set-valued map and it has a closed graph. But it is not necessarily upper semicontinuous.

Theorem 3.14 *Let X and Y be metric spaces and $F : X \rightrightarrows Y$ be a closed set-valued map. If Y is compact, then F is upper semicontinuous.*

Proof Assume to the contrary that F is not upper semicontinuous. Then there exists an open neighborhood $V_{F(x)}$ of $F(x)$ in Y such that for every open neighborhood U_x of x in X we have $F(U_x) \not\subset V_{F(x)}$.

We take $U_x = S_{1/n}(x)$, $n = 1, 2, \ldots$ Then for each n, we get a point $x_n \in S_{1/n}(x)$ such that $F(x_n) \not\subset V_{F(x)}$. For every n, let $y_n \in F(x_n)$ such that $y_n \notin V_{F(x_n)}$. Then we have $x_n \to x \in X$ and $\{y_n\} \subset Y$. Since Y is compact, we may assume without loss of generality that $y_n \to y \in Y$. We see that $y \notin V_{F(x)}$. Therefore for each n, we have $(x_n, y_n) \in \text{Graph}(F)$ and $\{(x_n, y_n)\} \to (x, y)$. So $(x, y) \in \text{Graph}(F)$ because $\text{Graph}(F)$ is a closed subset of $X \times Y$. So, $y \in F(x)$ and hence $y \in V_{F(x)}$. This contradicts the fact that $y \notin V_{F(x)}$ and the proof is complete. ■

If Y is not compact, then the conclusion of Theorem 3.14 does not hold.

Example 3.14 Let $X = \mathbb{R}$ be a metric space with the usual metric and $F : X \rightrightarrows X$ be a set-valued map defined as

$$F(x) = \begin{cases} \left\{\frac{1}{x}\right\}, & \text{when } x \neq 0, \\ \{0\}, & \text{when } x = 0. \end{cases}$$

Then F is closed, but not upper semicontinuous.

Exercise 3.2 Let X and Y be metric spaces. If $F_1 : X \rightrightarrows Y$ is a closed set-valued map and $F_2 : X \rightrightarrows Y$ is an upper semicontinuous set-valued map with compact values, then prove that $F_1 \cap F_2$ is upper semicontinuous.

The following examples show that the lower semicontinuity and closedness of a set-valued map are not equivalent.

Example 3.15 Let $X = Y = \mathbb{R}_+$ be the metric space with the usual metric and $F : X \rightrightarrows Y$ be defined by

$$F(x) = \begin{cases} [0, 2], & \text{when } x \neq 0, \\ [0, 1], & \text{when } x = 0. \end{cases}$$

Then F is lower semicontinuous, but not closed, even $F(x)$ is closed.

Example 3.16 Let $X = Y = \mathbb{R}_+$ be the metric space with the usual metric and $F : X \rightrightarrows Y$ be defined by

$$F(x) = \begin{cases} [0, 2], & \text{when } x \neq 0, \\ [0, 3], & \text{when } x = 0. \end{cases}$$

Then F is closed, but not lower semicontinuous, even $F(x)$ is closed.

Exercise 3.3 Let X and Y be metric spaces and $F : X \rightrightarrows Y$ be an upper semicontinuous set-valued map with compact values. If $x_n \to x$ and $y_n \in F(x_n)$ for all n, then show that there exists a subsequence $\{y_{n_i}\}$ of the sequence $\{y_n\}$ which converges to a point $y \in F(x)$.

Definition 3.10 Let X and Y be metric spaces. A set-valued map $F : X \rightrightarrows Y$ is said to be *bounded* if the image of a bounded set is bounded.

Theorem 3.15 *Let X, Y and Z be metric spaces, and $F_1 : X \rightrightarrows Y$ and $F_2 : Y \rightrightarrows Z$ be closed set-valued maps. If F_1 is bounded, then $F_2 \circ F_1$ is a closed set-valued map from X to Z.*

Proof Consider the sequences $\{x_n\}$ in X and $\{z_n\}$ in $(F_2 \circ F_1)(x_n)$ such that $x_n \to x$ and $z_n \to z$. Then we have to show that $z \in (F_2 \circ F_1)(x)$.

For every n, there exists an element $y_n \in F_1(x_n)$ such that $z_n \in F_2(y_n)$. Since every convergent sequence is bounded, $\{x_n\}$ is bounded and the boundedness of F_1 implies that the sequence $\{y_n\}$ is bounded. Hence, without loss of generality, we may assume that it converges to some point y. The closedness of F_1 and F_2 implies that $y \in F_1(x)$ and $z \in F_2(y)$, that is, $z \in (F_2 \circ F_1)(x)$. ■

3.3 Hausdorff Metric and Hausdorff Continuity for Set-valued Maps

Let X be a metric space. We use the following notations:

2^X = Set of all subsets of X;

2^X_{cl} = Set of all nonempty closed and bounded subsets of X;

2^X_b = Set of all nonempty bounded subsets of X.

2^X_q = Set of all nonempty compact subsets of X.

Evidently, we have $2^X_q \subseteq 2^X_{cl}$.

Throughout this section, unless otherwise specified, we assume that X is a metric space with the metric d. For any two nonempty elements M and N of 2^X, we define the number $d(M, N)$ as follows: let $m \in M$ and set

$$d(m, N) = \inf_{n \in N} d(m, n)$$

and

$$d(M, N) = \sup_{m \in M} d(m, N) = \sup_{m \in M} \inf_{n \in N} d(m, n). \tag{3.3}$$

By convention, $\inf \emptyset = \infty$.

Proposition 3.1 *Let M, N, Q be nonempty bounded subsets of X. Then,*

(a) *$d(M, N) = 0$ if and only if $M \subseteq \overline{N}$;*

(b) *$d(M, N) \leq d(M, Q) + d(Q, N)$.*

Proof (a) Let $M, N \in 2^X_b$ be such that $d(M, N) = 0$. Then,

$$\begin{aligned}
\sup_{m \in M} d(m, N) = 0 &\Leftrightarrow d(m, N) = 0 \text{ for all } m \in M \\
&\Leftrightarrow m \in \overline{N} \text{ for all } m \in M \quad \text{(By Theorem 1.2)} \\
&\Leftrightarrow M \subseteq \overline{N}.
\end{aligned}$$

(b) From the triangle inequality, we have

$$d(m,n) \leq d(m,q) + d(q,n), \quad \text{for all } m \in M,\ n \in N,\ q \in Q.$$

This implies that

$$\inf_{n\in N} d(m,n) \leq d(m,q) + \inf_{n\in N} d(q,n), \quad \text{for all } q \in Q.$$

This gives further that

$$\inf_{n\in N} d(m,n) = d(m,N) \leq d(m,q) + d(q,N),$$

and since this is valid for all $q \in Q$, we get

$$d(m,N) \leq d(m,Q) + d(Q,N).$$

By taking $\sup_{m\in M}$, we obtain

$$d(M,N) \leq d(M,Q) + d(Q,N).$$

The proof is complete. ■

Exercise 3.4 Let M, N be nonempty bounded subsets of a metric space (X, d). If $M \subseteq N$, then prove that $d(x, N) \leq d(x, M)$ for all $x \in X$.

Proof Let $m \in M$ be arbitrary. Then, $m \in N$ as $M \subseteq N$, and therefore,

$$d(x,m) \geq \inf\{d(x,n) : n \in N\} = d(x,N), \quad \text{for all } x \in X.$$

Since $m \in M$ is arbitrary, we have $d(x,M) = \inf\{d(x,m) : m \in M\} \geq d(x,N)$ for all $x \in X$. ■

Exercise 3.5 Let M, N, Q be nonempty bounded subsets of a metric space (X, d). If $N \subseteq Q$, then prove that $d(M, Q) \leq d(M, N)$.

Proof Since $N \subseteq Q$, by Exercise 3.4, $d(m,Q) \leq d(m,N)$ for all $m \in M$. Therefore, $\sup\{d(m,Q) : m \in M\} \leq \sup\{d(m,N) : m \in M\}$ and the result follows. ■

Exercise 3.6 For any nonempty bounded subset M of a metric space (X, d), prove that

$$d(x,M) \leq d(x,y) + d(y,M), \quad \text{for all } x, y \in X.$$

Exercise 3.7 For any nonempty bounded subsets M, N, Q of a metric space (X, d), prove that

$$d(M \cup N, Q) = \max\{d(M,Q), d(N,Q)\}.$$

Proof We have

$$\begin{aligned} d(M \cup N, Q) &= \sup\{d(x,Q) : x \in M \cup N\} \\ &= \max\left\{\sup\left\{d(x,Q) : x \in M\right\}, \sup\left\{d(x,Q) : x \in N\right\}\right\} \\ &= \max\left\{d(M,Q), d(N,Q)\right\}. \end{aligned}$$

■

Exercise 3.8 Let (X,d) be a metric space, $M \in 2^X_q$ and $x \in X$. Prove that there exists $m_x \in M$ such that $d(x,M) = d(x,m_x)$.

Proof By the definition of infimum, we can let $\{m_k\}$ be a sequence in M such that $d(x,m_k) < d(x,M) + \frac{1}{k}$. Since M is compact, there exists a subsequence $\{m_{k_i}\}$ of $\{m_k\}$ that converges to a point $m_x \in M$. Then, we have

$$d(x,M) \le d(x,m_x) \le d(x,m_{k_i}) + d(m_{k_i},m_x) < d(x,M) + \frac{1}{k_i} + d(m_{k_i},m_x).$$

Since $\lim_{i\to\infty}\left(\frac{1}{k_i} + d(m_{k_i},m_x)\right) = 0$, we have $d(x,m_x) = d(x,M)$. ■

Exercise 3.9 Let (X,d) be a metric space and $M,N \in 2^X_q$. Prove that there exist $\tilde{m} \in M$ and $\tilde{n} \in N$ such that $d(M,N) = d(\tilde{m},\tilde{n})$.

Proof By the definition of supremum, we can let $\{m_k\}$ be a sequence in M such that $d(M,N)$ is the limit of $d(m_k,N)$. By Exercise 3.8, there exists a sequence $\{n_k\}$ in N such that $d(m_k,N) = d(m_k,n_k)$. Since M and N are compact, there exist sequences $\{m_{k_i}\}$ of $\{m_k\}$ and $\{n_{k_{i_j}}\}$ of $\{n_{k_i}\}$ which converge to $\tilde{m} \in M$ and $\tilde{n} \in N$, respectively. Therefore,

$$d(M,N) = \lim_{j\to\infty} d\left(m_{k_{i_j}}, N\right) = \lim_{j\to\infty} d\left(m_{k_{i_j}}, n_{k_{i_j}}\right) = d\left(\tilde{m},\tilde{n}\right).$$

The proof is complete. ■

The mapping d defined by (3.3) is not a metric on 2^X because of several problems. For example, in $\mathbb{R}$, what is the distance between $\{0\}$ and $[0,\infty)$? It is infinite. That is not allowed in the definition of a metric. Therefore, we will restrict the use of d to bounded sets. What is the distance $d(\varnothing,\{0\})$? Again, infinite. So we will restrict the use of d to nonempty sets. What is the distance $d((0,1),[0.1])$? Now the distance is zero, even though the two sets are not equal. Therefore, we will restrict the use of d to closed sets. In general,

$$d(M,N) \ne d(N,M).$$

To avoid the difficulty posed by the above remarks, we define *Hausdorff metric* $\mathcal{H}$ on the family of all nonempty closed and bounded subsets of a metric space (X,d) by

$$\mathcal{H}(M,N) = \max\left\{d(M,N), d(N,M)\right\}. \tag{3.4}$$

$\mathcal{H}(M,N)$ is the distance from the set M to the set N and it is called *Hausdorff distance* from the set M to the set N.

Example 3.17 Let $X = \mathbb{R}^2$ be a metric space with the usual metric, $M = \{(m_1, m_2) \in \mathbb{R}^2 : m_1^2 + m_2^2 \leq 1\}$ and $N = \{(n_1, n_2) \in \mathbb{R}^2 : 0 \leq n_1 \leq 4, 0 \leq n_2 \leq 1\}$ (see Figure 3.9).

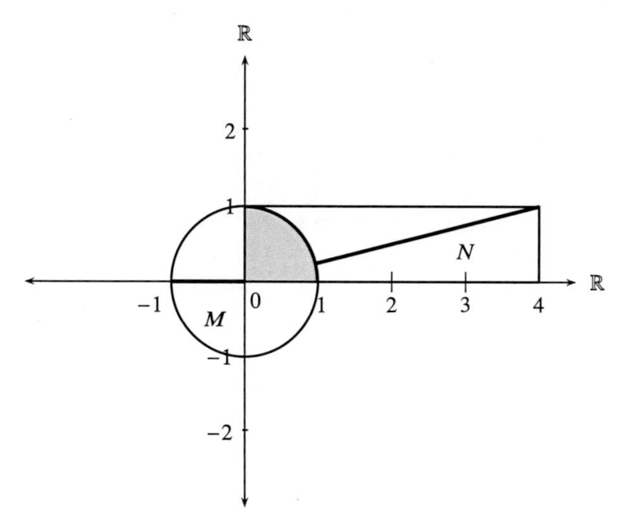

Figure 3.9 Distance between two intersecting sets

If $n \in M \cap N$, then $d(n, M) = 0$.

If $n \in N \setminus M$, then we suppose that the line joining n to the origin of the circle M intersects at the point $(x(n), y(n))$. Also, $d(n, M)$ is determined by taking infimum of all distances between n and $(x(n), y(n))$ for all $n \in N$. Then $d(N, M)$ is the largest distance from the upper right vertex of the rectangle at the point $(4, 1)$ to $(x(n), y(n))$. Therefore, $d(N, M)$ is equal to the distance from the point $\left(\frac{4}{\sqrt{17}}, \frac{1}{\sqrt{17}}\right)$ on the circle to the point $(4, 1)$. Hence,

$$d(N, M) = d\left(\left(\frac{4}{\sqrt{17}}, \frac{1}{\sqrt{17}}\right), (4, 1)\right) = \sqrt{17} - 1.$$

To find $d(M, N)$, we can use any point in the bottom lower left quadrant on the unit circle. Let us choose the point $(-1, 0)$. Then $d(M, N) = d((-1, 0), (0, 0)) = 1$, and therefore $\mathcal{H}(M, N) = \max\left\{1, \sqrt{17} - 1\right\} = \sqrt{17} - 1$.

In the following example we consider two sets which do not intersect.

Example 3.18 [24] Let $X = \mathbb{R}^2$ be a metric space with the usual metric, $M = \{(m_1, m_2) \in \mathbb{R}^2 : 0 \leq m_1 \leq 1, 0 \leq m_2 \leq 1\}$ and $N = \{(n_1, n_2) \in \mathbb{R}^2 : 3 \leq n_1 \leq 5, 0 \leq n_2 \leq 5\}$ (see Figure 3.10).

If $(m_1, m_2) \in M$, then $d((m_1, m_2), N) = d((m_1, m_2), (3, m_2)) = 3 - m_1$. Since $0 \leq m_1 \leq 1$, we find that $d(M, N) = 3$.

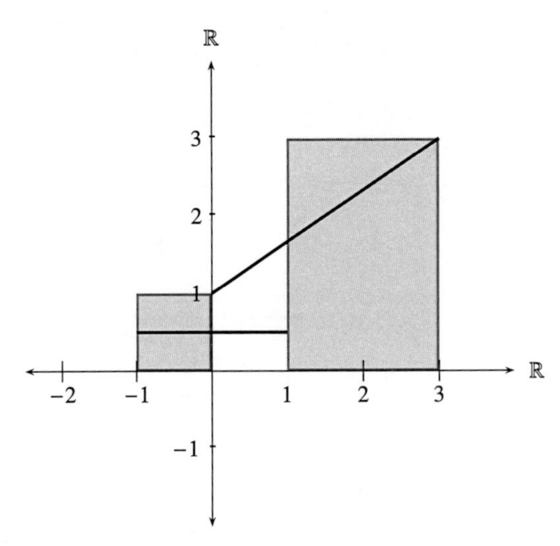

Figure 3.10 Distance between two nonintersecting sets

If $(n_1, n_2) \in N$, then $d((n_1, n_2), M) = d((n_1, n_2), (1, m_2)) = 3 - m_1$, where $0 \leq m_2 \leq 1$, which varies with our choice of (n_1, n_2). We find the point that maximizes $d((n_1, n_2), \text{M})$ is $(n_1, n_2) = (5, 4)$ such that $d(N, M) = d((5, 4), (1, 1)) = 5$. Therefore, $\mathcal{H}(M, N) = \max\{d(M, N), d(N, M)\} = 5$.

Theorem 3.16 *Let (X, d) be a metric space. Then, $\mathcal{H}(.,.)$ is a metric on 2^X_{cl}.*

Proof Let $M, N \in 2^X_{cl}$. Since M and N are closed and bounded, we have $\mathcal{H}(M, N) = d(m, n)$ for some $m \in M$ and $n \in N$. Hence $0 \leq \mathcal{H}(M, N) < \infty$.

Clearly, $\mathcal{H}(M, M) = \max\{d(M, M), d(M, M)\} = d(M, M) = \sup_{m \in M} d(m, M) = 0$.

Conversely, suppose that $\mathcal{H}(M, N) = 0$. If $m \in M$, then $d(m, N) = 0$. Since N is closed, by Theorem 1.2, $m \in N$. Therefore, $M \subseteq N$. Similarly, $N \subseteq M$, so $M = N$.

Clearly, $\mathcal{H}(M, N) = \mathcal{H}(N, M)$.

Finally, we prove the triangle inequality

$$\mathcal{H}(M, N) \leq \mathcal{H}(M, Q) + \mathcal{H}(Q, N), \quad \text{for all } M, N, Q \in 2^X_{cl}.$$

From Proposition 3.1 (b) we have

$$d(M, N) \leq d(M, Q) + d(Q, N) \quad \text{and} \quad d(N, M) \leq d(N, Q) + d(Q, M),$$

for all $M, N, Q \in 2^X_{cl}$. Hence,

$$\begin{aligned}
\mathcal{H}(M, N) &= \max\{d(M, N), d(N, M)\} \\
&\leq \max\{d(M, Q) + d(Q, N), d(N, Q) + d(Q, M)\} \\
&\leq \max\{d(M, Q), d(Q, M)\} + \max\{d(Q, N), d(N, Q)\} \\
&= \mathcal{H}(M, Q) + \mathcal{H}(Q, N).
\end{aligned}$$

This completes the proof. ∎

Remark 3.7 If $X = \mathbb{R}$ is a discrete metric space with discrete metric d_0 defined by

$$d_0(x,y) = \begin{cases} 0, & \text{if } x = y, \\ 1, & \text{if } x \neq y, \end{cases}$$

then $\left(2_q^X, \mathcal{H}_0\right)$ is a discrete metric space with the Hausdorff discrete metric $\mathcal{H}_0$ defined by

$$\mathcal{H}_0(M,N) = \begin{cases} 0, & \text{if } M = N, \\ 1, & \text{if } M \notin N. \end{cases}$$

Indeed, by Remark 1.12, the elements of 2_q^X are finite. Note that

$$d_0(x,N) = \inf\{d_0(x,n) : n \in N\} = \begin{cases} 0, & \text{if } x \in N, \\ 1, & \text{if } x \notin N. \end{cases}$$

Therefore,

$$d_0(M,N) = \max\{d_0(m,N) : m \in M\} = \begin{cases} 0, & \text{if } m \in N, \\ 1, & \text{if } m \notin N, \end{cases}$$

and hence,

$$\mathcal{H}_0(M,N) = \begin{cases} 0, & \text{if } M = N, \\ 1, & \text{if } M \notin N. \end{cases}$$

Exercise 3.10 Let $X = \mathbb{R}$ be a metric space with the usual metric. If $M = [0,10]$ and $N = [11,20]$, then prove that $\mathcal{H}(M,N) = 11$.

Proposition 3.2 *Let (X,d) be a metric space, $M, N \in 2_q^X$ and $r > 0$ be a given number. Then, $M \subseteq S_r[N]$ and $N \subseteq S_r[M]$ if and only if $\mathcal{H}(M,N) \leq r$.*

Proof We first show that $M \subseteq S_r[N]$ if and only if $d(M,N) \leq r$. Assume that $d(M,N) \leq r$. Since $d(M,N) = \sup_{m \in M} d(m,N) \leq r$, we obtain $d(m,N) \leq r$ for all $m \in M$. Since N is compact, there is a point $n \in N$ such that $d(m,n) \leq r$. Hence, each $m \in M$ belongs to a closed sphere of radius r with center at $n \in N$, that is, $m \in S_r[N]$ and hence $M \subseteq S_r[N]$.

Suppose that $M \subseteq S_r[N]$. Then for all $m \in M$, there is a point $n \in N$ such that $d(m,n) \leq r$. Thus for all $m \in M$, we have $d(m,N) \leq d(m,n) \leq r$. Since $d(M,N) = \sup_{m \in M} d(m,N)$, we conclude that $d(M,N) \leq r$.

By interchanging the sets M and N, we deduce that $N \subseteq S_r[M]$ if and only if $d(N,M) \leq r$.

Now $\mathcal{H}(M,N) = \max\left\{d(M,N), d(N,M)\right\} \leq r$ if and only if $d(M,N) \leq r$ and $d(N,M) \leq r$ if and only if $M \subseteq S_r[N]$ and $N \subseteq S_r[M]$. ■

In view of the above proposition, (3.4) can be rewritten as

$$\mathcal{H}(M,N) = \inf\left\{r > 0 : M \subseteq S_r[N] \text{ and } N \subseteq S_r[M]\right\}, \quad \text{for all } M, N \in 2_q^X.$$

Exercise 3.11 For any $M, N \in 2^X_{cl}$, prove that $d(m,N) \leq \mathcal{H}(M,N)$ for all $m \in M$.

Exercise 3.12 For any $M, N \in 2^X_{cl}$, prove that $d(x,M) \leq d(x,N) + \mathcal{H}(N,M)$ for all $x \in X$.

Exercise 3.13 Let (X,d) be a complete metric space. Prove the following statements.

(a) For each $m \in M \in 2^X_{cl}$ and $\varepsilon > 0$, there exists $n \in N \in 2^X_{cl}$ such that $d(m,n) \leq d(m,N) + \varepsilon$.
(b) For each $\varepsilon > 0$ and $m \in M \in 2^X_{cl}$, there exists $n \in N \in 2^X_{cl}$ such that $d(m,n) \leq \mathcal{H}(M,N) + \varepsilon$.

Exercise 3.14 For any $M, N \in 2^X_{cl}$, define $\mathscr{H}(M,N) = \frac{1}{2}\{d(M,N) + d(N,M)\}$.

(a) Prove that $\mathscr{H}$ is a metric on 2^X_{cl}.
(b) Prove that $\frac{1}{2}\mathcal{H}(M,N) \leq \mathscr{H}(M,N) \leq \mathcal{H}(M,N)$, and hence, $\mathcal{H}$ and $\mathscr{H}$ are equivalent metrics on 2^X_{cl}.
(c) Prove that $\mathscr{H} : 2^X_{cl} \times 2^X_{cl} \to \mathbb{R}$ is a continuous function.
(d) Give an example to show that two equivalent metrics on a metric space X may not generate equivalent Hausdorff metrics on 2^X_{cl}.

Exercise 3.15 Let (X,d) be a metric space and $\{M_k\}_{k\in\mathbb{N}}$ be a sequence of sets in 2^X_{cl}, and suppose that $\lim_{k\to\infty} \mathcal{H}(M_k, M_0) = 0$, where $M_0 \in 2^X_{cl}$. If $x_k \in M_k$ for all $k = 1, 2, \ldots$, and $\lim_{k\to\infty} x_k = x_0$, then prove that $x_0 \in M_0$.

To prove the completeness of the metric space $\left(2^X_q, \mathcal{H}\right)$, we need to show that the Cauchy sequence with respect to the metric $\mathcal{H}$ of nonempty compact sets in X converges with respect to the metric $\mathcal{H}$ to an element of 2^X_q. Before proving this, let us demonstrate how a sequence in the space $\left(2^X_q, \mathcal{H}\right)$ converges to an element of this space.

Example 3.19 Let $X = \mathbb{R}^2$ be a metric space with the usual metric. For each positive integer k, let $M_k = \{(r,\theta) : r = 1 + \frac{1}{k}\cos(k\theta),\ 0 \leq \theta \leq 2\pi\}$, and let M be the unit sphere in $\mathbb{R}^2$. Then we show that the sequence $\{M_k\}$ converges to the set M with respect to the Hausdorff metric.

For each k, we can see that M_k is nonempty compact set and hence the sequence $\{M_k\}$ lies in 2^X_q. We can also easily obtain that $\mathcal{H}(M_k, M) = \frac{1}{k}$.

Note that as k increases, the sequences $\{M_k\}$ converges to M. Therefore, $\{M_k\}$ is a Cauchy sequence that converges to $M \in 2^X_q$.

To prove that $\left(2^X_q, \mathcal{H}\right)$ is complete when the metric space (X,d) is complete, we first need to establish the following lemmas.

Lemma 3.1 [24] *Let (X,d) be a metric space, $\{M_k\}$ be a Cauchy sequence in the metric space $\left(2^X_q, \mathcal{H}\right)$, and $\{k_i\}$ be an increasing sequence of positive integers. If $\{m_{k_i}\}$ is a Cauchy sequence in X with $m_{k_i} \in M_{k_i}$ for all i, then there exists a Cauchy sequence $\{n_k\}$ in X such that $n_k \in M_k$ for all k and $n_{k_i} = m_{k_i}$ for all i.*

Proof Let $k_0 = 0$. For each k such that $k_{i-1} < k \leq k_i$, by Exercise 3.8, there exists $n_k \in M_k$ such that $d(m_{k_i}, M_k) = d(m_{k_i}, n_k)$. Then,

$$d(m_{k_i}, n_k) = d(m_{k_i}, M_k) \leq d(M_{k_i}, M_k) \leq \mathcal{H}(M_{k_i}, M_k).$$

Since $d(m_{k_i}, n_{k_i}) = d(m_{k_i}, M_{k_i}) = 0$ for all $m_{k_i} \in M_{k_i}$, it follows that $n_{n_i} = m_{k_i}$ for all i.

Let $\varepsilon > 0$. Since $\{m_{k_i}\}$ is a Cauchy sequence in X, there exists a positive integer $\mathcal{J}$ such that $d(m_{k_i}, m_{k_j}) < \frac{\varepsilon}{3}$ for all $i,j \geq \mathcal{J}$. Since $\{M_k\}$ is a Cauchy sequence in 2^X_q with respect to the metric $\mathcal{H}$, there exists a positive integer $\mathcal{N} \geq k_{\mathcal{J}}$ such that $\mathcal{H}(M_k, M_p) < \frac{\varepsilon}{3}$ for all $k,p \geq \mathcal{N}$. Suppose that $k,p \geq \mathcal{N}$, then there exist integers $i,j \geq \mathcal{J}$ such that $k_{i-1} < k \leq k_i$ and $k_{j-1} < p \leq k_j$, and we have

$$\begin{aligned}
d(n_k, n_p) &\leq d(n_k, m_{k_i}) + d(m_{k_i}, m_{k_j}) + d(m_{k_j}, n_p) \\
&= d(m_{k_i}, M_k) + d(m_{k_i}, m_{k_j}) + d(m_{k_j}, M_p) \\
&\leq \mathcal{H}(M_{k_i}, M_k) + d(m_{k_i}, m_{k_j}) + \mathcal{H}(M_{k_j}, M_p) \\
&< \frac{\varepsilon}{3} + \frac{\varepsilon}{3} + \frac{\varepsilon}{3} = \varepsilon.
\end{aligned}$$

Therefore, $\{n_k\}$ is a Cauchy sequence in X with $n_k \in M_k$ for all k and $n_{k_i} = m_{k_i}$ for all i. ∎

Lemma 3.2 [24] *Let (X, d) be a complete metric space and M be the set of all points $x \in X$ such that there is a sequence $\{m_k\}$ with $m_k \in M_k$ for all k that converges to x. If $\{M_k\}$ is a Cauchy sequence in $\left(2^X_q, \mathcal{H}\right)$, then the set M is nonempty and closed.*

Proof Since $\{M_k\}$ is a Cauchy sequence, there exists an integer J_1 such that $\mathcal{H}(M_k, M_j) < \frac{1}{2}$ for all $k,j \geq J_1$. Similarly, there exists $J_2 > J_1$ such that $\mathcal{H}(M_k, M_j) < \frac{1}{2^2}$ for all $k,j \geq J_2$. Continuing in this way, we obtain an increasing sequence $\{J_i\}$ such that $\mathcal{H}(M_k, M_j) < \frac{1}{2^i}$ for all $k,j \geq J_i$. Let m_{j_1} be a fixed point in M_{j_1}. Then we can choose $m_{j_2} \in M_{j_2}$ such that $d(m_{j_1}, m_{j_2}) = d(m_{j_1}, M_{j_2})$. Then, we have

$$d(m_{j_1}, m_{j_2}) = d(m_{j_1}, M_{j_2}) \leq d(M_{j_1}, M_{j_2}) \leq \mathcal{H}(M_{j_1}, M_{j_2}) < \frac{1}{2}.$$

Similarly, we can choose $m_{j_3} \in M_{j_3}$ such that

$$d(m_{j_2}, m_{j_3}) = d(m_{j_2}, M_{j_3}) \leq d(M_{j_2}, M_{j_3}) \leq \mathcal{H}(M_{j_2}, M_{j_3}) < \frac{1}{2^2}.$$

Continuing in this way, we obtain a sequence $\{m_{j_i}\}$ with $m_{j_i} \in M_{j_i}$ and for all i

$$d(m_{j_i}, m_{j_{i+1}}) = d(m_{j_i}, M_{j_{i+1}}) \leq d(M_{j_i}, M_{j_{i+1}}) \leq \mathcal{H}(M_{j_i}, M_{j_{i+1}}) < \frac{1}{2^i}.$$

Then, $\{m_{j_i}\}$ is a Cauchy sequence with $m_{j_i} \in M_{j_i}$ for all i. By Lemma 3.1, there exists a Cauchy sequence $\{n_k\}$ in X such that $n_k \in M_k$ for all k and $n_{k_i} = m_{k_i}$ for all i. Since X is complete, the Cauchy sequence $\{n_k\}$ converges to a point $n \in X$. Since $n_k \in M_k$ for all k and each M_k is compact, we have that $n \in M$. Therefore, M is nonempty.

To prove that the set M is closed, we assume that m is a limit point of M. Then, there exists a sequence $\{m_i\} \in M \setminus \{m\}$ that converges to m. Since each $m_i \in M$, there exists a sequence $\{n_k\}$ that converges to m_i with $n_k \in M_k$ for all k. Consequently, there exists an integer k_1 such that $m_{k_1} \in M_{k_1}$ and $d(m_{k_1}, m_1) < 1$. Similarly, there exist an integer $k_2 > k_1$ and a point $m_{k_2} \in M_{k_2}$ such that

$d(m_{k_2}, m_2) < \frac{1}{2}$. Continuing in this way, we obtain an increasing sequence $\{k_i\}$ of integers such that $d(m_{k_i}, m_i) < \frac{1}{i}$ for all i. Then

$$d(m_{k_i}, m) \leq d(m_{k_i}, m_i) + d(m_i, m) \rightarrow 0 \text{ as } i \rightarrow \infty,$$

that is, $\{m_{k_i}\}$ converges to m, and therefore it is Cauchy with $m_{k_i} \in M_{k_i}$ for all i. By Lemma 3.2, there exists a Cauchy sequence $\{n_k\}$ in X such that $n_k \in M_k$ for all k and $n_{k_i} = m_{k_i}$. Therefore, $m \in M$ and hence M is closed. ∎

Lemma 3.3 [24] *Let (X, d) be a metric space, $\{M_k\}$ be a sequence of totally bounded sets in X and M be a subset of X. If for each $\varepsilon > 0$, there exists a positive integer $\mathcal{N}$ such that $M \subseteq S_\varepsilon[M_\mathcal{N}]$, then M is totally bounded.*

Proof Let $\varepsilon > 0$. Choose a positive integer $\mathcal{N}$ such that $M \subseteq S_{\varepsilon/4}[M_\mathcal{N}]$. Since $M_\mathcal{N}$ is totally bounded, we can choose a finite set $\{m_1, m_2, \ldots, m_q\}$ of $M_\mathcal{N}$ such that $M_\mathcal{N} \subseteq \bigcup_{i=1}^{q} S_{\varepsilon/4}(m_i)$. By reordering the m_i's, we may assume that $S_{\varepsilon/2}(m_i) \cap M \neq \emptyset$ for $1 \leq i \leq p$ and $S_{\varepsilon/2}(m_i) \cap M = \emptyset$ for $i > p$. Then for each $1 \leq i \leq p$, let $n_i \in S_{\varepsilon/2}(x_i) \cap M$. We claim that $M \subseteq \bigcup_{i=1}^{p} S_\varepsilon(n_i)$. Let $m \in M$. Then $m \in S_{\varepsilon/4}[M_\mathcal{N}]$, and so $d(m, M_\mathcal{N}) \leq \frac{\varepsilon}{4}$. By Exercise 3.8, there exists $x \in M_\mathcal{N}$ such that $d(m, x) = d(m, M_\mathcal{N})$, and therefore

$$d(m, m_i) \leq d(m, x) + d(x, m_i) \leq \frac{\varepsilon}{4} + \frac{\varepsilon}{4} = \frac{\varepsilon}{2}.$$

Thus, $x \in S_{\varepsilon/2}[m_i]$ for some $1 \leq i \leq p$, and therefore $n_i \in S_{\varepsilon/2}(m_i) \cap M$ such that $d(m_i, n_i) < \frac{\varepsilon}{2}$. It follows that

$$d(m, n_i) \leq d(m, m_i) + d(m_i, n_i) < \frac{\varepsilon}{2} + \frac{\varepsilon}{2} = \varepsilon.$$

Since for each $m \in M$, we found n_i for $1 \leq i \leq p$ such that $m \in S_\varepsilon(n_i)$, it follows that $M \subseteq \bigcup_{i=1}^{p} S_\varepsilon(n_i)$. Hence M is totally bounded. ∎

Theorem 3.17 *If the metric space (X, d) is complete, then $\left(2_q^X, \mathcal{H}\right)$ is also complete.*

Proof Let $\{M_k\}$ be a Cauchy sequence in the metric space $\left(2_q^X, \mathcal{H}\right)$ and M be the set of all points $m \in X$ such that there exists a sequence $\{m_k\}$ with $m_k \in M_k$ for all k that converges to m. We need to prove that $M \in 2_q^X$ and $\{M_k\}$ converges to M with respect to the metric $\mathcal{H}$.

By Lemma 3.2, the set M is nonempty and closed. Let $\varepsilon > 0$. Since $\{M_k\}$ is a Cauchy sequence in the metric space $\left(2_q^X, \mathcal{H}\right)$, there exists a positive integer $\mathcal{N}$ such that

$$\mathcal{H}(M_k, M_p) < \varepsilon, \quad \text{for all } k, p \geq \mathcal{N}.$$

By Proposition 3.2, $M_p \subseteq S_\varepsilon[M_k]$ for all $p > k \geq \mathcal{N}$. Let $m \in M$. Then we show that $m \in S_\varepsilon[M_k]$. Fix $p \geq \mathcal{N}$. By the definition of M, there exists a sequence $\{m_q\}$ with $m_q \in M_q$ for all q converges to m. Since $S_\varepsilon[M_k]$ is closed and $m_q \in S_\varepsilon[M_k]$ for each q, it follows that $m \in S_\varepsilon[M_k]$, that is, $M \subseteq S_\varepsilon(M_k)$. By Lemma 3.3, M is totally bounded. Since M is a closed subset of a complete metric space, M is complete. Therefore, M is a nonempty complete and totally bounded subset of X, and hence M is compact.

Let $\varepsilon > 0$. Now we prove that $\{M_k\}$ converges to M with respect to the metric $\mathcal{H}$, that is, there exists a positive integer $\mathcal{N}$ such that $\mathcal{H}(M_k, M) < \varepsilon$ for all $k \geq \mathcal{N}$. In view of Proposition 3.2, we need to show that $M \subseteq S_\varepsilon[M_k]$ and $M_k \subseteq S_\varepsilon[M]$. From the first part of the proof, there exists $\mathcal{N}$ such that $M \subseteq S_\varepsilon[M_k]$ for all $k \geq \mathcal{N}$.

Now we prove that $M_k \subseteq S_\varepsilon[M]$. Let $\varepsilon > 0$. Since $\{M_k\}$ is a Cauchy sequence, there exists a positive integer $\mathcal{N}$ such that $\mathcal{H}(M_p, M_k) < \frac{\varepsilon}{2}$ for all $p, k \geq \mathcal{N}$. Since $\{M_k\}$ is a Cauchy sequence in 2^X_q, there exists a strictly increasing sequence $\{k_i\}$ of positive integers such that $k_1 > \mathcal{N}$ and $\mathcal{H}(M_p, M_k) < \frac{\varepsilon}{2^{i+1}}$ for all $p, k > k_i$. By Exercise 3.8, we obtain the following:

Since $M_k \subseteq S_{\varepsilon/2}[M_{k_1}]$, there exists $m_{k_1} \in M_{k_1}$ such that $d(n, m_{k_1}) \leq \frac{\varepsilon}{2}$.

Since $M_{k_1} \subseteq S_{\varepsilon/2^2}[M_{k_2}]$, there exists $m_{k_2} \in M_{k_2}$ such that $d(m_{k_1}, m_{k_2}) \leq \frac{\varepsilon}{2^2}$.

Since $M_{k_2} \subseteq S_{\varepsilon/2^3}[M_{k_3}]$, there exists $m_{k_3} \in M_{k_3}$ such that $d(m_{k_2}, m_{k_3}) \leq \frac{\varepsilon}{2^3}$.

Continuing in this way, we obtain a sequence $\{m_{k_i}\}$ such that for all positive integers i, $m_{k_i} \in M_{k_i}$ and $d(m_{k_i}, m_{k_{i+1}}) \leq \frac{\varepsilon}{2^{i+1}}$. Therefore, $\{m_{k_i}\}$ is a Cauchy sequence and thus by Lemma 3.1, its limit is m in M. We also have

$$\begin{aligned} d(n, m_{k_i}) &\leq d(n, m_{k_1}) + d(m_{k_1}, m_{k_2}) + \cdots + d(m_{k_{i-1}}, m_{k_i}) \\ &\leq \frac{\varepsilon}{2} + \frac{\varepsilon}{2^2} + \cdots + \frac{\varepsilon}{2^i} < \varepsilon. \end{aligned}$$

Since $d(n, m_{k_i}) \leq \varepsilon$ for all i, it follows that $d(n, m) \leq \varepsilon$, and therefore $n \in S_\varepsilon[M]$. Thus there exists $\mathcal{N}$ such that $M_k \subseteq S_\varepsilon[M]$ for all $k \geq \mathcal{N}$. Therefore, $\mathcal{H}(M_k, M) < \varepsilon$ for all $k \geq \mathcal{N}$, and hence $\{M_k\}$ converges to $M \in 2^X_q$. ■

Exercise 3.16 If the metric space (X, d) is complete, then prove that the metric space $\left(2^X_{cl}, \mathscr{H}\right)$ is also complete.

Note that the topology on 2^X_q derived from the Hausdorff distance $\mathcal{H}$ is not determined by the metric topology of (X, d). Two topologically equivalent metrics d and d' may lead to very different topologies on 2^X_q by the Hausdorff distance procedure. It follows from the example given below.

Example 3.20 Let $X = \mathbb{R}_+ = [0, +\infty)$,

$$d(x, y) = \left| \frac{x}{(1+x)} - \frac{y}{(1+y)} \right|,$$

and

$$d'(x, y) = \min\{1, |x - y|\}.$$

The metrics d and d' define the same topology on $\mathbb{R}_+$, but the topologies of the Hausdorff distance on 2^X_q are different, that is, the set $\mathbb{N}$ of natural numbers belongs to the closure of the set of all finite subsets of $\mathbb{N}$ in the first space but not in the second.

The above example shows that 2^X_q is not a closed subset of $\left(2^X_{cl}, \mathcal{H}\right)$ in general.

Exercise 3.17 If (X, d) is a complete metric space, then prove that 2^X_q is a closed subset of $\left(2^X_{cl}, \mathcal{H}\right)$.

Exercise 3.18 If the metric space (X, d) is complete, then prove that 2_q^X is a closed subset of $\left(2_{cl}^X, \mathscr{H}\right)$.

Theorem 3.18 *If the metric space (X, d) is totally bounded, then $\left(2_q^X, \mathcal{H}\right)$ is totally bounded.*

Proof Let $\varepsilon > 0$. Since X is totally bounded, there exists a finite subset $\{x_1, x_2, \dots, x_n\}$ of X such that $X \subseteq \bigcup_{i=1}^n S_{\varepsilon/3}(x_i)$. Let $C = \{C_k : 1 \leq k \leq 2^n - 1\}$ be the collection of all possible nonempty unions of the closure of these spheres. Since X is compact, the closure of each sphere is a compact set. Therefore, each C_k is a finite union of compact sets and hence compact; thus $C_k \in 2_q^X$.

Now we want to show that the sets $\mathbb{S}_\varepsilon(C_k)$ form a finite open cover of X, where $\mathbb{S}_\varepsilon(C_k)$ denotes the open sphere with center at C_k and radius $\varepsilon > 0$ in the metric space $\left(2_q^X, \mathcal{H}\right)$. It is sufficient to show that $X \subseteq \bigcup_{k=1}^{2^n-1} \mathbb{S}_\varepsilon(C_k)$.

Let $A \in 2_q^X$. Then we will show that A lies in one of $\mathbb{S}_\varepsilon(C_k)$, that is, $A \in \mathbb{S}_\varepsilon(C_k)$ for some k. Consider $D_A = \{x_i : A \cap \overline{S_\varepsilon(x_i)} \neq \emptyset\}$ and choose $C_j = \bigcup_{x_i \in D_A} \overline{S_{\varepsilon/3}(x_i)}$. Since $A \subseteq C_j$, by Proposition 3.1 (a), we have $d(A, C_j) = 0$. Let c be an element in C_j. Then there exist $x_i \in D_A$ and $a \in A$ such that $c, a \in \overline{S_{\varepsilon/3}(x_i)}$. This implies that $d(c, A) \leq \frac{2}{3}\varepsilon$. Since c was arbitrary, we obtain $d(C_j, A) \leq \frac{2}{3}\varepsilon$. Therefore, $\mathcal{H}(A, C_j) = d(C_j, A) < \varepsilon$, and hence, $A \subseteq \mathbb{S}_\varepsilon(C_j)$. Thus, 2_q^X is totally bounded. ∎

Theorem 3.19 *If the metric space (X, d) is compact, then $\left(2_q^X, \mathcal{H}\right)$ is compact.*

Proof Since every totally bounded and complete metric space is compact, by Theorem 3.18 and Theorem 3.17, we obtain the result. ∎

Theorem 3.20 *If (X, d) is a separable metric space, then $\left(2_q^X, \mathcal{H}\right)$ is also separable.*

Definition 3.11 Let (X, d_1) and (Y, d_2) be metric spaces. A set-valued map $F : X \to 2_q^Y$ is said to be *$\mathcal{H}$-continuous* if for any $\varepsilon > 0$, there exists a $\delta > 0$ such that for all $x, y \in X$,

$$\mathcal{H}(F(x), F(y)) < \varepsilon \quad \text{whenever} \quad d_1(x, y) < \delta.$$

Theorem 3.21 *Let X and Y be metric spaces. A mapping $F : X \to 2_q^Y$ is $\mathcal{H}$-continuous if and only if the set-valued map $F : X \rightrightarrows Y$ is compact-valued and upper semicontinuous as well as lower semicontinuous on X.*

Proof Let F be $\mathcal{H}$-continuous and G be an open subset of Y. We first show that F is upper semicontinuous by proving that the set $F_+^{-1}(G) = \{x \in X : F(x) \subseteq G\}$ is open in X.

Let $x_0 \in F_+^{-1}(G)$. Then $F(x_0) \subseteq G$. Since $F(x_0)$ is compact, there exists $\varepsilon > 0$ such that $S_\varepsilon(F(x_0)) \subseteq G$. Since F is $\mathcal{H}$-continuous, we can find a $\delta > 0$ such that for all $x \in S_\delta(x_0)$, we have

$$\mathcal{H}(F(x_0), F(x)) < \varepsilon.$$

This implies that $F(x) \subseteq S_\varepsilon(F(x_0)) \subseteq G$, and so $S_\delta(x_0) \subseteq F_+^{-1}(G)$. Thus $F_+^{-1}(G)$ is open in X.

Now we show that F is lower semicontinuous by proving that the set $F^{-1}(G) = \{x \in X : F(x) \cap G \neq \emptyset\}$ is open in X.

Let $x_0 \in F^{-1}(G)$. Then $F(x_0) \cap G \neq \emptyset$ and so we let $y_0 \in F(x_0) \cap G$. Since G is open and $y_0 \in G$, there exists $\varepsilon > 0$ such that $S_\varepsilon(y_0) \subseteq G$. Since F is $\mathcal{H}$-continuous, we can find a $\delta > 0$ such that for all $x \in S_\delta(x_0)$, we have

$$\mathcal{H}(F(x_0), F(x)) < \frac{\varepsilon}{2}.$$

We claim that $F(x) \cap S_\varepsilon(y_0) \neq \emptyset$. Assume to the contrary that $F(x) \cap S_\varepsilon(y_0) = \emptyset$. On the other hand $F(x_0) \subset S_{\varepsilon/2}(F(x))$. Therefore, $y_0 \in S_{\varepsilon/2}(F(x))$ and there exists $z_0 \in F(x)$ such that $d(y_0, z_0) < \varepsilon/2$. This implies that $z_0 \in S_\varepsilon(y_0)$ and we obtain a contradiction.

Conversely, assume that F is both upper semicontinuous as well as lower semicontinuous. Let $\varepsilon > 0$ and $x_0 \in X$. Let $U = S_\varepsilon(F(x_0))$. Then the sets $F^{-1}(U)$ and $F_+^{-1}(U)$ are open and $x_0 \in F^{-1}(U) \cap F_+^{-1}(U)$. Let $V = F^{-1}(U) \cap F_+^{-1}(U)$. Then V is an open neighborhood of x_0 such that $F(x) \subset S_\varepsilon(F(x_0))$ for all $x \in V$. We are looking for $\delta > 0$ such that $S_\delta(x_0) \subset V$ and $F(x_0) \subset S_\varepsilon(F(x))$ for every $x \in S_\delta(x_0)$.

To find such δ we cover the compact set by n open spheres $S_\varepsilon(y_i)$, $i = 1, 2, \dots, n$. Then $F(x_0) = \bigcup_{i=1}^n S_\varepsilon(y_i) \subset S_{\varepsilon/2}(F(x_0))$. Since F is lower semicontinuous, there are open spheres $S_{\delta_i}(x_0) \subset V$ such that

$$F(x) \cap S_{\varepsilon/2}(y_i) \neq \emptyset \quad \text{for all } x \in S_{\delta_i}(x_0).$$

Let $\delta = \min\{\delta_1, \delta_2, \dots, \delta_n\}$. Then $S_\delta(x_0) \subset V$ and any $y \in F(x_0)$ belongs to $S_{\varepsilon/2}(y_i)$ for some i. Furthermore, we know that for any $x \in S_\delta(x_0)$, $F(x) \cap S_{\varepsilon/2}(y_i) \neq \emptyset$ for all $i = 1, 2, \dots, n$. Thus for every $x \in S_\delta(x_0)$ and $y \in F(x_0)$, there exists $i = 1, 2, \dots, n$ such that

$$d(y, F(x)) \leq d(y, y_i) + d(y_i, F(x_i)) < \frac{\varepsilon}{2} + \frac{\varepsilon}{2} = \varepsilon.$$

Therefore for all $x \in S_\delta(x_0)$, we obtain

$$F(x_0) \subset S_\varepsilon(F(x)).$$

This completes the proof. ■

Note that if F is only upper semicontinuous or lower semicontinuous, then F is not $\mathcal{H}$-continuous in general.

Example 3.21 Let $X = [0, 1]$ be a metric space with the usual metric. Consider the set-valued map $F : X \rightrightarrows X$ defined by

$$F(x) = \begin{cases} \{1\}, & \text{if } x < \frac{1}{2}, \\ [0,1], & \text{if } x = \frac{1}{2}, \\ \{0\}, & \text{if } x > \frac{1}{2}. \end{cases}$$

Then, F is upper semicontinuous but not $\mathcal{H}$-continuous.

The following example shows that the only lower semicontinuity does not imply $\mathcal{H}$-continuity.

Example 3.22 Let $X = Y = [0, 1]$ be a metric space with the usual metric and $F : [0, 1] \to 2_q^{[0,1]}$ be defined by

$$F(x) = \begin{cases} [0,1], & \text{when } x \neq 0, \\ \{0\}, & \text{when } x = 0. \end{cases}$$

Then, F is lower semicontinuous with compact values but

$$\mathcal{H}(F(0), F(x)) = 1, \quad \text{for all } x \neq 0,$$

so, F is not $\mathcal{H}$-continuous.

Exercise 3.19 Let (X, d) be a metric space. If $F : X \to 2^X_{cl}$ is $\mathcal{H}$-continuous, then prove that the set-valued map $F : X \rightrightarrows X$ is lower semicontinuous.

Note that the $\mathcal{H}$-continuity of a set-valued map $F : X \to 2^X_{cl}$ does not imply upper semicontinuity.

Example 3.23 Let $X = \mathbb{R}$ be the usual metric space and $Y = \mathbb{R}^2$ be equipped with the bounded metric d^* defined by

$$d^*(x, y) = \frac{d(x, y)}{1 + d(x, y)},$$

where d is the usual metric on $\mathbb{R}^2$.

Consider the mapping $F : \mathbb{R} \to 2^{\mathbb{R}^2}_{cl}$ defined as

$$F(t) = \{(t, y) : y \in \mathbb{R}\}, \quad \text{for all } t \in \mathbb{R}.$$

Then, we have

$$\mathcal{H}(F(t), F(s)) \leq 2|t - s|, \quad \text{for all } t, s \in \mathbb{R},$$

and so, F is $\mathcal{H}$-continuous.

Let $U = \left\{(x, y) \in \mathbb{R}^2 : |y| < 1/x \text{ or } x = 0\right\}$. Then U is an open subset of $\mathbb{R}^2$ but $F_+^{-1}(U) = \{0\}$ is not open in $\mathbb{R}$. Consequently, F is not upper semicontinuous.

Exercise 3.20 Let X be a metric space and $F : X \to 2^X_{cl}$ be a $\mathcal{H}$-continuous set-valued mapping. Prove that the mapping $g : K \to [0, \infty)$ defined by $g(x) = d(x, F(x))$ is lower semicontinuous.

Proof Let $\{x_n\}$ be a sequence in X that converges to a point x. Then for any $z_n \in F(x_n)$, we have

$$\begin{aligned} g(x) = d(x, F(x)) &\leq d(x, x_n) + d(x_n, z_n) + d(z_n, F(x)) \\ &\leq d(x, x_n) + d(x_n, z) + \mathcal{H}(F(x), F(x_n)). \end{aligned}$$

Since $z_n \in F(x_n)$ is arbitrary, we have

$$g(x) \leq d(x, x_n) + d(x_n, F(x_n)) + \mathcal{H}(F(x), F(x_n)).$$

Taking the lim inf and using the $\mathcal{H}$-continuity of F, we obatin

$$g(x) \leq \liminf_{n \to \infty} g(x_n),$$

that is, g is lower semicontinuous. ∎

3.4 Fixed Points for Set-valued Maps

Definition 3.12 Let X be a nonempty set and $F : X \rightrightarrows X$ be a set-valued map with nonempty values. A point $x \in X$ is said to be a *fixed point* of F if $x \in F(x)$.

Definition 3.13 Let (X, d_1) and (Y, d_2) be two metric spaces. A mapping $F : X \to 2^Y_{cl}$ is said to be *set-valued Lipschitz map* if there exists a constant $\alpha > 0$ such that

$$\mathcal{H}(F(x), F(y)) \leq \alpha d_1(x, y), \quad \text{for all } x, y \in X.$$

The constant α is called a *Lipschitz constant* for F.
- If $\alpha < 1$, then F is called a *set-valued contraction map*.
- If $\alpha = 1$, then F is called *nonexpansive*.

Example 3.24 Let $X = [0, 1]$ be a metric space with the usual metric and let $f : X \to X$ be given by

$$f(x) = \begin{cases} \frac{1}{2}x + \frac{1}{2}, & \text{if } 0 \leq x \leq \frac{1}{2}, \\ -\frac{1}{2}x + 1, & \text{if } \frac{1}{2} \leq x \leq 1. \end{cases}$$

Define $F : X \rightrightarrows X$ by $F(x) = \{0\} \cup \{f(x)\}$ for all $x \in X$. Then, F is a set-valued contraction map and the fixed points of F are 0 and $\frac{2}{3}$.

Remark 3.8 Let $F, G : X \to 2^Y_{cl}$ be set-valued Lipschitz maps with Lipschitz constant α and β, respectively. Then, $F \cup G : X \to 2^Y_{cl}$ defined by $(F \cup G)(x) = F(x) \cup G(x)$, for all $x \in X$, is a set-valued Lipschitz map with Lipschitz constant $\max\{\alpha, \beta\}$.

But the intersection of two set-valued contraction maps need not be a contraction.

Example 3.25 Let $I^2 = \{(x, y) : 0 \leq x \leq 1 \text{ and } 0 \leq y \leq 1\}$, let $F : I^2 \to 2^{I^2}_{cl}$ be defined by

$$F(x, y) = \text{line segment in } I^2 \text{ from the point } \left(\frac{1}{2}x, 0\right) \text{ to the point } \left(\frac{1}{2}x, 1\right),$$
$$\text{for all } (x, y) \in I^2,$$

and let $G : I^2 \to 2^{I^2}_{cl}$ be defined by

$$G(x, y) = \text{line segment in } I^2 \text{ from the point } \left(\frac{1}{2}x, 0\right) \text{ to the point } \left(\frac{1}{3}x, 1\right),$$
$$\text{for all } (x, y) \in I^2.$$

Then, F and G are set-valued contraction maps and $F \cap G$, defined by

$$(F \cap G)(x, y) = \begin{cases} \left\{\left(\frac{1}{2}x, 0\right)\right\}, & \text{if } x \neq 0, \\ \{(x, y) \in I^2 : x = 0\}, & \text{if } x = 0, \end{cases}$$

for all $(x, y) \in I^2$, is not a contraction.

In 1969, Nadler [131] established the following analogue of the Banach contraction principle for set-valued maps.

Theorem 3.22 (Nadler's Theorem) *Let (X,d) be a complete metric space. If $F : X \to 2^X_{cl}$ is a set-valued contraction map, then F has a fixed point.*

Proof Let α be a Lipschitz constant for F, and let $x_0 \in X$. Choose $x_1 \in F(x_0)$. Since $F(x_0)$, $F(x_1) \in 2^X_{cl}$ and $x_1 \in F(x_0)$, there is a point $x_2 \in F(x_1)$ such that

$$d(x_1,x_2) \leq \mathcal{H}(F(x_0),F(x_1)) + \alpha.$$

Since $F(x_1)$, $F(x_2) \in 2^X_{cl}$ and $x_2 \in F(x_1)$, there is a point $x_3 \in F(x_2)$ such that

$$d(x_2,x_3) \leq \mathcal{H}\big(F(x_1),F(x_2)\big) + \alpha^2.$$

Continuing in this fashion, we produce a sequence $\{x_n\}_{n=1}^{\infty}$ of points of X such that $x_{n+1} \in F(x_n)$ and

$$d\left(x_n,x_{n+1}\right) \leq \mathcal{H}\big(F(x_{n-1}),F(x_n)\big) + \alpha^n, \quad \text{for all } n \geq 1.$$

Note that

$$\begin{aligned} d\left(x_n,x_{n+1}\right) &\leq \mathcal{H}\big(F(x_{n-1}),F(x_n)\big) + \alpha^n \\ &\leq \alpha d\left(x_{n-1},x_n\right) + \alpha^n \\ &\leq \alpha\left[\mathcal{H}\big(F(x_{n-2}),F(x_{n-1})\big) + \alpha^{n-1}\right] + \alpha^n \\ &\leq \alpha^2 d\left(x_{n-2},x_{n-1}\right) + 2\alpha^n \\ &\;\vdots \\ &\leq \alpha^n d\left(x_0,x_1\right) + n\alpha^n, \end{aligned}$$

for all $n \geq 1$. This implies that for any n and m, we have

$$\begin{aligned} d\left(x_n,x_{n+m}\right) &\leq d\left(x_n,x_{n+1}\right) + d\left(x_{n+1},x_{n+2}\right) + \cdots + d\left(x_{n+m-1},x_{n+m}\right) \\ &\leq \alpha^n d\left(x_0,x_1\right) + n\alpha^n + \alpha^{n+1} d\left(x_0,x_1\right) + (n+1)\alpha^{n+1} + \\ &\quad \cdots + \alpha^{n+m-1} d\left(x_0,x_1\right) + (n+m-1)\alpha^{n+m-1} \\ &= \sum_{k=n}^{n+m-1} \alpha^k d\left(x_0,x_1\right) + \sum_{k=n}^{n+m-1} k\alpha^k. \end{aligned}$$

It follows that the sequence $\{x_n\}$ is a Cauchy sequence. Since (X,d) is complete, the sequence $\{x_n\}$ converges to some point $x \in X$. Since F is $\mathcal{H}$-continuous, we have

$$\lim_{n\to\infty} \mathcal{H}(F(x_n),F(x)) = 0.$$

Also, since $x_n \in F(x_{n-1})$, we have

$$\lim_{n\to\infty} d(x_n,F(x)) = \lim_{n\to\infty}\left(\inf_{y\in F(x)} d(x_n,y)\right) = 0.$$

This implies that

$$d(x, F(x)) = \inf\{d(x, y) : y \in F(x)\} = 0,$$

and since $F(x)$ is closed, by Theorem 1.2, $x \in F(x)$. ∎

Theorem 3.23 [147] *Let (X, d) be a complete metric space and $F : X \to 2^X_q$ be a set-valued contraction map. Then, the set $Fix(F) = \{x \in X : x \in F(x)\}$ of all fixed points of F is compact.*

Proof Assume that $\text{Fix}(F)$ is not compact. Since it is complete, by Theorem 1.10, it cannot be totally bounded. So, there exist some $\delta > 0$ and some sequence $\{x_n\}$ in $\text{Fix}(F)$ such that $d(x_n, x_m) \geq \delta$ for any two different n and m. Put

$$\rho := \inf\{r : \exists x \in X \text{ such that } S_r(x) \text{ contains infinitely many } x_n\text{'s}\}.$$

Since for every $x \in X$, $S_{\delta/2}(x)$ contains at most one x_n, one has $\rho \geq \delta/2 > 0$. Fix ε such that

$$0 < \varepsilon < \rho\left(\frac{1-\alpha}{1+\alpha}\right),$$

where α is the contractivity constant, and choose $x \in X$ such that $J := \{n : x_n \in S_{\rho+\varepsilon}(x)\}$ is infinite. For each $n \in J$, we have

$$d(x_n, F(x)) \leq \mathcal{H}(F(x_n), F(x)) \leq \alpha d(x_n, x) < \alpha(\rho + \varepsilon),$$

and we can choose some $y_n \in F(x)$ such that $d(x_n, y_n) < \alpha(\rho + \varepsilon)$. By compactness of $F(x)$, there is a $y \in F(x)$ such that $J' := \{n \in J : d(y_m, y) < \varepsilon\}$ is infinite. Then for each $n \in J'$,

$$d(x_n, y) < \alpha(\rho + \varepsilon) + \varepsilon = \alpha\rho + \varepsilon(1 + \alpha) = r < \alpha\rho + \rho(1 - \alpha) = \rho,$$

which contradicts the definition of ρ since $S_r(y)$ contains infinitely many x_n's. ∎

Definition 3.14 Let (X, d) be a metric space. A set-valued mapping $F : X \to 2^X_{cl}$ is said to be *contractive* if

$$\mathcal{H}(F(x), F(y)) < d(x, y), \quad \text{for all } x, y \in X \text{ with } x \neq y.$$

Remark 3.9 It can be easily seen that if $F : X \to 2^X_{cl}$ is a contractive set-valued mapping and $y_1 \in F(x_1)$, then there exists $y_2 \in F(x_2)$ such that $d(y_1, y_2) < d(x_1, x_2)$.

Definition 3.15 Let (X, d) be a metric space and $x \in X$. An *orbit* $\mathcal{O}(x)$, also called *trajectory*, of a set-valued mapping $F : X \to 2^X_{cl}$ at the point x is a sequence $\{x_n : x_n \in F(x_{n-1})\}$ with $x_0 = x$.

The orbit $\mathcal{O}(x)$ is called *regular* if

$$d(x_{n+1}, x_{n+2}) \leq d(x_n, x_{n+1}) \quad \text{and} \quad d(x_{n+1}, x_{n+2}) \leq \mathcal{H}(F(x_n), F(x_{n+1})).$$

Remark 3.10 Let $F : X \to 2^X_q$ be a contractive set-valued mapping and $x \in X$. Then, the orbit $\mathcal{O}(x)$ defined by choosing $x_n \in F(x_{n-1})$ such that

$$d(x_{n-1}, x_n) = d(x_{n-1}, F(x_{n-1})) = \inf\{d(x_{n-1}, y) : y \in F(x_{n-1})\},$$

is regular.

Exercise 3.21 Let X be a metric space and $F : X \to 2^X_{cl}$ be a contractive set-valued mapping. If $x_n \to x$ and $y_n \to y$ with $y_n \in F(x_n)$ for all n, then prove that $y \in F(x)$.

Theorem 3.24 [156] *Let (X, d) be a bounded metric space, $F : X \to 2^X_{cl}$ be a contractive set-valued mapping and $x \in X$. If there is a regular orbit $\mathcal{O}(x)$ for F which contains a convergent subsequence $\{x_{n_i}\}$ such that $x_{n_i} \to y_0$ and $x_{n_i+1} \to y_1$, then $y_1 = y_0$, that is, F has a fixed point.*

Proof Let $\mathcal{O}(x)$ be a regular orbit with $x_{n_i} \to y_0$, $x_{n_i+1} \to y_1$ and $y_1 \in F(x_0)$. Let Δ denote the set of all diagonal elements in $X \times X$ and $Y = X \times X \setminus \Delta$. Define a function $f : Y \to \mathbb{R}$ as

$$f(p,q) = \frac{\mathcal{H}(F(p), F(q))}{d(p,q)}, \quad \text{for all } (p,q) \in Y.$$

Then, f is a continuous function. Since F is contractive, we have $f(p,q) < 1$. Therefore, if $y_1 \neq y_0$, then there is an $\alpha \in (0,1)$ and an open set U of Y such that $(y_0, y_1) \in U$ and if $(p,q) \in U$, then $0 \leq f(p,q) < \alpha$. Now choose $\delta > 0$ such that $\delta < \frac{1}{3}d(y_0, y_1)$ and $S_\delta(y_0) \times S_\delta(y_1) \subset U$.

Since $x_{n_i} \to y_0$, $x_{n_i+1} \to y_1$, there exists a positive integer N such that $x_{n_i} \in S_\delta(y_0)$ and $x_{n_i+1} \in S_\delta(y_1)$ for all $i \geq N$.

From the definition of f and the choice of U, we have

$$\mathcal{H}(F(x_{n_i}), F(x_{n_i+1})) < \alpha d(x_{n_i}, x_{n_i+1}).$$

Since $\mathcal{O}(x)$ is regular, we get $d(x_{n_i+1}, x_{n_i+2}) < \alpha d(x_{n_i}, x_{n_i+1})$. Further, if $l > j > N$, then

$$d(x_{n_l}, x_{n_l+1}) \leq d(x_{n_{l-1}+1}, x_{n_{l-1}+2}) < \alpha d(x_{n_{l-1}}, x_{n_{l-1}+1}).$$

By repeating this argument, we get $d(x_{n_l}, x_{n_l+1}) < \alpha^{l-j} d(x_{n_j}, x_{n_j+1})$. But with fixed j, $\alpha^{l-j} \to 0$ as $l \to \infty$, which implies that $d(x_{n_l}, x_{n_l+1}) \to 0$ as $l \to \infty$. This contradicts $d(x_{n_l}, x_{n_l+1}) > \delta$ for all $l > N$. Thus we conclude that $y_0 = y_1$, and hence, F has a fixed point. ∎

Corollary 3.4 *Let (X, d) be a compact metric space and $F : X \to 2^X_{cl}$ be a contractive set-valued mapping. Then F has a fixed point.*

3.5 Fixed Point Results for Directional Contraction Set-valued Mappings

Let (X, d) be a metric space and K be a nonempty subset of X. For all $x, y \in X$, define

$$[x,y] = \{z \in X : d(x,z) + d(z,y) = d(x,y)\},$$

and $(x, y] = [x, y] \setminus \{x\}$, $(x, y) = (x, y] \setminus \{y\}$.

Definition 3.16 (Directional Contraction) A set-valued mapping $F : K \to 2^X_{cl}$ is said to be a *directional contraction* if there exists $\alpha \in [0, 1)$ such that for all $x \in K$ and $y \in T(x)$,

$$\mathcal{H}(F(z), F(x)) \leq \alpha d(z, x), \quad \text{for all } z \in [x, y] \cap K. \tag{3.5}$$

The constant α in (3.5) is called a *contraction constant* of F.

We need the following lemma to prove the existence of a fixed point of a directional contraction set-valued mapping.

Lemma 3.4 *Let K be a nonempty closed subset of a metric space X and $F : K \to 2^X_{cl}$ be a directional contraction set-valued mapping with contraction constant $\alpha \in [0, 1)$ such that the following conditions hold.*

(i) *For each $x \in K$ and $y \in F(x) \setminus K$, there exists $z \in (x, y) \cap K$ such that $F(z) \subseteq K$;*
(ii) *The mapping $g : K \to [0, \infty)$ defined by $g(x) = d(x, F(x))$ is lower semicontinuous.*

Then for any β, $\alpha < \beta < 1$, there exists a set-valued mapping $G : K \to 2^X_{cl}$ which satisfies the following properties:

(a) *For each $x \in K$, $G(x)$ is nonempty and $G(x) \subseteq F(x)$;*
(b) *If $y \in G(x)$, then $d(x, y) \leq (1 - \beta + \alpha)^{-1} d(x, F(x))$;*
(c) *If $G(x) \cap K = \emptyset$ for some $x \in K$, then there exist $y \in G(x)$ (depending on x) and $z \in (x, y) \cap K$ (depending on x, y) such that*

$$d(x, y) \leq d(x, F(x)) + (\beta - \alpha) d(x, z). \tag{3.6}$$

Proof Define a set-valued mapping $G : K \to 2^X_{cl}$ by

$$G(x) = \{y \in F(x) : d(x, y) \leq (1 - \beta + \alpha)^{-1} d(x, F(x))\}.$$

Since $(1 - \beta + \alpha) < 1$, $G(x) \neq \emptyset$ for all $x \in K$ and satisfies (a) and (b).

Suppose that $G(x) \cap K = \emptyset$ for some $x \in K$. Choose a sequence $\{y_n\} \subseteq F(x)$ such that

$$d(x, y_n) \to d(x, F(x)). \tag{3.7}$$

Since the sequence $\{y_n\}$ is eventually in $G(x)$, we may assume that the sequence $\{y_n\} \subseteq G(x)$. Then it follows by the supposition that for each $n \in \mathbb{N}$, $y_n \in F(x) \setminus K$, and consequently by (i), for each $n \in \mathbb{N}$, there exists z_n such that

$$z_n \in (x, y_n) \cap K \quad \text{and} \quad F(z_n) \subseteq K. \tag{3.8}$$

Since $d(x, z_n) \leq d(x, y_n)$, it follows from (3.7) that there is a subsequence $\{z_{n_k}\}$ of the sequence $\{z_n\}$ and a real number $\lambda \geq 0$ such that

$$d(x, z_{n_k}) \to \lambda. \tag{3.9}$$

We claim that $\lambda > 0$. Suppose that $\lambda = 0$. Then the sequence $\{z_{n_k}\}$ converges to x. Moreover, since $y_n \in F(x)$, it follows by the definition of F and (3.8) that

$$\mathcal{H}(F(x), F(z_{n_k})) \leq \alpha d(x, z_{n_k}) \to 0 \text{ as } k \to \infty. \tag{3.10}$$

Now we prove that $F(x) \subseteq K$. If y is an arbitrary element of $F(x)$, then by Exercise 3.13 (b) for each $k \in \mathbb{N}$, there is a $w_k \in F(z_{n_k})$ such that

$$d(y, w_k) \leq \mathcal{H}(F(x), F(z_{n_k})) + \frac{1}{k} \to 0 \text{ as } k \to \infty.$$

Since $\{w_k\} \subseteq K$ and K is closed, it follows that $y \in F(x)$ and hence $F(x) \subseteq K$.

However, this contradicts the supposition that $G(x) \cap K = \emptyset$. Thus $\lambda > 0$.

Now choose an $\varepsilon > 0$ such that $\delta = (\beta - \alpha)\lambda - \varepsilon > 0$. Then by (3.9), $(\beta - \alpha)d(x, z_{n_k}) \geq \delta$ eventually and hence by (3.7) and the last inequality, we have that

$$d(x, y_{n_k}) \leq d(x, F(x)) + \delta \leq d(x, F(x)) + (\beta - \alpha)d(x, z_{n_k})$$

eventually. Thus, there exists $y = y_{n_k}$ and the corresponding $z = z_{n_k}$ satisfying (3.8) such that (3.6) holds. ∎

Theorem 3.25 [151] *Let K be a nonempty closed subset of a metric space X and $F : K \to 2^X_{cl}$ be a directional contraction set-valued mapping with contraction constant $\alpha \in [0, 1)$ such that the following conditions hold.*

(i) *For each $x \in K$ and $y \in F(x) \setminus K$, there exists $z \in (x, y) \cap K$ such that $F(z) \subseteq K$;*

(ii) *The mapping $g : K \to [0, \infty)$ defined by $g(x) = d(x, F(x))$ is lower semicontinuous.*

Then, F has a fixed point.

Proof Let G be the same as defined in the previous lemma. Define a mapping $f : K \to K$ as follows: For $x \in K$, let $f(x)$ be any element of $G(x) \cap K$ if $G(x) \cap K \neq \emptyset$; and if $G(x) \cap K = \emptyset$, then by the above lemma, there exist elements $y \in G(x)$ (depending on x) and $z \in (x, y) \cap K$ (depending on x, y) satisfying (3.6), let $f(x) = z$ in this case.

Note that for any $x \in K$,

$$\mathcal{H}(F(x), F(f(x))) \leq \alpha d(x, f(x)). \tag{3.11}$$

This is obvious if $G(x) \cap K = \emptyset$ and if $G(x) \cap K \neq \emptyset$, then since $f(x) \in F(x)$ and $f(x) \in [x, f(x)] \cap K$, the definition of F implies (3.11).

Set $\varphi(x) = (1 - \beta)^{-1} g(x)$ for all $x \in K$. Then by condition (ii), φ is lower semicontinuous on K. We show that φ satisfies the following Caristi's condition:

$$d(x, f(x)) \leq \varphi(x) - \varphi(f(x)), \quad \text{for all } x \in K. \tag{3.12}$$

We consider the following two cases:

Case I: When $A(x) \cap K \neq \emptyset$, then $f(x) \in G(x)$ and hence by Lemma 3.4 (b), we have

$$d(x, f(x)) \leq (1 - \beta + \alpha)^{-1} d(x, F(x)).$$

This implies that

$$\alpha(1 - \beta)^{-1} d(x, f(x)) \leq \varphi(x) - d(x, f(x)).$$

Since $d(m, N) \leq \mathcal{H}(M, N)$ for all $m \in M$ and all $M, N \in 2^X_{cl}$, by the last inequality and (3.11), we have

$$\begin{aligned} \varphi(f(x)) &= (1 - \beta)^{-1} g(f(x)) \\ &\leq (1 - \beta)^{-1} \mathcal{H}(F(x), F(f(x))) \\ &\leq \varphi(x) - d(x, f(x)). \end{aligned}$$

Thus (3.12) holds.

CASE II: When $G(x) \cap K = \emptyset$, there is a $y \in F(x)$ (depending on x) such that $f(x) \in (x, y)$ and satisfies (3.6). Thus by (3.6), we obtain

$$\begin{aligned} d(f(x), F(x)) &\leq d(f(x), y) \\ &= d(x, y) - d(x, f(x)) \\ &\leq d(x, F(x)) - (1 - \beta + \alpha) d(x, f(x)). \end{aligned}$$

Since $d(x, N) \leq d(x, M) + \mathcal{H}(M, N)$ for all $x \in X$ and all $M, N \in 2^X_{cl}$, by the last inequality and (3.11), we have

$$\begin{aligned} (1 - \beta)\varphi(f(x)) &= g(f(x)) \\ &\leq d(f(x), F(x)) + \mathcal{H}(F(x), F(f(x))) \\ &\leq d(x, F(x)) - (1 - \beta) d(x, f(x)), \end{aligned}$$

that is, $d(x, f(x)) \leq \varphi(x) - \varphi(f(x))$. Thus, (3.12) holds.

Consequently by Caristi's Theorem 2.14, there exists $x \in K$ such that $f(x) = x$. This implies that $x \in F(x)$ for otherwise $f(x) \notin G(x) \cap K$ and hence by the definition of f, $G(x) \cap K = \emptyset$. Thus, $f(x) \in (x, y)$ for some $y \in G(x)$ (y depends on x) which contradicts $x \neq f(x)$. Consequently, $x \in F(x)$. ∎

Remark 3.11 Sehgal [151] established the above theorem and lemma in the setting of 2^X_b. The above theorem is further improved by Park [136, 137].

Remark 3.12 In view of Exercise 3.20, the condition (ii) in Theorem 3.25 holds if F is $\mathcal{H}$-continuous.

3.6 Fixed Point Results for Dissipative Set-valued Maps

We first give the following fixed point theorem for set-valued maps by using Caristi's Theorem 2.14.

Theorem 3.26 (Caristi–Kirk Fixed Point Theorem) *Let (X, d) be a complete metric space and $\varphi : X \to (-\infty, +\infty]$ be a proper, bounded below, and lower semicontinuous functional on X. Let $F : X \rightrightarrows X$ be a set-valued map such that for each $x \in X$, there exists $y \in F(x)$ satisfying*

$$\varphi(y) + d(x, y) \leq \varphi(x).$$

Then, there exists a fixed point $\bar{x} \in X$ of F, that is, $\bar{x} \in F(\bar{x})$ and $\varphi(\bar{x}) < +\infty$.

Proof Let y be an element in X such that $y \in F(x)$ and $\varphi(y) + d(x, y) \leq \varphi(x)$. For each $x \in X$, let $f(x) = y$. Then, f is a mapping from X into itself and satisfying

$$\varphi(f(x)) + d(x, f(x)) \leq \varphi(x), \quad \text{for all } x \in X.$$

By Caristi's Theorem 2.14, there exists $\bar{x} \in X$ such that $\bar{x} = f(\bar{x}) \in F(\bar{x})$ and $\varphi(\bar{x}) < +\infty$. ∎

Definition 3.17 Let (X, d) be a metric space and $F : X \rightrightarrows X$ be a set-valued map with nonempty values.

(a) A function $\varphi : X \to [0, +\infty)$ is said to be a *weak entropy* of F if

$$\text{for all } x \in X, \text{ there exists } y \in F(x) \text{ such that } \varphi(y) + d(x, y) \leq \varphi(x).$$

(b) A function $\varphi : X \to [0, +\infty)$ is said to be an *entropy* of F if

$$\text{for all } x \in X \text{ and all } y \in F(x), \text{ we have } \varphi(y) + d(x, y) \leq \varphi(x).$$

(c) The set-valued map $F : X \rightrightarrows X$ is said to be *weakly dissipative* if there exists a weak entropy of F.

(d) The set-valued map $F : X \rightrightarrows X$ is said to be *dissipative* if there exists an entropy of F.

(e) A sequence $\{x_n\}_{n=0}^{\infty}$ in X is said to be a *trajectory*, also known as *orbit*, starting at x if $x_0 = x$ and $x_{n+1} \in F(x_n)$ for $n = 0, 1, 2, \ldots$ The set of all such trajectories is denoted by $\mathcal{O}(F, x)$.

Aubin and Siegel [19] established the following fixed point theorem for weakly dissipative set-valued maps.

Theorem 3.27 *Let (X, d) be a complete metric space. Then any weakly dissipative closed set-valued map $F : X \rightrightarrows X$ with nonempty values has a fixed point. Furthermore, for any $x_0 \in X$, there exists a trajectory starting at x_0 and converging to a fixed point of F.*

Proof Let φ be a weak entropy of F. By Definition 3.17 (a), for any given $x \in X$, we construct the sequence $\{x_n\} \in \mathcal{O}(F, x)$ such that

$$d(x_p, x_q) \leq \sum_{n=p}^{q-1} d(x_n, x_{n+1}) \leq \varphi(x_p) - \varphi(x_q).$$

Since the sequence $\{\varphi(x_n)\}$ is nonincreasing and bounded below, $\{x_n\}$ is a Cauchy sequence.

Let $x_n \to \bar{x}$ as $n \to \infty$. Since $x_{n+1} \in F(x_n)$, we have that $(\bar{x}, \bar{x})$ is in the closure of the graph of F. Since F is closed, its graph is also closed, and therefore $(\bar{x}, \bar{x}) \in \text{Graph}(F)$, that is, $\bar{x} \in F(\bar{x})$. ■

Aubin and Siegel [19] gave the following partial characterization of those set-valued maps which admit a weak entropy, and hence are weakly dissipative.

Proposition 3.3 *Let (X, d) be a complete metric space and $F : X \rightrightarrows X$ be a weakly dissipative set-valued map with nonempty values. If $\varphi : X \to [0, \infty)$ is a weak entropy of F, then*

$$\varphi_0(x) \leq \varphi(x) < +\infty, \quad \textit{for all } x \in X, \tag{3.13}$$

where $\varphi_0 : X \to [0, +\infty]$ is a function defined by

$$\varphi_0(x) = \inf \left\{ \sum_{n=0}^{\infty} d(x_n, x_{n+1}) : \{x_n\} \in \mathcal{O}(F, x) \right\}, \quad \textit{for all } x \in X. \tag{3.14}$$

Proof Let $x \in X$ and $\{x_n\} \in \mathcal{O}(F, x)$. As in the proof of Theorem 3.27, we have

$$\sum_{n=p}^{q-1} d(x_n, x_{n+1}) \leq \varphi(x_p) - \varphi(x_q).$$

For $p = 0$, we have

$$\sum_{n=0}^{q-1} d(x_n, x_{n+1}) \leq \varphi(x_0) - \varphi(x_q) \leq \varphi(x_0).$$

Letting limit as $q \to \infty$, we get

$$\sum_{n=0}^{\infty} d(x_n, x_{n+1}) \leq \varphi(x_0).$$

Thus,

$$\inf\left\{\sum_{n=0}^{\infty} d(x_n, x_{n+1}) : \{x_n\} \in \mathcal{O}(F, x)\right\} \leq \varphi(x_0),$$

and therefore,

$$\varphi_0(x) \leq \varphi(x_0).$$

■

As a partial converse of the above proposition, we have the following result.

Theorem 3.28 *Let (X, d) be a complete metric space and $F : X \rightrightarrows X$ be an upper semicontinuous set-valued map with nonempty compact values. If $\varphi_0(x) < +\infty$ for all $x \in X$, then F is weakly dissipative and, in particular, φ_0 is the smallest weak entropy of F.*

Proof For any $\varepsilon > 0$, define the function $\varphi_\varepsilon : X \to [0, \infty)$ by

$$\varphi_\varepsilon(x) = \inf\left\{\sum_{n=0}^{\infty} d(x_n, x_{n+1}) : x_0 = x \text{ and } \{x_{n+1}\} \in S_\varepsilon(F(x_n))\right\}, \quad \text{for all } x \in X, \tag{3.15}$$

where $S_\varepsilon(F(x_n)) = \{y \in X : d(y, F(x_n)) < \varepsilon\}$.

We may define a function $\varphi : X \to [0, \infty)$ by

$$\varphi(x) = \lim_{\varepsilon \to 0} \varphi_\varepsilon(x), \quad \text{for all } x \in X. \tag{3.16}$$

Properties (3.15) and (3.16) are well-defined for if $\varepsilon_2 < \varepsilon_1$, then $\varphi_{\varepsilon_1} \leq \varphi_{\varepsilon_2} \leq \varphi \leq \varphi_0$.

Now, for each $\varepsilon > 0$, select $x_\varepsilon \in S_\varepsilon(F(x))$ such that

$$\varphi_\varepsilon(x_\varepsilon) + d(x, x_\varepsilon) \leq \varphi_\varepsilon(x) + \varepsilon. \tag{3.17}$$

Since $F(x)$ is compact, we may select a subsequence $\{x_k\} \in S_{1/k}(F(x_n))$ converging to some $x^* \in F(x)$. Also, since F is upper semicontinuous, for any $\delta > 0$, there exists k_0 with $\delta > 1/k_0$ such that $S_{1/k}(F(x_{1/k})) \subseteq S_\delta(F(x^*))$ for all $k \geq k_0$. Thus, we have

$$\varphi_\delta(x^*) \leq d(x^*, x_k) + \varphi_k(x_k), \quad \text{for all } k \geq k_0. \tag{3.18}$$

Combining (3.17) and (3.18), we obtain

$$\varphi_\delta(x^*) - d(x^*, x_k) + d(x, x_k) \le \varphi_k(x) + \frac{1}{k}, \quad \text{for all } k \ge k_0. \tag{3.19}$$

Letting as $k \to \infty$ in (3.15), we get

$$\varphi_\delta(x^*) + d(x, x^*) \le \varphi(x). \tag{3.20}$$

Since δ was arbitrary, we finally have

$$\varphi(x^*) + d(x, x^*) \le \varphi(x). \tag{3.21}$$

Hence, φ is a weak entropy for F. Since $\varphi(x) \le \varphi_0(x)$ for all x, Proposition 3.3 implies that $\varphi(x) = \varphi_0(x)$ for all x, that is, $\varphi \equiv \varphi_0$. ∎

We have the following characterization of the dissipative set-valued maps.

Theorem 3.29 [19] *Let (X, d) be a complete metric space. A set-valued map $F : X \rightrightarrows X$ with nonempty values is dissipative if and only if $\varphi_*(x) < +\infty$ for all $x \in X$, where $\varphi_* : X \to [0, \infty)$ is a function defined by*

$$\varphi_*(x) = \sup\left\{\sum_{n=0}^{\infty} d(x_n, x_{n+1}) : \{x_n\} \in \mathcal{O}(F, x)\right\}, \quad \textit{for all } x \in X. \tag{3.22}$$

In this case, φ_* is the smallest entropy of F.

Proof We may re-write (3.22) as

$$\varphi_*(x) = \sup\{d(x, y) + \varphi_*(y) : y \in F(x)\}. \quad \text{for all } x \in X. \tag{3.23}$$

Then, the result follows from this observation. ∎

From Theorem 3.27 and Theorem 3.28, we derive the Nadler's Theorem 3.22.

Theorem 3.30 *Let (X, d) be a metric space and $F : X \to 2^X_q$ be a contraction set-valued map with constant $\alpha \in (0, 1)$. Then, T is weakly dissipative. Furthermore, if X is complete, then T has a fixed point.*

Proof Since a set-valued contraction map is upper semicontinuous, it suffices to see that $\varphi_0(x) < +\infty$ for all x. Let $x \in X$. Since F has compact values, we may select a trajectory $\{x_n\}$ such that $d(x_n, x_{n+1}) = d(x_n, F(x_n))$. Since F is contraction, we have

$$\varphi_0(x) \le \sum_{n=0}^{\infty} d(x_n, x_{n+1}) \le \sum_{n=0}^{\infty} \alpha^n d(x_0, x_1) < +\infty.$$

This completes the proof. ∎

3.7 Fixed Point Results for Ψ-contraction Set-valued Mappings

Corresponding to the concept of ψ-contraction for single-valued mappings, Mizoguchi and Takahashi [129] gave the following definition of a Ψ-contraction set-valued mapping.

Definition 3.18 Let (X,d) be a metric space and $\Psi : (0,\infty) \to [0,1)$ be a function such that $\limsup_{r\to t^+} \Psi(r) < 1$ for all $t \in [0,\infty)$. A set-valued mapping $F : X \to 2^X_{cl}$ is said to be Ψ-*contraction* if

$$\mathcal{H}(F(x),F(y)) \leq \Psi\left(d(x,y)\right) d(x,y), \quad \text{for all } x,y \in X,\ x \neq y. \tag{3.24}$$

Mizoguchi and Takahashi [129] established the following fixed point theorem. However, Daffer and Kaneko [65] gave an alternative proof of this theorem. But we include the proof given by Suzuki [166] which seems to be simpler than the proof given in [65, 129].

Theorem 3.31 (Mizoguchi–Takahashi Fixed Point Theorem) *Let (X,d) be a complete metric space and $F : X \to 2^X_{cl}$ be a Ψ-contraction set-valued mapping, where $\Psi : (0,\infty) \to [0,1)$ is a function such that $\limsup_{r\to t^+} \Psi(r) < 1$ for all $t \in [0,\infty)$. Then, F has a fixed point $\bar{x} \in X$.*

Proof Define a function $\beta : [0,\infty) \to [0,1)$ by $\beta(t) = \frac{\Psi(t)+1}{2}$ for all $t \in [0,\infty)$. Then the following assertions hold:

- $\limsup_{s\to t+0} \beta(s) < 1$ for all $t \in [0,\infty)$;
- For $x,y \in X$ and $u \in F(x)$, there exists an element $v \in F(y)$ such that $d(u,v) \leq \beta\left(d(x,y)\right) d(x,y)$.

Putting $u = y$, we obtain the following:

- For $x \in X$ and $y \in F(x)$, there exists an element $v \in F(y)$ such that $d(y,v) \leq \beta\left(d(x,y)\right) d(x,y)$.

Thus, we can define a sequence $\{x_n\} \in X$ in the following way:

$$x_{n+1} \in F(x_n) \quad \text{and} \quad d(x_{n+1},x_{n+2}) \leq \beta\left(d(x_n,x_{n+1})\right) d(x_n,x_{n+1}), \quad \text{for } n \in \mathbb{N}.$$

Since $\beta(t) < 1$ for all $t \in [0,\infty)$, $\{d(x_n,x_{n+1})\}$ is a nonincreasing sequence in $\mathbb{R}$. Hence, $\{d(x_n,x_{n+1})\}$ converges to some nonnegative real number τ. Since $\limsup_{s\to\tau+0} \beta(s) < 1$ and $\beta(\tau) < 1$, there exist $r \in [0,1)$ and $\varepsilon > 0$ such that $\beta(s) \leq r$ for all $s \in [\tau,\tau+\varepsilon]$. We can take $N \in \mathbb{N}$ such that $\tau \leq d(x_n,x_{n+1}) \leq \tau+\varepsilon$ for all $n \in \mathbb{N}$ with $n \geq N$. Since

$$d(x_{n+1},x_{n+2}) \leq \beta\left(d(x_n,x_{n+1})\right) d(x_n,x_{n+1}) \leq r\, d(x_n,x_{n+1}), \quad \text{for } n \in \mathbb{N} \text{ with } n \geq N,$$

we have

$$\sum_{n=1}^{\infty} d(x_n,x_{n+1}) \leq \sum_{n=1}^{N} d(x_n,x_{n+1}) + \sum_{n=1}^{\infty} r^n d(x_N,x_{N+1}) < \infty,$$

and hence, $\{x_n\}$ is a Cauchy sequence. Since X is complete, $\{x_n\}$ converges to some point $\bar{x} \in X$. Since

$$\begin{aligned} d(\bar{x},F(\bar{x})) &= \lim_{n\to\infty} d(x_{n+1},F(\bar{x})) \leq \lim_{n\to\infty} \mathcal{H}\left(F(x_n),F(\bar{x})\right) \\ &\leq \lim_{n\to\infty} \beta\left(d(x_n,\bar{x})\right) \leq \lim_{n\to\infty} d(x_n,\bar{x}) = 0, \end{aligned}$$

and $F(\bar{x})$ is closed, we obtain $\bar{x} \in F(\bar{x})$. ∎

Corollary 3.5 *Let (X, d) be a complete metric space and $F : X \to 2^X_{cl}$ be a set-valued mapping. If $\Psi : (0, \infty) \to [0, 1)$ is a monotone increasing function such that $0 \leq \Psi(t) < 1$ for all $t \in [0, \infty)$ and if*

$$\mathcal{H}(F(x), F(y)) \leq \Psi\left(d(x, y)\right) d(x, y), \quad \text{for all } x, y \in X,\ x \neq y,$$

then F has a fixed point $\bar{x} \in X$.

For further discussion and generalizations, we refer to [152, 171].

3.8 Fixed Point Results for Weakly Contraction Set-valued Mappings

Definition 3.19 Let (X, d) be a metric space. A set-valued mapping $F : X \to 2^X_{cl}$ is said to be *weakly contraction* if

$$\mathcal{H}(F(x), F(y)) \leq d(x, y) - \varphi\left(d(x, y)\right), \quad \text{for all } x, y \in X, \tag{3.25}$$

where $\varphi : [0, \infty) \to [0, \infty)$ is a function such that $\varphi(t) > 0$ for all $t \in (0, \infty)$ and $\varphi(0) = 0$.

Remark 3.13 **(a)** If F is single-valued and φ is continuous and nondecreasing, then the definition of weakly contraction mapping is same as appeared in [2].

(b) If $\varphi(t) = (1 - \alpha)t$, where $\alpha \in (0, 1)$ is contractivity constant, then a weakly contraction mapping is a contraction.

(c) If $\Psi(t) = 1 - \varphi(t)/t$ for $t > 0$ and $\Psi(0) = 0$, then condition (3.25) reduces to the condition (3.24).

(d) If φ is lower semicontinuous from the right, then $\Psi(t) = t - \varphi(t)$ is upper semicontinuous from the right and in this case condition (3.25) reduces to the following condition:

$$\mathcal{H}(F(x), F(y)) \leq \Psi\left(d(x, y)\right), \quad \text{for all } x, y \in X. \tag{3.26}$$

This condition can be seen as an extension of condition (2.26) for set-valued mappings.

Exercise 3.22 Let (X, d) be a complete metric space and $F : X \to 2^X_{cl}$ be a weakly contraction mapping. Then prove that the graph of F is closed.

Theorem 3.32 [20] *Let (X, d) be a complete metric space and $F : X \to 2^X_{cl}$ be a weakly contraction mapping for which φ is lower semicontinuous from the right and $\limsup\limits_{t \to 0+}(t/\varphi(t)) < \infty$. Then, F has a fixed point in X.*

Proof Let $M = \{(x, y) : x \in X,\ y \in F(x)\}$ be the graph of F. Then by Exercise 3.22, M is closed. Define ρ on M by

$$\rho((x, z), (u, v)) = \max\{d(x, u), d(z, v)\}.$$

Then, (M, ρ) is a complete metric space.

Now define $\phi : M \to [0, \infty)$ by

$$\phi(x, z) = d(x, z),$$

and $c : [0, \infty) \to [0, \infty)$ by

$$c(x) = \begin{cases} \frac{2t}{\varphi(t)}, & \text{if } t > 0, \\ \limsup\limits_{t \to 0+} \dfrac{2t}{\varphi(t)}, & \text{if } t = 0. \end{cases}$$

Note that ϕ is continuous and c is upper semicontinuous from the right since φ is lower semicontinuous from the right. Assume the contrary that F has no fixed point. Then, for each $(x, z) \in M$, we see that $x \neq z$. Since $z \in F(x)$, we can choose $v \in F(z)$ such that

$$d(z, v) \leq d(x, z) - \frac{1}{2}\varphi(d(x, z)). \tag{3.27}$$

Observe that

$$\rho((x, z), (z, v)) = \max\{d(x, z), d(z, v)\} = d(x, z) = \phi(x, z).$$

From (3.27), we have

$$\rho((x, z), (z, v)) \leq \frac{2\phi(x, z)}{\varphi(\phi(x, z))}[\phi(x, z) - \phi(z, v)].$$

Now define $g : M \to M$ by $g(x, z) = (z, v)$. Then, g has no fixed point and satisfies

$$\rho((x, z), g(z, v)) \leq c(\phi(x, z))[\phi(x, z) - \phi(g(x, z))].$$

Then, by Theorem 2.20, g should have a fixed point, which is a contradiction. Therefore, we conclude that F has a fixed point. ∎

3.9 Stationary Points for Set-valued Maps

Definition 3.20 Let X be a nonempty set and $F : X \rightrightarrows X$ be a set-valued map with nonempty values. A point $x \in X$ is said to be a *stationary point* or *endpoint* of F if $F(x) = \{x\}$.

Clearly, every stationary point of a set-valued map is a fixed point, but the converse need not be true.

The following stationary point theorem is established by Dancs, Hegedűs, and Medvegyev [67]; hence, it is called the DHM theorem.

Theorem 3.33 (DHM Theorem) *Let (X, d) be a complete metric space and $F : X \rightrightarrows X$ be a set-valued map such that the following conditions hold.*

- **(i)** *For all $x \in X$, $F(x)$ is closed;*
- **(ii)** *For all $x \in X$, $x \in F(x)$;*
- **(iii)** *For all $x, y \in X$, $y \in F(x)$ implies $F(y) \subseteq F(x)$;*
- **(iv)** *For any sequence $\{x_n\}$ in X such that*

$$x_2 \in F(x_1),\ x_3 \in F(x_2),\ \dots,\ x_n \in F(x_{n-1}), \dots, \tag{3.28}$$

the distance $d(x_n, x_{n+1})$ tends to zero as $n \to +\infty$.

Then, the set-valued map F has a stationary point $\bar{x} \in X$, that is, $F(\bar{x}) = \{\bar{x}\}$.

The iterate (3.28) is called the *generalized Picard iteration*.

Proof Since d satisfies condition (iv), the equivalent metric $d' = \dfrac{d}{(1+d)}$ also satisfies the condition (iv). So, we can suppose that d is bounded on X. Since $F(x) \neq \emptyset$ for all $x \in X$, we can construct a generalized Picard iteration such that $x_1 = \tilde{x}$, $x_n \in F(x_{n-1})$ and

$$d(x_n, x_{n-1}) \geq \frac{\operatorname{diam}\left(F(x_{n-1})\right)}{2} - \frac{1}{2^{n-1}}.$$

Hence, from conditions (iii) and (iv), we have $F(x_{n-1}) \supseteq F(x_n)$ and the diameter of $F(x_n)$, $\operatorname{diam}\left(F(x_n)\right) \to 0$ as $n \to +\infty$. Since X is complete, $\{F(x_n)\}$ is a decreasing sequence of nonempty closed sets such that $\operatorname{diam}\left(F(x_n)\right) \to 0$ as $n \to +\infty$, by Cantor's Intersection Theorem 1.4, we have

$$\bigcap_{n=1}^{\infty} F(x_n) = \{\bar{x}\}.$$

The point $\bar{x}$ is a stationary point, since on the other hand $\bar{x} \in F(x_n)$ and (iii) imply $F(\bar{x}) \subseteq \bigcap_{n=1}^{\infty} F(x_n) = \{\bar{x}\}$. From condition (ii), we have $\{\bar{x}\} \subseteq F(\bar{x})$ and hence $F(\bar{x}) = \bar{x}$. ■

Jachymski [100] replaced the condition (iv) of Theorem 3.33 by the following condition:

$$\text{for all } x \in X \text{ and for any } \varepsilon > 0, \text{ there exists } y \in F(x) \text{ such that } \operatorname{diam}(F(y)) < \varepsilon,$$

and established the following result.

Theorem 3.34 *Let (X, d) be a complete metric space and $F : X \rightrightarrows X$ be a set-valued map such that the following conditions hold.*

- **(i)** *For all $x \in X$, $F(x)$ is closed;*
- **(ii)** *For all $x \in X$, $x \in F(x)$;*
- **(iii)** *For all $x, y \in X$, $y \in F(x)$ implies $F(y) \subseteq F(x)$;*
- **(iv)** *For all $x \in X$ and for any $\varepsilon > 0$, there exists $y \in F(x)$ such that* $\operatorname{diam}(F(y)) < \varepsilon$.

Then, the set-valued map F has a stationary point $\bar{x} \in X$, that is, $F(\bar{x}) = \{\bar{x}\}$.

Proof Let $x_0 \in X$ be an arbitrary fixed element. Then by (iv), there is $x_1 \in F(x_0)$ such that $\operatorname{diam}(F(x_1)) < 1$. By (iii), $F(x_1) \subseteq F(x_0)$. Similarly, there is $x_2 \in F(x_1)$ such that $\operatorname{diam}(F(x_2)) < \frac{1}{2}$, and again by (iii) $F(x_2) \subseteq F(x_1)$. By induction, we obtain a sequence $\{x_n\}_{n=0}^{\infty}$ in X such that $\operatorname{diam}(F(x_n)) < \frac{1}{n}$ and $F(x_n) \subseteq F(x_{n-1})$. That is, $\{F(x_n)\}$ is a decreasing sequence of nonempty closed sets in a complete metric space X such that $\operatorname{diam}\left(F(x_n)\right) \to 0$ as $n \to +\infty$. By Cantor's Intersection Theorem 1.4, there exists $\bar{x} \in X$ such that

$$\bigcap_{n \in \mathbb{N}} F(x_n) = \{\bar{x}\}.$$

Since $\bar{x} \in F(x_n)$ for all $n \in \mathbb{N}$, by (iii), we have $F(\bar{x}) \subseteq F(x_n)$, that is,

$$F(\bar{x}) \subseteq \bigcap_{n \in \mathbb{N}} F(x_n) \subseteq \{\bar{x}\}.$$

Hence, $F(\bar{x}) = \{\bar{x}\}$ because $F(\bar{x})$ is nonempty. ■

The following result gives the characterization of the completeness of metric spaces.

Theorem 3.35 *Let (X, d) be a metric space and $F : X \rightrightarrows X$ be a set-valued map which satisfies conditions (i)–(iv) of Theorem 3.33. If F has no stationary point, then (X, d) is complete.*

Proof Assume contrary that (X, d) is not complete. By the converse of Cantor's Intersection Theorem 1.4, there exists a decreasing sequence $\{A_n\}$ (that is, $X = A_0 \supseteq A_1 \supseteq \cdots \supseteq A_n \supseteq \cdots$) of nonempty closed subsets of X such that $\text{diam}(A_n) \to 0$ as $n \to \infty$ and $\bigcap_{n=1}^{\infty} A_n = \emptyset$. For each n, set $B_n := X \setminus A_n$. Then, $X = \bigcup_{n=1}^{\infty} B_n$. For $x \in X$, define

$$n(x) = \min\{n \in \mathbb{N} : x \in B_n\}.$$

Note that for $x \in X$, $x \in B_{n(x)} = X \setminus A_{n(x)}$. Thus, if $n(x) > 1$, then $x \in A_{n(x)-1}$. Define the set-valued map $F : X \rightrightarrows X$ by

$$F(x) = A_{n(x)}, \quad \text{for all } x \in X.$$

Then, $F(x)$ is a closed set for all $x \in X$ and F has no fixed point. First we show that F satisfies the condition (iii) of Theorem 3.34. For this, let $x \in X$ and $y \in F(x)$. Since $y \in F(x) = A_{n(x)}$, it follows that $y \notin X \setminus A_{n(x)} = B_{n(x)}$. Since $\{B_n\}$ is ascending sequence, we see that $n(y) > n(x)$. Hence, $A_{n(y)} \subseteq A_{n(x)}$, that is, $F(y) \subseteq F(x)$.

Now we show that F satisfies the condition (iv) of Theorem 3.34. Let $x \in X$ and $\varepsilon > 0$. Note that $\text{diam}(A_n) \to 0$ as $n \to \infty$. Then there exists $m \in \mathbb{N}$ such that $m > n(x)$ and $\text{diam}(A_n) < \varepsilon$. Since $A_m \subset A_{m-1}$, there exists $y \in A_m \setminus A_{m-1}$. Since $m - 1 \geq n(x)$ and $n(y) = m$, we have $y \in A_{n(x)}$ and $F(y) = A_m$. Thus, $y \in F(x)$ and $\text{diam}(F(y)) < \varepsilon$. It means that condition (iv) of Theorem 3.34 holds. ■

The above theorem can be re-stated as follows:

Theorem 3.36 *If the metric space (X, d) is noncomplete, then there is a set-valued map $F : X \rightrightarrows X$ which satisfies conditions (i)–(iv) of Theorem 3.33 but has no stationary point.*

For a set-valued mapping $F : X \rightrightarrows X$ with nonempty values, we define a set-valued map $H_F : X \rightrightarrows X$ by

$$H_F(x) = \bigcup_{\{x_n\} \in \mathcal{O}(F,x)} \{x_n\}, \quad \text{for all } x \in X, \tag{3.29}$$

and define $L_F : \mathcal{O}(F, x) \rightrightarrows X$ by

$$L_F(\{x_n\}) = \bigcap_{n \geq 0} \overline{H_F(x_n)}. \tag{3.30}$$

$L_F(\{x_n\})$ is called the *limit set* of the trajectory.

Proposition 3.4 *Let (X, d) be a metric space and $x \in X$ be any element. Then, the following assertions hold.*

(a) $x \in H_F(x)$.

(b) *For all $y \in H_F(x)$, $H_F(y) \subseteq H_F(x)$.*

(c) $F(H_F(x)) \subseteq H_F(x)$.

(d) *If either $H_F(x)$, for all x, is closed, or F is lower semicontinuous with compact values, then $F(L_F(\{x_n\})) \subseteq L_F(\{x_n\})$.*

Theorem 3.37 [19] *Let X be a complete metric space and $F : X \rightrightarrows X$ be dissipative with entropy φ. Then for any $x \in X$, there exists $\{x_n\} \in \mathcal{O}(F, x)$ converging to some $\bar{x} \in X$ such that $L_F(\{x_n\}) = \{\bar{x}\}$.*

Proof One can first check that H_F is dissipative, and φ is an entropy for H_F. Let

$$\rho(x) = \inf\{\varphi(y) : y \in H_F(x)\}. \tag{3.31}$$

Then, the diameter $\operatorname{diam}(H_F(x)) = \sup\{d(y,z) : y, z \in H_F(x)\}$ satisfies the inequality

$$\operatorname{diam}(H_F(x)) \leq 2[\varphi(x) - \rho(x)], \quad \text{for all } x \in X, \tag{3.32}$$

since

$$\begin{aligned} \operatorname{diam}(H_F(x)) &\leq \sup_{y,z \in H_F(x)} \{d(y,x) + d(x,z)\} \\ &\leq \sup_{y \in H_F(x)} \{d(y,x)\} \leq 2[\varphi(x) - \rho(x)]. \end{aligned}$$

Choose $\{x_n\} \in \mathcal{O}(H_F, x)$ such that $\varphi(x_{n+1}) \leq \rho(x_n) + \frac{1}{2^n}$. By Proposition 3.4 (b), $\rho(x_n) \leq \rho(x_{n+1})$; thus by (3.32), we have

$$\operatorname{diam}(H_F(x_{n+1})) \leq 2[\varphi(x_{n+1}) - \rho(x_{n+1})] \leq \frac{1}{2^{n-1}}.$$

Thus, again by Proposition 3.4 (b), $\{x_n\}$ is Cauchy and

$$L_F(\{x_n\}) = \bigcap \overline{H_F(x_n)} = \bar{x} = \lim_{n \to \infty} x_n.$$

This completes the proof. ■

As an application of Proposition 3.4 (d) and Theorem 3.37, we give the following stationary point theorems due to Aubin and Siegel [19].

Theorem 3.38 *Let X be a complete metric space and $F : X \rightrightarrows X$ be a dissipative set-valued map. Further, let either F be lower semicontinuous with compact values or H_F be closed. Then, F has a stationary point.*

The following result is the Caristi's Fixed Point Theorem 2.14.

Theorem 3.39 *Let X be a complete metric space and $F : X \rightrightarrows X$ be a dissipative set-valued map with lower semicontinuous entropy φ. Then, T has a stationary point.*

Proof Let $\Phi : X \rightrightarrows X$ be defined by

$$\Phi(x) = \{y \in X : d(x,y) \leq \varphi(x) - \varphi(y)\}, \quad \text{for all } x \in X.$$

Then, $\Phi = H_\Phi$ is closed, and $F(x) \subseteq \Phi(x)$ for all $x \in X$. Therefore by Theorem 3.38, Φ has a stationary point, and hence so does $F \subseteq \Phi$. ■

Finally, we give the application of Theorem 3.39 to derive a theorem of Ekeland [75].

Theorem 3.40 *Let (X,d) be a complete metric space and $\varphi : X \to [0,\infty)$ be lower semicontinuous. For any $\varepsilon > 0$, let $y_\varepsilon \in X$ be such that $\varphi(y_\varepsilon) \le \inf_{y\in X} \varphi(y)+\varepsilon$. Then there exists $x_\varepsilon \in X$ such that*

$$d(x_\varepsilon, y_\varepsilon) \le \sqrt{\varepsilon} \quad \text{and} \quad \varphi(x_\varepsilon) = \min_{x\in X} \left(\varphi(x) + \sqrt{\varepsilon} d(x_\varepsilon, x)\right). \tag{3.33}$$

Proof Let

$$F(x) = H_F(x) = \left\{y \in X : d(x,y) \le \varphi(x)/\sqrt{\varepsilon} - \varphi(y)/\sqrt{\varepsilon}\right\}.$$

By Theorem 3.39, F has a stationary point x_ε. The result follows from the definition of F. ∎

Chapter 4

Variational Principles and Their Applications

In 1972, Ekeland [74] (see also [75, 76]) established a theorem on the existence of an approximate minimizer of a bounded below and lower semicontinuous function. This theorem is now known as Ekeland's variational principle (in short, EVP). It is one of the most applicable results from nonlinear analysis and used as a tool to study the problems from fixed point theory, optimization, optimal control theory, game theory, nonlinear equations, dynamical systems, etc.; see, for example, [15, 17, 18, 40, 74–77, 70, 135, 141] and the references therein. Several well-known fixed point results, namely, Banach contraction principle, Clarke's fixed point theorem, Caristi's fixed point theorem, Nadler's fixed point theorem, etc., can be easily derived by using EVP. Later, it was found that several well-known results, namely, the Caristi-Kirk fixed point theorem [52, 53], Takahashi's minimization theorem [168], the Petal theorem [141], and the Daneš drop theorem [68], from nonlinear analysis are equivalent to Ekeland's variational principle. Since the discovery of EVP, there have also appeared many of its extensions or equivalent formulations; see, for example, [4, 9, 12, 18, 23, 41, 66, 80, 85, 90–93, 105, 116–119, 127, 128, 135, 137–139, 141, 159, 161, 162, 165, 168, 169, 178, 181, 182] and the references therein. Sullivan [161] established that the conditions of EVP on a metric space (X, d) imply the completeness of the metric space (X, d). In 1982, McLinden [127] showed how EVP, or more precisely the augmented form of it provided by Rockafellor [149], can be adapted to extremum problems of minimax type. For further detail on the equivalence of these results, we refer to [4, 69, 85, 91, 138] and the references therein.

In this chapter, we present several forms of Ekeland's variational principle, its equivalence with Takahashi's minimization theorem and the Caristi-Kirk fixed point theorem, and some applications to fixed point theory and weak sharp minima. A variational principle given by Borwein and Preiss [39] and further revised by Li and Shi [116] is discussed. By using EVP, a fixed point theorem due to Uderzo [171] for $\mathcal{H}$-continuous and directional Ψ-contraction set-valued mappings is also presented.

4.1 Ekeland's Variational Principle and Its Applications

Let X be a nonempty set. A function $f : X \to \mathbb{R} \cup \{+\infty\}$ is said to be *proper* if $f(x) \neq +\infty$ for all $x \in X$. The domain of f, denoted by $\operatorname{dom}(f)$, is defined by

$$\operatorname{dom}(f) = \{x \in X : f(x) < +\infty\}.$$

Let (X, d) be a metric space, K be a nonempty subset of X, and $f : X \to \mathbb{R} \cup \{+\infty\}$ be a proper function. The well-known Weierstrass's theorem states that if K is compact and f is lower semicontinuous, then the following constrained minimization problem (in short, CMP)

$$\inf_{x \in K} f(x)$$

has a solution. Not only this, the solution set of CMP is compact.

Definition 4.1 For a given $\varepsilon > 0$, an element x_ε is said to be an *approximate ε-solution* of the following minimization problem (in short, MP)

$$\inf_{x \in X} f(x),$$

if

$$\inf_X f \leq f(x_\varepsilon) \leq \inf_X f + \varepsilon,$$

where $\inf_X f := \inf_{x \in X} f(x)$.

The following Ekeland's variational principle is the basic tool to establish the existence of solutions, or approximate ε-solutions, for minimization problems that fail to satisfy the compactness requirement of Weierstrass's theorem.

Theorem 4.1 (Strong Form of Ekeland's Variational Principle) *Let (X, d) be a complete metric space and $f : X \to \mathbb{R} \cup \{+\infty\}$ be a proper, bounded below, and lower semicontinuous function. Let $\varepsilon > 0$ and $\hat{x} \in X$ be given such that*

$$f(\hat{x}) \leq \inf_X f + \varepsilon.$$

Then, for a given $\lambda > 0$, there exists $\bar{x} \in X$ such that

(a) $f(\bar{x}) \leq f(\hat{x})$;

(b) $d(\hat{x}, \bar{x}) \leq \lambda$;

(c) $f(\bar{x}) < f(x) + \frac{\varepsilon}{\lambda} d(x, \bar{x})$, *for all $x \in X \setminus \{\bar{x}\}$.*

Proof For the sake of convenience, we set $d_\lambda(u, v) = (1/\lambda)d(u, v)$. Then, d_λ is equivalent to d and (X, d_λ) is complete. Let us define a partial ordering $\preccurlyeq$ on X by

$$x \preccurlyeq y \quad \text{if and only if} \quad f(x) \leq f(y) - \varepsilon d_\lambda(x, y).$$

It is easy to see that this ordering is (i) reflexive, that is, for all $x \in X$, $x \preccurlyeq x$; (ii) antisymmetric, that is, for all $x, y \in X$, $x \preccurlyeq y$ and $y \preccurlyeq x$ imply $x = y$; (iii) transitive, that is, for all $x, y, z \in X$, $x \preccurlyeq y$ and $y \preccurlyeq z$ imply $x \preccurlyeq z$.

We define a sequence $\{S_n\}$ of subsets of X as follows: Start with $x_1 = \hat{x}$ ($\hat{x}$ is the same as given in the statement of the theorem) and define

$$S_1 = \{x \in X : x \preccurlyeq x_1\}; \quad x_2 \in S_1 \text{ such that } f(x_2) \leq \inf_{S_1} f + \frac{\varepsilon}{2},$$

$$S_2 = \{x \in X : x \preccurlyeq x_2\}; \quad x_3 \in S_2 \text{ such that } f(x_3) \leq \inf_{S_2} f + \frac{\varepsilon}{2^2},$$

and inductively

$$S_n = \{x \in X : x \preccurlyeq x_n\}; \quad x_{n+1} \in S_n \text{ such that } f(x_{n+1}) \leq \inf_{S_n} f + \frac{\varepsilon}{2^n}. \tag{4.1}$$

Clearly, $S_1 \supset S_2 \supset S_3 \supset ....$ We claim that each S_n is closed.

Indeed, let $u_j \in S_n$ with $u_j \to u \in X$. Then $u_j \preccurlyeq x_n$ and so $f(u_j) \leq f(x_n) - \varepsilon d_\lambda(u_j, x_n)$. By taking limit as $j \to \infty$ and using the lower semicontinuity of f and the continuity of d and so the continuity of d_λ, we conclude that $u \in S_n$.

Now, we prove that the diameter of these sets S_n, $\operatorname{diam}(S_n) \to 0$ as $n \to \infty$.

Indeed, take an arbitrary point $u \in S_n$. Then, on one hand, $u \preccurlyeq x_n$ implies that

$$f(u) \leq f(x_n) - \varepsilon d_\lambda(u, x_n) \tag{4.2}$$

and, on other hand, we observe that u also belongs to S_{n-1}. So it is one of the points which entered into the competition when we picked x_n. Therefore, from (4.1), we have

$$f(x_n) \leq f(u) + \frac{\varepsilon}{2^{n-1}}. \tag{4.3}$$

From (4.2) and (4.3), we obtain

$$d_\lambda(u, x_n) \leq 2^{-n+1}, \quad \text{for all } u \in S_n$$

which gives $\operatorname{diam}(S_n) \leq 2^{-n}$, and hence $\operatorname{diam}(S_n) \to 0$ as $n \to \infty$.

Since (X, d_λ) is complete and $\{S_n\}$ is a decreasing sequence of closed sets, by Cantor's Intersection Theorem 1.4, we infer that

$$\bigcap_{n=1}^{\infty} S_n = \{\bar{x}\}.$$

We still have to prove that this unique point $\bar{x}$ satisfies conditions (a)–(c).

Since $\bar{x} \in S_1$, we have

$$\bar{x} \preccurlyeq x_1 = \hat{x} \quad \text{if and only if} \quad f(\bar{x}) \leq f(\hat{x}) - \varepsilon d_\lambda(\bar{x}, \hat{x})$$

and so, $f(\bar{x}) \leq f(\hat{x})$. Hence (a) is proved.

Now let $x \neq \bar{x}$. We cannot have $x \preccurlyeq \bar{x}$, because otherwise x would belong to $\bigcap_{n=1}^{\infty} S_n$. So $x \not\preccurlyeq \bar{x}$, which means that

$$f(x) > f(\bar{x}) - \varepsilon d_\lambda(x, \bar{x}),$$

and hence (c) is proved.

Finally, by writing

$$d_\lambda(\hat{x}, x_n) = d_\lambda(x_1, x_n) \leq \sum_{j=1}^{n-1} d_\lambda(x_j, x_{j+1}) \leq \sum_{j=1}^{n-1} 2^{-j},$$

and taking limit as $n \to \infty$, we obtain $d_\lambda(\hat{x}, \bar{x}) \leq 1$ and so $d(\hat{x}, \bar{x}) \leq \lambda$. This proves (b). ■

Remark 4.1 (a) Let $S_r[x]$ be the closed sphere with center at x and radius $r > 0$. In this context, condition (b) of Theorem 4.1 can be written as $\bar{x} \in S_\lambda[\hat{x}]$.

(b) Strong form of Ekeland's variational principle says that for $\lambda, \varepsilon > 0$ and $\hat{x}$ an ε-approximate solution of an optimization problem, there exists a new point $\bar{x}$ that is not worse than $\hat{x}$ and belongs to a λ-neighborhood of $\hat{x}$ and, especially, $\bar{x}$ satisfies (c). Relation (c) says, in fact, that $\bar{x}$ minimizes globally $f(\cdot) + \left(\frac{\varepsilon}{\lambda}\right) d(\cdot, \bar{x})$, which is a Lipschitz perturbation of f.

(c) Ha [90] presented examples to illustrate that the conclusion of Ekelend's variational principle may not hold at all or it holds not for all positive ε when f is lower semicontinuous only on its domain.

Remark 4.2 Theorem 2.17 implies Theorem 4.1.

Proof Applying Theorem 2.17 with the metric $(\varepsilon/\lambda)d$ and $\tilde{x} = \hat{x}$, we have a point $\bar{x}$ such that

$$(\varepsilon/\lambda)d(\bar{x}, x) > \varphi(\bar{x}) - \varphi(x), \quad \text{for all } x \in X \setminus \{\bar{x}\},$$

and

$$(\varepsilon/\lambda)d(\tilde{x}, \bar{x}) \leq \varphi(\tilde{x}) - \varphi(\bar{x}).$$

Hence, we get (a) and (c). The inequality $\varphi(\tilde{x}) \leq \inf_{x \in X} \varphi(x) + \varepsilon$ implies that $\varphi(\tilde{x}) - \varphi(\bar{x}) \leq \varepsilon$. Thus, $(\varepsilon/\lambda)d(\tilde{x}, \bar{x}) \leq \varepsilon$, which gives (b). ■

Aubin and Frankowska [18] established the following form of Ekeland's variational principle which is equivalent to Theorem 4.1.

Theorem 4.2 (Strong Form of Ekeland's Variational Principle) *Let (X, d) be a complete metric space and $f : X \to \mathbb{R} \cup \{+\infty\}$ be a proper, bounded below, and lower semicontinuous function. Let $\hat{x} \in \mathrm{dom}(f)$ and $e > 0$ be fixed. Then, there exists $\bar{x} \in X$ such that*

(aa) $f(\bar{x}) - f(\hat{x}) + ed(\hat{x}, \bar{x}) \leq 0$;

(bb) $f(\bar{x}) < f(x) + ed(x, \bar{x})$, *for all $x \in X \setminus \{\bar{x}\}$.*

Proof For the sake of convenience, we set $d_e(u, v) = ed(u, v)$. Then, d_e is equivalent to d and (X, d_e) is complete. For all $x \in X$, define

$$S(x) = \{y \in X : f(y) - f(x) + d_e(x, y) \leq 0\}.$$

We assume that $f(x) \neq +\infty$ for all $x \in X$. For all $x \in X$, we have $x \in S(x)$, and therefore, $S(x)$ is nonempty. So, we can let $y \in S(x)$. Then,

$$f(y) - f(x) + d_e(x, y) \leq 0, \tag{4.4}$$

and also let $z \in S(y)$, then

$$f(z) - f(y) + d_e(y, z) \leq 0. \tag{4.5}$$

Adding (4.4) and (4.5), we obtain

$$f(z) - f(x) + d_e(x, z) \leq 0.$$

Therefore, $z \in S(x)$ which implies that $S(y) \subseteq S(x)$.

We claim that for every $x \in X$, $S(x)$ is closed. Indeed, let $\{x_n\}$ be a sequence in $S(x)$ such that $x_n \to x^* \in X$. Then,

$$f(x_n) - f(x) + d_e(x, x_n) \leq 0.$$

By using the lower semicontinuity of f and continuity of d and hence continuity of d_e, we have

$$f(x^*) - f(x) + d_e(x, x^*) \leq \liminf_{n\to\infty} f(x_n) - f(x) + \liminf_{n\to\infty} d_e(x, x_n) \leq 0.$$

Therefore, $x^* \in S(x)$ and so $S(x)$ is closed for every $x \in X$.

For all $x \in \text{dom}(f)$, define $\mathcal{V}(x)$ by

$$\mathcal{V}(x) := \inf_{z\in S(x)} f(z).$$

For every $z \in S(x)$,

$$d_e(x, z) \leq f(x) - f(z) \leq f(x) - \mathcal{V}(x),$$

so that the diameter of $S(x)$, $\text{diam}\,(S(x)) = \sup_{y,z\in S(x)} d_e(y, z)$, is not greater than $2(f(x) - \mathcal{V}(x))$.

Define a sequence in the following manner: Fix $x_0 = \hat{x} \in X$; take $x_1 \in S(x_0)$ such that

$$f(x_1) \leq \mathcal{V}(x_0) + 2^{-1}.$$

Denote by x_2 any point in $S(x_1)$ such that

$$f(x_2) \leq \mathcal{V}(x_1) + 2^{-2}.$$

Continue in this manner, we obtain a sequence $\{x_n\}$ such that

$$x_{n+1} \in S(x_n) \quad \text{and} \quad f(x_{n+1}) \leq \mathcal{V}(x_n) + 2^{-(n+1)}.$$

Since $S(x_{n+1}) \subseteq S(x_n)$, we deduce that

$$\mathcal{V}(x_{n+1}) = \inf_{z\in S(x_{n+1})} f(z) \geq \inf_{z\in S(x_n)} f(z) = \mathcal{V}(x_n),$$

and thus,

$$\mathcal{V}(x_n) \leq \mathcal{V}(x_{n+1}).$$

On the other hand, inequality $\mathcal{V}(y) \leq f(y)$ implies that

$$\mathcal{V}(x_{n+1}) \leq f(x_{n+1}) \leq \mathcal{V}(x_n) + 2^{-(n+1)} \leq \mathcal{V}(x_{n+1}) + 2^{-(n+1)},$$

and therefore,

$$0 \leq 2\left(f(x_{n+1}) - \mathcal{V}(x_{n+1})\right) \leq 2^{-n}.$$

It follows that

$$\text{diam}\,(S(x_n)) \leq 2^{-n} \to 0 \quad \text{as } n \to \infty.$$

By Cantor's Intersection Theorem 1.4, there exists exactly one point $\bar{x} \in X$ such that

$$\bigcap_{n=0}^{\infty} S(x_n) = \{\bar{x}\}.$$

This implies that $\bar{x} \in S(x_0) = S(\hat{x})$, that is,

$$f(\bar{x}) - f(\hat{x}) + d_e(\hat{x}, \bar{x}) \leq 0$$

and so (aa) holds.

Moreover, $\bar{x}$ also belongs to all $S(x_n)$ and, since $S(\bar{x}) \subseteq S(x_n)$ for all n, we have

$$S(\bar{x}) = \{\bar{x}\}.$$

It follows that $x \notin S(\bar{x})$ whenever $x \neq \bar{x}$ implying that

$$f(x) - f(\bar{x}) + d_e(\bar{x}, x) > 0,$$

that is, (bb) holds. ■

Remark 4.3 Theorem 4.1 and Theorem 4.2 are equivalent.

Proof Assume that Theorem 4.2 and the hypothesis of Theorem 4.1 hold. By (aa), we have $f(\bar{x}) \leq f(\hat{x}) - ed(\hat{x}, \bar{x})$. Since $e > 0$, we obtain $f(\bar{x}) \leq f(\hat{x})$ and so (a) holds.

By the hypothesis of Theorem 4.1, $f(\hat{x}) \leq \inf_X f + \varepsilon$, that is, $f(\hat{x}) - \inf_X f \leq \varepsilon$. Therefore, in particular, $f(\hat{x}) - f(\bar{x}) \leq \varepsilon$. Set $e = \frac{\varepsilon}{\lambda}$. Then by (aa) of Theorem 4.2, we have

$$d(\hat{x}, \bar{x}) \leq \frac{\lambda}{\varepsilon} \left[f(\hat{x}) - f(\bar{x}) \right] \leq \frac{\lambda}{\varepsilon} \varepsilon = \lambda,$$

and (b) of Theorem 4.1 follows.

Conversely, suppose that Theorem 4.1 and the hypothesis of Theorem 4.2 hold. Set $\varepsilon = e\lambda$ and let $\hat{x} \in \mathrm{dom}(f)$ and $e > 0$ be given. Take $f(\hat{x}) - \inf_X f \leq \varepsilon$, and consider

$$S_1 := \{x \in X : f(x) - f(\hat{x}) \leq -\varepsilon\}$$

and

$$S_2 := \{x \in X : f(x) - f(\hat{x}) + ed(\hat{x}, x) \leq 0\}.$$

The lower semicontinuity of f implies that S_1 and S_2 are closed. Furthermore, $S_2 \neq \emptyset$ as $\hat{x} \in S_2$. By applying Theorem 4.1 for ε, λ and $S_1 \cup S_2$ instead of X, we obtain $\bar{x} \in S_1 \cup S_2$ such that (a), (b), and

(cc) $$f(x) - f(\bar{x}) + \frac{\varepsilon}{\lambda} d(x, \bar{x}) > 0 \quad \text{for all } x \in S_1 \cup S_2 \setminus \{\bar{x}\}$$

hold.

If $\bar{x} \in S_2$, then (aa) holds. If $\bar{x} \in S_1$, then we have

$$f(\bar{x}) \leq f(\hat{x}) - \varepsilon.$$

Adding this inequality and (b) multiplied by e, we obtain

$$f(\bar{x}) - f(\hat{x}) + ed(\hat{x}, \bar{x}) \leq 0,$$

that is, $\bar{x} \in S_2$. So (aa) holds.

To show (bb), it is sufficient to check (cc) is satisfied also for $x \notin S_1 \cup S_2$. By the definition of S_2, $x \notin S_2$ means $f(x) - f(\hat{x}) + ed(\hat{x}, x) > 0$. Adding this inequality and (aa), we get

$$f(x) - f(\bar{x}) + ed(x, \bar{x}) \geq 0,$$

that is, (bb) holds. ∎

The following result is known as the weak formulation of Ekeland's variational principle.

Corollary 4.1 (Weak Form of Ekeland's Variational Principle) *Let (X, d) be a complete metric space and $f : X \to \mathbb{R} \cup \{+\infty\}$ be a proper, bounded below, and lower semicontinuous function. Then, for any given $\varepsilon > 0$, there exists $\bar{x} \in X$ such that*

$$f(\bar{x}) \leq \inf_X f + \varepsilon$$

and

$$f(\bar{x}) < f(x) + \varepsilon d(x, \bar{x}), \quad \textit{for all } x \in X \textit{ with } x \neq \bar{x}. \tag{4.6}$$

Proof It follows from the fact that there always exists some point $\hat{x} \in X$ such that

$$f(\hat{x}) \leq \inf_X f + \varepsilon.$$

Then by Theorem 4.1, there exists $\bar{x} \in X$ (by taking $\lambda = 1$) such that

$$f(\bar{x}) \leq f(\hat{x}) \leq \inf_X f + \varepsilon$$

and

$$f(\bar{x}) < f(x) + \varepsilon d(x, \bar{x}), \quad \text{for all } x \in X \text{ with } x \neq \bar{x}.$$ ∎

Remark 4.4 As in [58], the lower semicontinuity of the function involved in Ekeland's variational principle and Caristi's fixed point theorem can be replaced by lower semicontinuity from above.

The property of Ekeland's variational principle for proper but extended real-valued lower semicontinuous and bounded below functions on a metric space characterizes completeness of the metric space.

Theorem 4.3 (Converse of Ekeland's Variational Principle) *A metric space (X, d) is complete if for every function $f : X \to \mathbb{R} \cup \{+\infty\}$ which is proper, bounded below, and lower semicontinuous on X and for every given $\varepsilon > 0$, there exists $\bar{x} \in X$ such that*

$$f(\bar{x}) \leq \inf_X f + \varepsilon$$

and

$$f(\bar{x}) \leq f(x) + \varepsilon d(x, \bar{x}), \quad \textit{for all } x \in X.$$

Proof Consider a Cauchy sequence $\{x_n\}$ in X and define a function $f : X \to [0, +\infty)$ by

$$f(x) = \lim_{n\to\infty} d(x_n, x), \quad \text{for all } x \in X.$$

Then, f is well-defined, nonnegative, and continuous. Since $\{x_n\}$ is Cauchy, we have

$$f(x_m) = \lim_{n\to\infty} d(x_n, x_m) \to 0, \quad \text{as } m \to \infty.$$

Therefore, $\inf\limits_X f = 0$.

For given $0 < \varepsilon < 1$, choose $\bar{x} \in X$ such that $f(\bar{x}) \leq \varepsilon$ and

$$f(\bar{x}) \leq f(x) + \varepsilon d(x, \bar{x}), \quad \text{for all } x \in X. \tag{4.7}$$

Letting $x = x_n$ in (4.7) and taking limit as $n \to \infty$, we obtain $f(\bar{x}) \leq \varepsilon f(\bar{x})$, and so, $f(\bar{x}) = 0$. This implies that $\{x_n\}$ converges to $\bar{x}$. ■

Zhong [181] extended Ekeland's variational principle and derived a minimax theorem under a weak "compactness condition."

Theorem 4.4 [181] *Let (X, d) be a complete metric space, x_0 be a fixed element of X and $f : X \to \mathbb{R} \cup \{+\infty\}$ be a proper, bounded below, and lower semicontinuous function. Let $h : [0, \infty) \to [0, \infty)$ be a continuous nondecreasing function such that*

$$\int_0^\infty \frac{1}{(1 + h(r))}\, dr = +\infty. \tag{4.8}$$

Let $\varepsilon > 0$ and $\hat{x} \in X$ be given such that

$$f(\hat{x}) \leq \inf_X f + \varepsilon. \tag{4.9}$$

Then, for a given $\lambda > 0$, there exists $\bar{x} \in X$ such that

(a) $f(\bar{x}) \leq f(\hat{x})$*;*

(b) $d(\bar{x}, x_0) \leq r_0 + \bar{r}$*;*

(c) $f(x) \geq f(\bar{x}) - \frac{\varepsilon}{\lambda(1+h(d(x_0,\bar{x})))} d(x, \bar{x})$*, for all $x \in X$,*

where $r_0 = d(x_0, \hat{x})$ and $\bar{r}$ is a positive constant such that

$$\int_{r_0}^{r_0+\bar{r}} \frac{1}{1 + h(r)}\, dr \geq \lambda. \tag{4.10}$$

If $h(r) = 0$ for all r and $x_0 = \hat{x}$, then Theorem 4.4 reduces to EVP (Theorem 4.1).

Proof Define inductively a sequence $\{x_n\}$ in X as follows: Take $x_1 = \hat{x}$. If x_n is known, then x_n is such that either

(i) $f(x) \geq f(x_n) - \dfrac{\varepsilon}{\lambda(1 + h(d(x_0, x_n)))} d(x, x_n)$, for all $x \in X$,

or

(ii) $E_n := \left\{ x \in X : f(x) < f(x_n) - \dfrac{\varepsilon}{\lambda(1 + h(d(x_0, x_n)))} d(x, x_n) \right\} \neq \emptyset$.

If case (i) holds, then we take $x_{n+1} = x_n$; and if case (ii) holds, then we choose $x_{n+1} \in E_n$ such that

$$f(x_{n+1}) < \inf_{E_n} f + \frac{1}{n+1}. \tag{4.11}$$

Thus, we obtain a sequence $\{x_n\}$ in X such that

$$f(x_{n+1}) \leq f(x_n) - \frac{\varepsilon}{\lambda(1 + h(d(x_0, x_n)))} d(x_n, x_{n+1}), \quad \text{for all } n \in \mathbb{N}. \tag{4.12}$$

We claim that

$$d(x_0, x_n) < r_0 + \tilde{r}, \quad \text{for all } n \in \mathbb{N}. \tag{4.13}$$

It is easy to see that (4.13) holds for $n = 1$. Now assume that (4.13) holds for some k with $1 \leq n < k$, but $d(x_0, x_k) \geq r_0 + \tilde{r}$. This implies that $x_2, \ldots, x_k$ are defined as case (ii). Hence,

$$\begin{aligned} f(x_k) &< f(x_{k-1}) - \frac{\varepsilon}{\lambda(1 + h(d(x_0, x_{k-1})))} d(x_{k-1}, x_k) \\ &< f(x_1) - \sum_{n=1}^{k-1} \frac{\varepsilon}{\lambda(1 + h(d(x_0, x_n)))} d(x_n, x_{n+1}). \end{aligned}$$

From the above inequality and (4.9), we obatin

$$\sum_{n=1}^{k-1} \frac{\varepsilon}{\lambda(1 + h(d(x_0, x_n)))} d(x_n, x_{n+1}) < f(x_1) - f(x_k) \leq f(\hat{x}) - \inf_X f < \varepsilon,$$

which implies that

$$\sum_{n=1}^{k-1} \frac{d(x_n, x_{n+1})}{1 + h(d(x_0, x_n))} < \lambda. \tag{4.14}$$

Without loss of generality, we can assume that $d(x_0, x_n)$, $n = 1, 2, \ldots, k$, is nondecreasing. In fact, if there is some n, $1 \leq n \leq k - 1$, such that

$$d(x_0, x_{n+1}) < d(x_0, x_n),$$

which implies that $n \leq k - 2$ since $d(x_0, x_k) \geq r_0 + \tilde{r} > d(x_0, x_n)$ for every $n < k$; then by the monotonicity of h, we get

$$\frac{1}{1 + h(d(x_0, x_n))} \leq \frac{1}{1 + h(d(x_0, x_{n+1}))}.$$

By the triangle inequality, we have

$$\begin{aligned}\frac{d(x_n, x_{n+2})}{1 + h(d(x_0, x_n))} &\le \frac{d(x_n, x_{n+1}) + d(x_{n+1}, x_{n+2})}{1 + h(d(x_0, x_n))} \\ &\le \frac{d(x_n, x_{n+1})}{1 + h(d(x_0, x_n))} + \frac{d(x_{n+1}, x_{n+2})}{1 + h(d(x_0, x_{n+1}))}.\end{aligned}$$

Therefore, (4.14) still holds if we take

$$\frac{1}{1 + h(d(x_0, x_n))} d(x_n, x_{n+2})$$

instead of

$$\frac{d(x_n, x_{n+1})}{1 + h(d(x_0, x_n))} + \frac{d(x_{n+1}, x_{n+2})}{1 + h(d(x_0, x_{n+1}))}.$$

Thus, we can delete x_{n+1} from $\{x_i\}_{i=1}^k$. Since h and $d(x_0, x_n)$, $n = 1, 2, \dots, k$, are nondecreasing, we have

$$\begin{aligned}\sum_{n=1}^{k-1} \frac{d(x_n, x_{n+1})}{1 + h(d(x_0, x_n))} &\ge \sum_{n=1}^{k-1} \frac{d(x_{n+1}, x_0) - d(x_0, x_n)}{1 + h(d(x_0, x_n))} \\ &\ge \sum_{n=1}^{k-1} \int_{d(x_0, x_n)}^{d(x_0, x_{n+1})} \frac{1}{1 + h(r)}\, dr \\ &= \int_{d(x_0, x_1)}^{d(x_0, x_k)} \frac{1}{1 + h(r)}\, dr \\ &\ge \int_{r_0}^{r_0 + \bar{r}} \frac{1}{1 + h(r)}\, dr \ge \lambda,\end{aligned}$$

which contradicts (4.14). Thus, (4.13) holds.

We now prove that $\{x_n\}$ converges to some point $\bar{x} \in X$ for which (a)–(c) hold. By the definition of $\{x_n\}$, we know that if there exists some x_k defined as in case (i), then $x_n = x_k$ for any $n \ge k$ and x_k satisfies (a)–(c). Therefore, without loss of generality, we can assume that all of x_n are defined as in case (ii). By (4.9) and (4.12), we get

$$\sum_{k=1}^{n} \frac{\varepsilon}{\lambda(1 + h(d(x_0, x_k)))} d(x_k, x_{k+1}) \le f(x_1) - f(x_{n+1}) \le f(\hat{x}) - \inf_X f < \varepsilon.$$

Letting limit as $n \to \infty$, we obtain

$$\sum_{k=1}^{\infty} \frac{\varepsilon}{\lambda(1 + h(d(x_0, x_k)))} d(x_k, x_{k+1}) \le \lambda. \tag{4.15}$$

Take a constant c such that $c \geq 1 + h(r_0 + \bar{r})$. Then, from (4.13), we obtain

$$\frac{c}{1 + h(d(x_0, x_n))} \geq 1, \quad \text{for all } n \in \mathbb{N}. \tag{4.16}$$

Hence,

$$\begin{aligned} d(x_{n+p}, x_n) &\leq \sum_{k=n}^{n+p-1} d(x_k, x_{k+1}) \\ &\leq c \sum_{k=n}^{n+p-1} \frac{1}{1 + h(d(x_0, x_k))} d(x_k, x_{k+1}) \\ &\leq c \sum_{k=n}^{\infty} \frac{1}{1 + h(d(x_0, x_k))} d(x_k, x_{k+1}). \end{aligned}$$

From (4.15), we know that $\{x_n\}$ is a Cauchy sequence. Hence, $\{x_n\}$ converges to a point $\bar{x} \in X$. It follows from (4.13) that $\bar{x} \in X$ satisfies (b). Using (4.12) and the lower semicontinuity of f, we get

$$f(\bar{x}) \leq \liminf_{n \to \infty} f(x_n) \leq f(x_n) \leq f(x_1).$$

This shows that (a) holds. Now if $\bar{x}$ does not satisfy (c), then there exists some $x^* \in X$ such that

$$f(x^*) < f(\bar{x}) - \frac{\varepsilon}{\lambda(1 + h(d(x_0, \bar{x})))} d(\bar{x}, x^*). \tag{4.17}$$

Since $\lim_{n \to \infty} x_n = \bar{x}$ and $f(\bar{x}) \leq f(x_n)$, there exists some $n_0 \in \mathbb{N}$ such that

$$f(x^*) < f(x_n) - \frac{\varepsilon}{\lambda(1 + h(d(x_0, x_n)))} d(x_n, x^*), \quad \text{for all } n \geq n_0. \tag{4.18}$$

This shows that $x^* \in E_n$ as $n \geq n_0$. From (4.11) and (4.18), we obtain

$$\begin{aligned} \inf_{E_n} f + \frac{1}{n+1} &\geq f(x_{n+1}) \\ &\geq f(x^*) + \frac{\varepsilon}{\lambda(1 + h(d(x_0, x_{n+1})))} d(x_{n+1}, x^*) \\ &\geq \inf_{E_n} f + \frac{\varepsilon}{\lambda(1 + h(d(x_0, x_{n+1})))} d(x_{n+1}, x^*). \end{aligned}$$

Combining this with (4.16), we get

$$d(x_{n+1}, x^*) \leq \frac{\lambda}{\varepsilon} \frac{c}{n+1}.$$

Letting limit as $n \to \infty$, we obtain $\lim_{n \to \infty} x_n = x^*$. It follows that $\bar{x} = x^*$, which contradicts (4.17) and this completes the proof. ∎

4.1.1 Applications to Fixed Point Theorems

As a first application of Ekeland's variational principle, we prove the well-known Banach contraction principle.

Theorem 4.5 (Banach Contraction Principle) *Let (X, d) be a complete metric space and $T : X \to X$ be a contraction mapping. Then, T has a unique fixed point in X.*

Proof Consider the function $f : X \to [0, \infty)$ defined by

$$f(x) = d(x, T(x)), \quad \text{for all } x \in X.$$

Then, f is bounded below and continuous on X. Choose ε such that $0 < \varepsilon < 1 - \alpha$, where α is the Lipschitz constant. By Corollary 4.1, there exists $\bar{x} \in X$ (depending on ε) such that

$$f(\bar{x}) \leq f(x) + \varepsilon d(x, \bar{x}), \quad \text{for all } x \in X.$$

Putting $x = T(\bar{x})$, we have

$$\begin{aligned} d(\bar{x}, T(\bar{x})) &\leq d\left(T(\bar{x}), T(T(\bar{x}))\right) + \varepsilon d(\bar{x}, T(\bar{x})) \\ &\leq \alpha d(\bar{x}, T(\bar{x})) + \varepsilon d(\bar{x}, T(\bar{x})) \\ &= (\alpha + \varepsilon) d(\bar{x}, T(\bar{x})). \end{aligned}$$

If $\bar{x} \neq T(\bar{x})$, then we obtain $1 \leq \alpha + \varepsilon$, which contradicts our assumption that $\alpha + \varepsilon < 1$. Therefore, we have $\bar{x} = T(\bar{x})$. The uniqueness of $\bar{x}$ can be proved as in the original proof of the Banach contraction principle. ■

As a second application of Ekeland's variational principle, we prove Caristi's fixed point theorem.

Theorem 4.6 (Caristi's Fixed Point Theorem) [52] *Let (X, d) be a complete metric space and $T : X \to X$ be a mapping such that*

$$d(x, T(x)) + \varphi(T(x)) \leq \varphi(x), \quad \text{for all } x \in X, \tag{4.19}$$

where $\varphi : X \to \mathbb{R} \cup \{+\infty\}$ is a proper, lower semicontinuous, and bounded below function. Then, there exists $\bar{x} \in X$ such that $T(\bar{x}) = \bar{x}$ and $\varphi(\bar{x}) < \infty$.

Proof By using Corollary 4.1 with $\varepsilon = 1$, we obtain $\bar{x} \in X$ such that

$$\varphi(\bar{x}) < \varphi(x) + d(x, \bar{x}), \quad \text{for all } x \in X \setminus \{\bar{x}\}. \tag{4.20}$$

We claim that $\bar{x} = T(\bar{x})$. Otherwise all $y = T(\bar{x}) \in X$ are such that $y \neq \bar{x}$. Then from (4.19) and (4.20), we have

$$d(y, \bar{x}) + \varphi(y) \leq \varphi(\bar{x}) \quad \text{and} \quad \varphi(\bar{x}) < \varphi(y) + d(y, \bar{x}),$$

which cannot hold simultaneously. ■

Definition 4.2 Let (X, d) be a metric space. For any $x, y \in X$, the *segment* joining x and y is defined by

$$[x, y] = \{z \in X : d(x, z) + d(z, y) = d(x, y)\}. \tag{4.21}$$

Definition 4.3 Let (X, d) be a metric space. A mapping $T : X \to X$ is said to be a *directional contraction* if

(i) T is continuous, and

(ii) there exists $\alpha \in (0, 1)$ such that for any $x \in X$ with $T(x) \neq x$, there exists $z \in [x, T(x)] \setminus \{x\}$ such that

$$d(T(x), T(z)) \leq \alpha d(x, z).$$

Theorem 4.7 (Clarke's Fixed Point Theorem) *Let (X, d) be a complete metric space and $T : X \to X$ be a directional contraction mapping. Then, T has a fixed point.*

Proof Define the function $f : X \to [0, +\infty)$ by

$$f(x) = d(x, T(x)), \quad \text{for all } x \in X.$$

Then, f is continuous and bounded below. Applying Corollary 4.1 to f with $\varepsilon \in (0, 1 - \alpha)$, we conclude that there exists $\bar{x} \in X$ such that

$$f(\bar{x}) \leq f(x) + \varepsilon d(x, \bar{x}), \quad \text{for all } x \in X. \tag{4.22}$$

If $T(\bar{x}) = \bar{x}$, then we are done. Otherwise, by the definition of a directional contraction mapping, there exists a point $z \neq \bar{x}$ with $z \in [\bar{x}, T(\bar{x})]$, that is,

$$d(\bar{x}, z) + d(z, T(\bar{x})) = d(\bar{x}, T(\bar{x})) = f(\bar{x}) \tag{4.23}$$

satisfying

$$d\left(T(\bar{x}), T(z)\right) \leq \alpha d(\bar{x}, z). \tag{4.24}$$

Setting $x = z$ in (4.22), we get

$$f(\bar{x}) \leq f(z) + \varepsilon d(z, \bar{x}).$$

By using the above inequality and (4.23), we obtain

$$d(\bar{x}, z) + d(z, T(\bar{x})) \leq d(z, T(z)) + \varepsilon d(z, \bar{x}),$$

that is,

$$d(\bar{x}, z) \leq d(z, T(z)) - d(z, T(\bar{x})) + \varepsilon d(z, \bar{x}). \tag{4.25}$$

By the triangle inequality and (4.24), we obtain

$$d(z, T(z)) - d(z, T(\bar{x})) \leq d(T(\bar{x}), T(z)) \leq \alpha d(\bar{x}, z). \tag{4.26}$$

By combining (4.25) and (4.26), we get

$$d(\bar{x}, z) \leq (\alpha + \varepsilon) d(\bar{x}, z),$$

a contradiction. ∎

Remark 4.5 It is clear that every contraction mapping is a directional contraction. Therefore, Clarke's fixed point theorem is more general than the Banach contraction principle.

In the following example, we show that Theorem 4.7 is applicable but Theorem 4.5 is not.

Example 4.1 Let $X = \mathbb{R}^2$ be a metric space with the metric

$$d(x,y) = |x_1 - y_1| + |x_2 - y_2|, \quad \text{for all } x = (x_1,x_2),\ y = (y_1,y_2) \in \mathbb{R}^2.$$

A segment between two points (a_1,a_2) and (b_1,b_2) consists of the closed rectangle having the two points as diagonally opposite corners. Define

$$T(x) = \left(\frac{3x_1}{2} - \frac{x_2}{3},\ x_1 + \frac{x_2}{3}\right), \quad \text{for all } x = (x_1,x_2) \in \mathbb{R}^2.$$

Then, T is a directional contraction mapping. Indeed, if $y = (y_1,y_2) = T(x) \neq x = (x_1,x_2)$. Then, $y_2 \neq x_2$ (for otherwise we will also have $y_1 = x_1$). Now the set $[x,y]$ contains points of the form (x_1,t) with t arbitrarily close to x_2 but not equal to x_2. For such points, we have

$$d\left(T(x_1,t), T(x_1,x_2)\right) = \frac{2}{3} d\left((x_1,t),(x_1,x_2)\right).$$

Hence, T is a directional contraction mapping. We can easily check that the fixed points of T are all points of the form $(x, 3x/2)$. Since T has more than one fixed point, clearly then the Banach contraction theorem does not apply to this mapping.

Daneš [69] proved that the Daneš drop theorem [68], Krasnoselskii-Zabreiko renorming theorem [179], Browder's generalization of the Bishop–Phelps theorem [37, 44], Caristi's fixed point thoerem, and Ekeland's variational principle are all equivalent in the sense that each result can be derived by using the other.

By using Ekeland's variational principle, Mizoguchi and Takahashi [129] derived the following Caristi–Kirk theorem [53], which is the set-valued version of Theorem 4.6.

Theorem 4.8 (Caristi–Kirk Fixed Point Theorem) *Let (X,d) be a complete metric space and $T : X \rightrightarrows X$ be a set-valued map with nonempty values such that for each $x \in X$, there exists $y \in T(x)$ satisfying*

$$d(x,y) + \varphi(y) \leq \varphi(x), \tag{4.27}$$

where $\varphi : X \to \mathbb{R} \cup \{+\infty\}$ is a proper, lower semicontinuous, and bounded below function. Then, T has a fixed point, that is, there exists $\bar{x} \in X$ such that $\bar{x} \in T(\bar{x})$.

Proof For each $x \in X$, we put $f(x) = y$, where $y \in X$ such that $y \in T(x)$ and $d(x,y) + \varphi(y) \leq \varphi(x)$. Then, f defines a mapping from X into itself such that

$$d(x,f(x)) + \varphi(f(x)) \leq \varphi(x), \quad \text{for all } x \in X.$$

Since φ is proper, there exists $u \in X$ with $\varphi(u) < +\infty$. So, let

$$\mathcal{X} = \left\{x \in X : \varphi(x) \leq \varphi(u) - d(u,x)\right\}.$$

Then, $\mathcal{X}$ is nonempty and closed. We can also see that $\mathcal{X}$ is invariant under the mapping f. In fact, for each $x \in \mathcal{X}$, we have

$$d(x,f(x)) + \varphi(f(x)) \le \varphi(x) \le \varphi(u) - d(u,x),$$

and hence,

$$\begin{aligned}\varphi(f(x)) &\le \varphi(u) - \{d(u,x) + d(x,f(x))\}\\ &\le \varphi(u) - d(u,f(x)).\end{aligned}$$

This implies that $f(x) \in \mathcal{X}$. Now, from Caristi's fixed point theorem, we obtain $z \in \mathcal{X}$ such that $z = f(z) \in T(z)$. ∎

Now we prove Ekeland's variational principle (Theorem 4.1) by using the Caristi–Kirk Theorem 4.8.

Proof of Ekeland's Variational Principle by using Theorem 4.8 [129] Without loss of generality, we may assume that $\lambda = 1$. Let $\varepsilon > 0$ be given and choose $\hat{x} \in X$ such that

$$f(\hat{x}) \le \inf_X f + \varepsilon.$$

Let $\mathcal{X} := \{x \in X : f(x) \le f(\hat{x}) - \varepsilon d(\hat{x}, x)\}$. Then, $\mathcal{X}$ is nonempty. By lower semicontinuity of f, $\mathcal{X}$ is a closed subset of a complete metric space, and hence, a complete metric space. For each $x \in \mathcal{X}$, let

$$S(x) = \{y \in X : y \ne x, f(y) \le f(x) - \varepsilon d(x,y)\},$$

and then define

$$T(x) = \begin{cases} x, & \text{if } S(x) = \varnothing,\\ S(x), & \text{if } S(x) \ne \varnothing.\end{cases}$$

Then, T is a set-valued map from $\mathcal{X}$ to itself with nonempty values. Indeed, $T(x) = x \in \mathcal{X}$ if $S(x) = \varnothing$. Since $T(x) = S(x)$ if not, we have for all $y \in T(x)$,

$$\begin{aligned}\varepsilon d(\hat{x},y) &\le \varepsilon d(\hat{x},x) + \varepsilon d(x,y)\\ &\le f(\hat{x}) - f(x) + f(x) - f(y)\\ &= f(\hat{x}) - f(y),\end{aligned}$$

and hence, $y \in \mathcal{X}$. For all $x \in \mathcal{X}$ and $y \in T(x)$, we have

$$\frac{1}{\varepsilon} f(y) + d(x,y) \le \frac{1}{\varepsilon} f(x).$$

From Theorem 4.8, T has a fixed point $\bar{x} \in \mathcal{X}$. Consequently, $S(\bar{x}) = \varnothing$, that is, $f(x) > f(\bar{x}) - \varepsilon d(\bar{x}, x)$ for all $x \in X$ with $x \ne \bar{x}$. Since $\bar{x} \in \mathcal{X}$, we have

$$f(\bar{x}) \le f(\hat{x}) - \varepsilon d(\hat{x}, \bar{x}) \le f(\hat{x}).$$

Further, we have

$$\begin{aligned}\varepsilon d(\hat{x},\bar{x}) &\leq f(\hat{x}) - f(\bar{x})\\ &\leq f(\hat{x}) - \inf_{x\in X} f(x)\\ &\leq \varepsilon,\end{aligned}$$

and hence, $d(\hat{x},\bar{x}) \leq 1$. ■

By using Ekeland's variational principle, Mizoguchi and Takahashi [129] also derived several fixed point theorems.

By using Theorem 4.4, Zhong et al. [183] derived the following Caristi's type fixed point theorem.

Theorem 4.9 *Let (X, d) be a complete metric space and x_0 be a fixed element of X. Let $\varphi : X \to [0,\infty)$ be a lower semicontinuous function and $h : [0,\infty) \to [0,\infty)$ be the same as in Theorem 4.4. Let $T : X \to X$ be a mapping such that*

$$\frac{d(x,T(x))}{1+h(d(x_0,x))} \leq \varphi(x) - \varphi(T(x)), \quad \textit{for all } x \in X. \tag{4.28}$$

Then, T has a fixed point.

Proof Applying Theorem 4.4 to the function $\varphi : X \to [0,\infty)$ with $\varepsilon = 1/2$ and $\lambda = 1$, there exists $\bar{x} \in X$ such that

$$\varphi(x) \geq \varphi(\bar{x}) - \frac{\varepsilon}{1+h(d(x_0,\bar{x}))} d(x,\bar{x}), \quad \text{for all } x \in X. \tag{4.29}$$

Taking $T(\bar{x})$ instead of x in (4.29), we get

$$\varphi(\bar{x}) - \varphi(T(\bar{x})) \leq \frac{\varepsilon}{1+h(d(x_0,\bar{x}))} d(T(\bar{x}),\bar{x}).$$

From (4.28), we get

$$\frac{\varepsilon}{1+h(d(x_0,\bar{x}))} d(T(\bar{x}),\bar{x}) \geq \varphi(\bar{x}) - \varphi(T(\bar{x})) \geq \frac{\varepsilon}{1+h(d(x_0,\bar{x}))} d(T(\bar{x}),\bar{x}),$$

which implies that $T(\bar{x}) = \bar{x}$. ■

Zhong et al. [183] also derived the following fixed point theorem for set-valued maps by using Theorem 4.4.

Theorem 4.10 *Let (X, d) be a complete metric space, x_0 be a given element of X, and $\sigma \in (0,1]$ be a constant. Let $h : [0,\infty) \to [0,\infty)$ be the same as in Theorem 4.4. Let $F : X \to 2^X_{cl}$ be a $\mathcal{H}$-continuous set-valued map such that for any $x \in X$, if $x \notin F(x)$, then for any $y \in F(x)$,*

$$\mathcal{H}(F(y),F(x)) \leq \left(1 - \frac{\sigma}{1+h(d(x_0,x))}\right) d(x,y). \tag{4.30}$$

Then, T has a fixed point.

Proof Since F is $\mathcal{H}$-continuous, by Exercise 3.20, the function $f : X \to [0, \infty)$ defined by $f(x) = d(x, F(x))$ is lower semicontinuous. By using Theorem 4.4 with $\varepsilon = \sigma/2$ and $\lambda = 1$, we obtain some $\bar{x} \in X$ such that

$$d(x, F(x)) \geq d(\bar{x}, F(\bar{x})) - \frac{\sigma}{2(1 + h(d(x_0, \bar{x})))} d(x, \bar{x}), \quad \text{for all } x \in X. \tag{4.31}$$

If $\bar{x} \notin F(\bar{x})$, then by (4.30), for any $y \in F(\bar{x})$, we have

$$\mathcal{H}(F(y), F(\bar{x})) \leq \left(1 - \frac{\sigma}{1 + h(d(x_0, \bar{x}))}\right) d(\bar{x}, y). \tag{4.32}$$

Taking y instead of x in (4.31) and then combining the resultant with (4.32), we obtain

$$\begin{aligned}
d(\bar{x}, F(\bar{x})) &\leq d(y, F(y)) + \frac{\sigma}{2(1 + h(d(x_0, \bar{x})))} d(y, \bar{x}) \\
&\leq \mathcal{H}(F(y), F(\bar{x})) + \frac{\sigma}{2(1 + h(d(x_0, \bar{x})))} d(y, \bar{x}) \\
&\leq \left(1 - \frac{\sigma}{1 + h(d(x_0, \bar{x}))}\right) d(y, \bar{x}) + \frac{\sigma}{2(1 + h(d(x_0, \bar{x})))} d(y, \bar{x}) \\
&= \left(1 - \frac{\sigma}{2(1 + h(d(x_0, \bar{x})))}\right) d(y, \bar{x}).
\end{aligned}$$

This implies that

$$d(\bar{x}, F(\bar{x})) \leq \left(1 - \frac{\sigma}{2(1 + h(d(x_0, \bar{x})))}\right) d(\bar{x}, F(\bar{x})),$$

which is impossible. Therefore, $\bar{x} \in F(\bar{x})$. ■

Remark 4.6 Zhong et al. [183] considered $\mathscr{H}(F(y), F(x)) = \sup_{v \in F(y)} d(v, F(x)) = \sup_{v \in F(y)} \inf_{u \in F(x)} d(v, u)$ in place of $\mathcal{H}(F(y), F(x))$ in (4.30).

Example 4.2 [183] Let $X = \{(x, y) \in \mathbb{R} \times \mathbb{R} : 0 \leq y \leq x\}$ and the metric d on X be defined by $d((x, y), (u, v)) = \max\{|x - u|, |y - v|\}$ for all $(x, y), (u, v) \in X$. Let $h : [0, \infty) \to [0, \infty)$ be the same as in Theorem 4.10 and $F(x, y) = \left(x, \left(1 - \frac{\sigma}{1+h(|(x,y)|)}\right) y\right)$, where $0 < \sigma < 1$ is a constant and $|(x, y)| = |(0, 0) - (x, y)|$. Then, F is not a contraction mapping. Indeed, if we take $(x, y), (u, v) \in X$ such that $|x - u| > 2\sigma y$, then

$$d\left(F((x, y)), F((u, y))\right) = |F((x, y)) - F((u, y))| = |x - u| = |(x, y) - (u, y)|,$$

which shows that F is not a contraction mapping.

However, F satisfies the conditions of Theorem 4.10. In fact, for any $(x, y) \in X$, we have

$$\begin{aligned} d\left(F^2((x,y)), F((x,y))\right) &= \left|F^2((x,y)) - F((x,y))\right| \\ &= \left|\left(x, \left(1 - \frac{\sigma}{1+h(x)}\right)^2 y\right) - \left(x, \left(1 - \frac{\sigma}{1+h(x)}\right) y\right)\right| \\ &= \left(1 - \frac{\sigma}{1+h(x)}\right)\left(\frac{\sigma}{1+h(x)}\right) y \\ &= \left(1 - \frac{\sigma}{1+h(x)}\right) \left|(x,y) - F((x,y))\right|. \end{aligned}$$

This shows that the condition (4.30) is satisfied and by Theorem 4.10, F has a fixed point. In addition, F is continuous and any $(x, 0) \in X$ is a fixed point of F.

It also shows that the Banach contraction principle and Nadler's theorem are not applicable.

The following example shows that condition (4.8) is important in Theorem 4.10.

Example 4.3 [183] Let $X = [0, \infty)$ be a metric space with the usual metric and $T : X \to X$ be defined by $T(x) = \sqrt{x^2 + 1}$. Then,

$$\begin{aligned} \left|T^2(x) - T(x)\right| &= |T(x) - x| \frac{\sqrt{x^2+1} + x}{\sqrt{x^2+2} + \sqrt{x^2+1}} \\ &\leq |T(x) - x| \left|1 - \frac{1}{4x^2 + (\sqrt{2}+1)x + 3}\right|. \end{aligned} \tag{4.33}$$

Let $h(r) = 4r^2 + \left(\sqrt{2} + 1\right) r + 2$ and $x_0 = 0$. Then, $h(r)$ is a continuous nondecreasing function and $d(x_0, x) = x$. From (4.33), we have

$$\left|T^2(x) - T(x)\right| \leq \left(1 - \frac{1}{1 + h(d(x_0, x))}\right) |T(x) - x|,$$

that is, T and h satisfy condition (4.30). But T has no fixed point and

$$\int_0^\infty \frac{dr}{1 + h(r)} = \int_0^\infty \frac{dr}{4r^2 + (\sqrt{2}+1)r + 3} < +\infty.$$

This shows that the condition (4.8) is important in Theorem 4.10.

4.1.2 Directional Ψ-contraction Set-valued Mappings

Uderzo [171] introduced the following concept of directional Ψ-contraction for set-valued mappings.

Definition 4.4 Let (X, d) be a metric space and K be a nonempty subset of X. A set-valued mapping $F : K \to 2^X_{cl}$ is said to be *directional* Ψ*-contraction* if there exist $\alpha \in (0, 1]$, $\psi : (0, \infty) \to [\alpha, 1]$ and $\Psi : (0, \infty) \to [0, 1)$ such that for each $x \in K$ with $x \notin F(x)$, there is a $y \in K \setminus \{x\}$ satisfying

$$\psi(d(x, y))\, d(x, y) + d(y, F(x)) \leq d(x, F(x)) \tag{4.34}$$

and

$$\mathcal{H}(F(y), F(x)) \leq \Psi(d(x,y))\, d(x,y). \tag{4.35}$$

Remark 4.7 Uderzo [171] considered

$$\mathscr{H}(F(y), F(x)) = \sup_{v \in F(y)} d(v, F(x)) = \sup_{v \in F(y)} \inf_{u \in F(x)} d(v,u)$$

in place of $\mathcal{H}(F(y), F(x))$ in (4.35).

Example 4.4 [171] Let $X = [0,1] \times [0,1]$ and let $\mathbb{Q}$ denote the set of all rational numbers. Consider the usual metric space (X, d) and a set-valued mapping $F : X \to 2^X_{cl}$ defined by

$$F(x_1, x_2) = \begin{cases} \{(1,1)\}, & \text{if } (x_1, x_2) \in X \cap (\mathbb{Q} \times \mathbb{Q}), \\ \{(0,0)\}, & \text{otherwise.} \end{cases}$$

Then, F is directional Ψ-contraction with $\alpha = 1$, $\psi \equiv 1$ and $\Psi \equiv 0$. Indeed, it is easy to see that for any point $x = (x_1, x_2) \in X \cap (\mathbb{Q} \times \mathbb{Q}) \setminus \{(1,1)\}$, the line segment connecting $x = (x_1, x_2)$ with $z = (1,1) = F(x_1, x_2)$ contains a point $y = (y_1, y_2) \in X \cap (\mathbb{Q} \times \mathbb{Q}) \setminus \{(x_1, x_2), (1,1)\}$ such that $d(x,y) + d(y,z) = d(x,z)$ and $d(F(y), F(x)) = 0$.

Analogously, for any point $x = (x_1, x_2) \in X \setminus (\mathbb{Q} \times \mathbb{Q})$, the line segment joining (x_1, x_2) with $w = (0,0) = F(x_1, x_2)$ contains a point $y = (y_1, y_2) \in X \setminus (\mathbb{Q} \times \mathbb{Q}) \cup \{(x_1, x_2)\}$ such that $d(x,y) + d(y,w) = d(x,w)$ and $d(F(y), F(x)) = 0$. Nonetheless, F fails to be a contraction on X.

Uderzo [171] established the following fixed point theorem for directional Ψ-contraction set-valued mappings. This result can be seen as a generalization of a result due to Song [158].

Theorem 4.11 *Let K be a nonempty closed subset of a complete metric space (X, d) and $F : K \to 2^X_{cl}$ be $\mathcal{H}$-continuous and directional Ψ-contraction. Assume that there exist $x_0 \in K$ and $\delta > 0$ such that $d(x_0, F(x_0)) \leq \alpha\delta$ and $\sup_{t \in (0,\delta]} \Psi(t) < \inf_{t \in (0,\delta]} \psi(t)$, where $\alpha \in (0,1]$, ψ and Ψ are functions occurring in the definition of directional Ψ-contraction. Then, F has a fixed point.*

Proof By hypothesis, there exist $\beta > 0$ and $\delta > 0$ such that

$$\sup_{t \in (0,\delta]} [\Psi(t) - \psi(t)] \leq \sup_{t \in (0,\delta]} \Psi(t) - \inf_{t \in (0,\delta]} \psi(t) \leq -\beta. \tag{4.36}$$

Since F is $\mathcal{H}$-continuous, by Exercise 3.20, the displacement function $f(x) = d(x, F(x))$ is lower semicontinuous. Since K is complete if equipped with the metric induced by d, and $f(x_0) \leq \alpha\delta$, then it is possible to apply Ekelend's variational principle around x_0, to get for any $\lambda > 0$ the existence of $x_\lambda \in K$ such that $f(x_\lambda) \leq f(x_0)$ and

$$f(x_\lambda) < f(x) + \frac{\alpha\delta}{\lambda} d(x_\lambda, x), \quad \text{for all } x \in K \setminus \{x_\lambda\}. \tag{4.37}$$

Suppose ab absurdo that $f(x_\lambda) > 0$ for every $\lambda > 0$. Take $\lambda = \frac{2\alpha\delta}{\beta}$. By the definition of directional Ψ-contraction, there exists $y \in K \setminus \{x_\lambda\}$ such that

$$\psi(d(x_\lambda, y))\, d(x_\lambda, y) + d(y, F(x_\lambda)) \leq f(x_\lambda), \tag{4.38}$$

and

$$\mathcal{H}(F(y), F(x_\lambda)) \le \Psi(d(x_\lambda, y))\, d(x_\lambda, y). \tag{4.39}$$

Since $\alpha \le \psi(d(x_\lambda, y))$, it follows from (4.38) that

$$d(x_\lambda, y) \le \frac{1}{\alpha} f(x_\lambda) \le \frac{1}{\alpha} f(x_0),$$

so $0 < d(x_\lambda, y) \le \delta$. Moreover, we also have

$$d(y, F(x_\lambda)) \le f(x_\lambda) - \psi(d(x_\lambda, y))\, d(x_\lambda, y).$$

Since for any $M, N \in 2^X_{cl}$ and any $x \in X$, we have $d(x, N) \le d(x, M) + \mathcal{H}(M, N)$, from the above inequality we obtain

$$\begin{aligned} f(y) &\le d(y, F(x_\lambda)) + \mathcal{H}(F(y), F(x_\lambda)) \\ &\le f(x_\lambda) + [\Psi(d(x_\lambda, y)) - \psi(d(x_\lambda, y))] d(x_\lambda, y). \end{aligned}$$

Putting $x = y$ in inequality (4.37) and taking into account inequality (4.36) along with $0 < d(x_\lambda, y) \le \delta$, we have

$$\begin{aligned} f(x_\lambda) &< f(y) + \frac{\alpha\delta}{\frac{2\alpha\delta}{\beta}} d(x_\lambda, y) \\ &\le f(x_\lambda) + \left\{ [\Psi(d(x_\lambda, y)) - \psi(d(x_\lambda, y))] + \frac{\beta}{2} \right\} d(x_\lambda, y) \\ &\le f(x_\lambda) - \frac{\beta}{2} d(x_\lambda, y) < f(x_\lambda), \end{aligned}$$

which yields an absurdum. Therefore, it must be $f(x_\lambda) = 0$ and this completes the proof. ■

Definition 4.5 Let K be a nonempty subset of a metric space (X, d). A set-valued mapping $F : K \to 2^X_{cl}$ is said to have the *almost fixed point property* in K if $\inf_{x \in K} d(x, F(x)) = 0$.

Corollary 4.2 *Let K be a nonempty closed subset of a complete metric space (X, d) and $F : K \to 2^X_{cl}$ be $\mathcal{H}$-continuous and directional Ψ-contraction such that*

$$\limsup_{s \to 0^+} \Psi(s) < \liminf_{s \to 0^+} \psi(s). \tag{4.40}$$

If F has the almost fixed point property in K, then it admits a fixed point in K.

Proof Observe that (4.40) implies the existence of $\delta_1 > 0$ and $\delta_2 > 0$ such that

$$\sup_{s \in (0, \delta_1]} \Psi(s) < \inf_{s \in (0, \delta_2]} \psi(s),$$

so that it suffices to take $\delta = \min\{\delta_1, \delta_2\}$ to get satisfied the related inequality in the hypothesis of Theorem 4.11. Besides, since $\inf_{x \in K} f(x) = 0$, there exists $x_0 \in K$ such that $f(x_0) \le \alpha\delta$. ■

4.2 Borwein–Preiss Variational Principle

To deal with differentiability problems of convex functions, Borwein and Preiss [39] considered the following form of the variational principle which was further revised by Li and Shi [116].

Definition 4.6 Let (X, d) be a complete metric space. A continuous function $\rho : X \times X \to [0, \infty]$ is said to be a *gauge-type function* on X if

(i) $\rho(x, x) = 0$ for all $x \in X$,
(ii) for all $\{x_n\}, \{y_n\} \in X$, $\rho(x_n, y_n) \to 0$ implies $d(x_n, y_n) \to 0$,
(iii) for all $y \in X$, the function $x \mapsto \rho(x, y)$ is lower semicontinuous.

Theorem 4.12 (Borwein–Preiss Variational Principle) *Let (X, d) be a complete metric space and $f : X \to \mathbb{R} \cup \{+\infty\}$ be a proper, bounded below, and lower semicontinuous function. Let $\rho : X \times X \to [0, \infty]$ be a gauge-type function and $\{\alpha_n\}_{n=0}^{\infty}$ be a sequence of positive numbers. Let $\varepsilon > 0$ and $\hat{x} \in X$ be given such that*

$$f(\hat{x}) \le \inf_X f + \varepsilon.$$

Then, there exist $\bar{x} \in X$ and a sequence $\{x_n\}$ such that

(a) $\rho(\hat{x}, \bar{x}) \le \dfrac{\varepsilon}{\alpha_0}$ *and* $\rho(x_n, \bar{x}) \le \dfrac{\varepsilon}{2^n \alpha_0}$,

(b) $f(\bar{x}) + \sum_{n=0}^{\infty} \alpha_n \rho(\bar{x}, x_n) \le f(\hat{x})$,

(c) $f(x) + \sum_{n=0}^{\infty} \alpha_n \rho(x, x_n) > f(\bar{x}) + \sum_{n=0}^{\infty} \alpha_n \rho(\bar{x}, x_n)$, *for all* $x \in X \setminus \{\bar{x}\}$.

Proof We construct a sequence $\{x_n\}$ in X and a sequence $\{S_n\}$ of subsets of X inductively as follows: Start with $x_0 = \hat{x}$ ($\hat{x}$ is the same as given in the hypothesis) and define

$$S_0 := \{x \in X : f(x) + \alpha_0 \rho(x, x_0) \le f(x_0)\}. \tag{4.41}$$

Then, S_0 is nonempty as $x_0 \in S_0$. Since f and $\rho(\cdot, x_0)$ are lower semicontinuous, the set S_0 is closed. We also have that, for all $x \in S_0$,

$$\alpha_0 \rho(x, x_0) \le f(x_0) - f(x) \le f(\hat{x}) - \inf_X f \le \varepsilon. \tag{4.42}$$

Take $x_1 \in S_0$ such that

$$f(x_1) + \alpha_0 \rho(x_1, x_0) \le \inf_{x \in S_0} \left[f(x) + \alpha_0 \rho(x, x_0) \right] + \frac{\alpha_1 \varepsilon}{2\alpha_0}, \tag{4.43}$$

and define similarly

$$S_1 := \left\{ x \in S_0 : f(x) + \sum_{k=0}^{1} \alpha_k \rho(x, x_k) \le f(x_1) + \alpha_0 \rho(x_1, x_0) \right\}. \tag{4.44}$$

Continuing in this way, we obtain x_i and S_i for $i = 0, 1, 2, \dots, n-1$ such that

$$f(x_i) + \sum_{k=0}^{i-1} \alpha_k \rho(x_i, x_k) \le \inf_{x \in S_{i-1}} \left[f(x) + \sum_{k=0}^{i-1} \alpha_k \rho(x, x_k) \right] + \frac{\alpha_i \varepsilon}{2^i \alpha_0} \tag{4.45}$$

and

$$S_i := \left\{ x \in S_{i-1} : f(x) + \sum_{k=0}^{i} \alpha_k \rho(x, x_k) \le f(x_i) + \sum_{k=0}^{i-1} \alpha_k \rho(x_i, x_k) \right\}. \tag{4.46}$$

We choose $x_n \in S_{n-1}$ such that

$$f(x_n) + \sum_{k=0}^{n-1} \alpha_k \rho(x_n, x_k) \le \inf_{x \in S_{n-1}} \left[f(x) + \sum_{k=0}^{n-1} \alpha_k \rho(x, x_k) \right] + \frac{\alpha_n \varepsilon}{2^n \alpha_0} \tag{4.47}$$

and define

$$S_n := \left\{ x \in S_{n-1} : f(x) + \sum_{k=0}^{n} \alpha_k \rho(x, x_k) \le f(x_n) + \sum_{k=0}^{n-1} \alpha_k \rho(x_n, x_k) \right\}. \tag{4.48}$$

Note that for every $n = 1, 2, \dots$, S_n is nonempty and closed. From (4.47) and (4.48), we have, for all $x \in S_n$, that

$$\begin{aligned}
\alpha_n \rho(x, x_n) &\le \left[f(x_n) + \sum_{k=0}^{n-1} \alpha_k \rho(x_n, x_k) \right] - \left[f(x) + \sum_{k=0}^{n-1} \alpha_k \rho(x, x_k) \right] \\
&\le \left[f(x_n) + \sum_{k=0}^{n-1} \alpha_k \rho(x_n, x_k) \right] - \inf_{x \in S_{n-1}} \left[f(x) + \sum_{k=0}^{n-1} \alpha_k \rho(x, x_k) \right] \\
&\le \frac{\alpha_n \varepsilon}{2^n \alpha_0},
\end{aligned}$$

that is,

$$\rho(x, x_n) \le \frac{\varepsilon}{2^n \alpha_0}, \quad \text{for all } x \in S_n. \tag{4.49}$$

Since ρ is a gauge-type function, from (4.49), we have that $d(x, x_n) \to 0$. Therefore, $\operatorname{diam}(S_n) \to 0$. Since X is complete, by Cantor's Intersection Theorem 1.4, there exists a unique $\bar{x} \in \bigcap_{n=0}^{\infty} S_n$ which satisfies (a) by (4.42) and (4.49). Obviously, we have $x_n \to \bar{x}$. For any $x \neq \bar{x}$, we have that $x \notin \bigcap_{n=0}^{\infty} S_n$, and therefore for some i, we have

$$f(x) + \sum_{k=0}^{\infty} \alpha_k \rho(x, x_k) \ge f(x) + \sum_{k=0}^{i} \alpha_k \rho(x, x_k) > f(x_i) + \sum_{k=0}^{i-1} \alpha_k \rho(x_i, x_k). \tag{4.50}$$

On the other hand, it follows from (4.41), (4.48), and $\bar{x} \in \bigcap_{n=0}^{\infty} S_n$ that, for all $q \geq i$,

$$\begin{aligned} f(x_0) &\geq f(x_i) + \sum_{k=0}^{i-1} \alpha_k \rho(x_i, x_k) \\ &\geq f(x_q) + \sum_{k=0}^{q-1} \alpha_k \rho(x_q, x_k) \\ &\geq f(\bar{x}) + \sum_{k=0}^{q} \alpha_k \rho(\bar{x}, x_k). \end{aligned} \tag{4.51}$$

Taking limit in (4.51) as $q \to \infty$, we obtain

$$\begin{aligned} f(\hat{x}) = f(x_0) &\geq f(x_i) + \sum_{k=0}^{i-1} \alpha_k \rho(x_i, x_k) \\ &\geq f(\bar{x}) + \sum_{k=0}^{\infty} \alpha_k \rho(\bar{x}, x_k), \end{aligned} \tag{4.52}$$

which verifies (b). Combining (4.50) and (4.52) yields (c). ∎

By adopting the proof of Theorem 4.12 for nonnegative sequence $\{\alpha_n\}_{n=0}^{\infty}$, $\alpha_0 > 0$, we can derive the following generalization for both Ekeland's variational principle and the Borwein–Preiss variational principle.

Theorem 4.13 *Let (X, d) be a complete metric space and $f : X \to \mathbb{R} \cup \{+\infty\}$ be a proper, bounded below, and lower semicontinuous function. Let $\rho : X \times X \to [0, \infty]$ be a gauge-type function and $\{\alpha_n\}_{n=0}^{\infty}$ be a sequence of nonnegative numbers with $\alpha_0 > 0$. Let $\varepsilon > 0$ and $\hat{x} \in X$ be given such that*

$$f(\hat{x}) \leq \inf_X f + \varepsilon.$$

Then, there exists a sequence $\{x_n\}$ converging to some $\bar{x} \in X$ such that

(a) $\rho(\hat{x}, \bar{x}) \leq \dfrac{\varepsilon}{\alpha_0}$,

(b) $f(\bar{x}) + \sum_{n=0}^{\infty} \alpha_n \rho(\bar{x}, x_n) \leq f(\hat{x})$,

(c) $f(x) + \sum_{n=0}^{\infty} \alpha_n \rho(x, x_n) > f(\bar{x}) + \sum_{n=0}^{\infty} \alpha_n \rho(\bar{x}, x_n)$, *for all $x \in X \setminus \{\bar{x}\}$.*

Moreover, if $\alpha_k > 0$ and $\alpha_l = 0$ for all $l > k \geq 0$, then the following assertion holds:

(d) *For all $x \in X \setminus \{\bar{x}\}$, there exists $j \geq k$ such that*

$$f(x) + \sum_{n=0}^{k-1} \alpha_n \rho(x, x_n) + \alpha_k \rho(x, x_j) > f(\bar{x}) + \sum_{n=0}^{k-1} \alpha_n \rho(\bar{x}, x_n) + \alpha_k \rho(\bar{x}, x_j).$$

Proof Set $x_0 = \hat{x}$. There are two cases for $\{\alpha_n\}_{n=0}^{\infty}$: (i) infinitely many $\alpha_n > 0$ and (ii) only finitely many $\alpha_n > 0$.

For the first case, without loss of generality, we can assume that $\alpha_n > 0$ for all $n = 0, 1, 2, ...$ In this case, (a)–(c) holds by Theorem 4.12.

Assume that $\alpha_k > 0$ for some $k \geq 0$ and $\alpha_l = 0$ for all $l > k$. Without loss of generality, we assume that $\alpha_i > 0$ for all $i \leq k$. Thus, when $n \leq k$, we take the same x_n and S_n as in Theorem 4.12. When $n > k$, we take $x_n \in S_{n-1}$ such that

$$f(x_n) + \sum_{i=0}^{k-1} \alpha_i \rho(x_n, x_i) \leq \inf_{x \in S_{n-1}} \left(f(x) + \sum_{i=0}^{k-1} \alpha_i \rho(x, x_i) \right) + \frac{\varepsilon \alpha_k}{2^n \alpha_0}.$$

Set

$$S_n := \left\{ x \in S_{n-1} : f(x) + \sum_{i=0}^{k-1} \alpha_i \rho(x, x_i) + \alpha_k \rho(x, x_n) \leq f(x_n) + \sum_{i=0}^{k-1} \alpha_i \rho(x_n, x_i) \right\}.$$

Then, by the same deduction as in Theorem 4.12, (a)–(c) also hold. But when $x \neq \bar{x}$, it may follow that there exists an $m > k$ such that

$$\begin{aligned} f(x) + \sum_{i=0}^{k-1} \alpha_i \rho(x, x_i) + \alpha_k \rho(x, x_m) &> f(x_m) + \sum_{i=0}^{k-1} \alpha_i \rho(x_m, x_i) \\ &\geq f(\bar{x}) + \sum_{i=0}^{k-1} \alpha_i \rho(\bar{x}, x_i) + \alpha_k \rho(\bar{x}, x_m). \end{aligned}$$

Thus, (d) holds. ■

4.3 Takahashi's Minimization Principle

We derive the following existence result for a solution of an optimization problem without compactness and convexity assumptions on the underlying space.

Theorem 4.14 (Takahashi's Minimization Theorem) *Let (X, d) be a complete metric space and $f : X \to \mathbb{R} \cup \{+\infty\}$ be a proper, bounded below, and lower semicontinuous function. Suppose that, for each $\hat{x} \in X$ with $\inf_X f < f(\hat{x})$, there exists $z \in X$ such that $z \neq \hat{x}$ and*

$$f(z) + d(\hat{x}, z) \leq f(\hat{x}).$$

Then, there exists $\bar{x} \in X$ such that $f(\bar{x}) = \inf_{x \in X} f(x)$.

Proof Assume to the contrary that $\inf_{x \in X} f(x) < f(y)$ for all $y \in X$ and let $\hat{x} \in X$ with $f(\hat{x}) < +\infty$. We define inductively a sequence $\{x_n\}$ in X, starting with $x_1 = \hat{x}$. Suppose that $x_n \in X$ is known. Then, choose $x_{n+1} \in S_{n+1}$ such that

$$S_{n+1} = \{x \in X : f(x) \leq f(x_n) - d(x_n, x)\}$$

and

$$f(x_{n+1}) \leq \frac{1}{2}\left\{ \inf_{x \in S_{n+1}} f(x) + f(x_n) \right\}. \tag{4.53}$$

We claim that $\{x_n\}$ is a Cauchy sequence. Indeed, if $m > n$, then

$$\begin{aligned} d(x_n, x_m) &\leq \sum_{j=n}^{m-1} d(x_j, x_{j+1}) \\ &\leq \sum_{j=n}^{m-1} \{f(x_j) - f(x_{j+1})\} \\ &= f(x_n) - f(x_m). \end{aligned} \tag{4.54}$$

Since $\{f(x_n)\}$ is a decreasing sequence and the function f is bounded below, there exists $\varepsilon > 0$ such that

$$f(x_n) - f(x_m) < \varepsilon, \quad \text{for all } m > n.$$

Therefore, from (4.54), we have

$$d(x_n, x_m) < \varepsilon, \quad \text{for all } m > n,$$

and hence, $\{x_n\}$ is a Cauchy sequence in X. Since X is a complete metric space, there exists $\tilde{x} \in X$ such that $x_n \to \tilde{x}$. Then, if $m \to \infty$ in (4.54), the lower semicontinuity of f and continuity of d imply that

$$d(x_n, \tilde{x}) \leq f(x_n) - \lim_{m \to \infty} f(x_m) \leq f(x_n) - f(\tilde{x}). \tag{4.55}$$

On the other hand, by hypothesis, there exists a $z \in X$ such that $z \neq \tilde{x}$ and

$$f(z) + d(\tilde{x}, z) \leq f(\tilde{x}). \tag{4.56}$$

By using (4.55) and (4.56), we have

$$\begin{aligned} f(z) &\leq f(\tilde{x}) - d(\tilde{x}, z) \\ &\leq f(\tilde{x}) - d(\tilde{x}, z) + f(x_n) - f(\tilde{x}) - d(x_n, \tilde{x}) \\ &= f(x_n) - \{d(x_n, \tilde{x}) + d(\tilde{x}, z)\} \\ &\leq f(x_n) - d(x_n, z). \end{aligned}$$

Consequently, $z \in S_{n+1}$ for all $n \in \mathbb{N}$. Using (4.53), we obtain

$$2f(x_{n+1}) - f(x_n) \leq \inf_{x \in S_{n+1}} f(x) \leq f(z).$$

Hence,

$$f(\tilde{x}) \leq \lim_{n \to \infty} f(x_n) \leq f(z) \leq f(\tilde{x}) - d(\tilde{x}, z) < f(\tilde{x}),$$

which is a contradiction. Therefore, there exists $\bar{x} \in X$ such that $f(\bar{x}) = \inf_{x \in X} f(x)$. ■

Remark 4.8 Takahashi's Minimization Theorem 4.14 and Ekeland's Variational Principle (Theorem 4.1) are equivalent in the sense that one can be derived by using the other.

Proof We first prove Theorem 4.1 by using Theorem 4.14. Let

$$X_0 = \{x \in X : f(x) \leq f(\hat{x}) - \varepsilon d_\lambda(\hat{x}, x)\}.$$

Since $\hat{x} \in X_0$, X_0 is nonempty. By lower semicontinuity of f and continuity of d_λ, X_0 is closed. Further, for each $x \in X_0$,

$$\varepsilon d_\lambda(\hat{x}, x) \leq f(\hat{x}) - f(x) \leq f(\hat{x}) - \inf_{y \in X} f(y) \leq \varepsilon,$$

and hence, $d_\lambda(\hat{x}, x) \leq 1$, and thus $d(\hat{x}, x) \leq \lambda$. We also have $f(x) \leq f(\hat{x})$.

Assume to the contrary of conclusion (c) in Theorem 4.1 that for every $x \in X_0$, there exists $y \in X$ such that $y \neq x$ and $f(y) \leq f(x) - \varepsilon d_\lambda(y, x)$. Then,

$$\begin{aligned} \varepsilon d_\lambda(\hat{x}, y) &\leq \varepsilon d_\lambda(\hat{x}, x) + \varepsilon d_\lambda(x, y) \\ &\leq f(\hat{x}) - f(x) + f(x) - f(y) \\ &= f(\hat{x}) - f(y), \end{aligned}$$

and hence, $y \in X_0$. Then, by Theorem 4.14, there exists $\bar{x} \in X$ such that $f(\bar{x}) = \inf_{x \in X_0} f(x)$. This is a contradiction of the hypothesis that there exists $y_0 \in X_0$ with $f(y_0) < f(\bar{x})$.

Now we prove Theorem 4.14 by using Theorem 4.1. By Theorem 4.1, for any given $\varepsilon > 0$, there exists $\bar{x} \in X$ such that

$$f(\bar{x}) < f(x) + \varepsilon d(x, \bar{x}), \quad \text{for all } x \in X \text{ with } x \neq \bar{x}. \tag{4.57}$$

We claim that $f(\bar{x}) = \inf_{x \in X} f(x)$.

Assume to the contrary that there exists $w \in X$ such that $f(w) > \inf_{x \in X} f(x)$. By hypothesis, there exists $z \in X$ such that $z \neq w$ and

$$f(z) + d(w, z) \leq f(w)$$

contradicting (4.57). Hence, $f(\bar{x}) = \inf_{x \in X} f(x)$. ■

Remark 4.9 Takahashi's Minimization Theorem 4.14 and Caristi's Fixed Point Theorem 2.13 are equivalent in the sense that one can be derived by using the other.

Proof We first prove Theorem 2.13 by using Theorem 4.14. Suppose that T does not have any fixed point, that is, for any $x \in X$, $T(x) \neq x$. Then for every $x \in X$, there exists $y \in X$ such that $x \neq y$ and $\varphi(y) + d(x, y) \leq \varphi(x)$. So from Theorem 4.14, there exists $\bar{x} \in X$ such that $\varphi(\bar{x}) = \inf_{x \in X} \varphi(x)$. For such $\bar{x} \in X$, we have

$$0 < d(\bar{x}, T(\bar{x})) \leq \varphi(\bar{x}) - \varphi(T(\bar{x})) \leq \varphi(T(\bar{x})) - \varphi(T(\bar{x})) = 0,$$

a contradiction.

Now we prove Theorem 2.13 by using Theorem 4.1. By Theorem 4.1, for any given $\varepsilon > 0$, there exists $\bar{x} \in X$ such that

$$f(\bar{x}) < f(x) + \varepsilon d(x, \bar{x}), \quad \text{for all } x \in X \text{ with } x \neq \bar{x}. \tag{4.58}$$

We claim that $f(\bar{x}) = \inf_{x \in X} f(x)$.

Assume to the contrary that there exists $w \in X$ such that $f(w) > \inf_{x \in X} f(x)$. By hypothesis, there exists $z \in X$ such that $z \neq w$ and

$$f(z) + d(w, z) \leq f(w)$$

contradicting (4.58). Hence, $f(\bar{x}) = \inf_{x \in X} f(x)$. ■

The following theorem due to Takahashi [168] characterizes the completeness of the underlying metric space X.

Theorem 4.15 *A metric space (X, d) is complete if for every uniformly continuous function $f : X \to \mathbb{R} \cup \{+\infty\}$ and every $\hat{x} \in X$ with $\inf_X f < f(\hat{x})$, there exists $z \in X$ such that $z \neq \hat{x}$ and*

$$f(z) + d(\hat{x}, z) \leq f(\hat{x}),$$

then there exists $\bar{x} \in X$ such that $f(\bar{x}) = \inf_{x \in X} f(x)$.

Proof Let $\{x_n\}$ be a Cauchy sequence in X. Consider the function $f : X \to [0, +\infty)$ defined by

$$f(x) = \lim_{n \to \infty} d(x_n, x), \quad \text{for all } x \in X.$$

Then, f is uniformly continuous and $\inf_{x \in X} f(x) = 0$. Let $f(\hat{x}) > 0$. Then, there exists an $x_m \in X$ such that

$$x_m \neq \hat{x}, \quad f(x_m) < \frac{1}{3} f(\hat{x}) \quad \text{and} \quad d(x_m, \hat{x}) - f(\hat{x}) < f(\hat{x}).$$

Thus, we have

$$3f(x_m) + d(x_m, \hat{x}) < f(\hat{x}) + 2f(\hat{x}) = 3f(\hat{x}).$$

Therefore, there exists an $\bar{x} \in X$ such that $f(\bar{x}) = \inf_{x \in X} f(x) = 0$, and so, $0 = f(\bar{x}) = \lim_{n \to \infty} d(x_n, \bar{x})$. Thus, $\{x_n\}$ converges to $\bar{x}$, and hence, X is complete. ■

We give an application of Takahashi's Minimization Theorem 4.14 to prove the Nadler's fixed point theorem for set-valued maps [131].

Theorem 4.16 (Nadler's Fixed Point Theorem) *Let (X, d) be a complete metric space and $T : X \to 2^X_{cl}$ be a contraction set-valued mapping. Then, T has a fixed point, that is, there exists $\bar{x} \in X$ such that $\bar{x} \in T(\bar{x})$.*

Proof Assume that T does not have any fixed point. Then, $d(x, T(x)) > 0$ for all $x \in X$. Choose a number $\varepsilon > 0$ with $\varepsilon < \frac{1}{\alpha} - 1$. Then, for every $x \in X$, there exists $y \in X$ such that $y \in T(x)$ and

$$d(x, y) \leq (1 + \varepsilon) d(x, T(x)).$$

Since

$$d(y, T(y)) \leq H(T(x), T(y)) \leq \alpha d(x, y) \\ \leq \alpha(1 + \varepsilon)d(x, T(x)),$$

we have $\inf_{x \in X} d(x, T(x)) = 0$. Further, since

$$d(x, T(x)) - d(y, T(y)) \geq \frac{1}{1+\varepsilon} d(x, y) - \alpha d(x, y) \\ = \left(\frac{1}{1+\varepsilon} - \alpha\right) d(x, y),$$

we obtain

$$f(y) + d(x, y) \leq f(x),$$

where f is a continuous function defined by

$$f(x) = \left(\frac{1}{1+\varepsilon} - \alpha\right)^{-1} d(x, T(x)), \quad \text{for all } x \in X.$$

By Theorem 4.14, there exists $\bar{x} \in X$ such that $f(\bar{x}) = 0$, that is, $d(\bar{x}, T(\bar{x})) = 0$, a contradiction. Hence, T has a fixed point. ■

4.3.1 Applications to Weak Sharp Minima

Daffer et. al. [66] and Hamel [91] gave an application of Takahashi's minimization Theorem 4.14 to prove the existence of a weak sharp minima for a class of lower semicontinuous functions.

Let (X, d) be a complete metric space and $f : X \to \mathbb{R} \cup \{+\infty\}$ be a proper lower semicontinuous function. We define

$$m = \inf\{f(x) : x \in X\}$$

and

$$M = \{y \in X : f(y) = m\}. \tag{4.59}$$

We say that f has *weak sharp minima* if

$$d(x, M) \leq f(x) - m, \quad \text{for all } x \in X.$$

Theorem 4.17 [66] *Let (X, d) be a complete metric space and $f : X \to \mathbb{R} \cup \{+\infty\}$ be a proper, bounded below, and lower semicontinuous function. Suppose that, for each $\hat{x} \in X$ with $\inf_X f < f(\hat{x})$, there exists $z \in X$ such that $z \neq \hat{x}$ and*

$$d(\hat{x}, z) \leq f(\hat{x}) - f(z).$$

Then, M defined by (4.59) is nonempty and f has weak sharp minima.

Proof For $\hat{x} \in X$, define

$$S(\hat{x}) = \{z \in X : d(\hat{x}, z) \leq f(\hat{x}) - f(z)\}.$$

By the lower semicontinuity of f, $S(\hat{x})$ is closed. By Takahashi's Minimization Theorem 4.14, $M \neq \emptyset$.

We note that for all $z \in S(\hat{x}), f(z) \leq f(\hat{x})$. Assume contrary that there exists $x_0 \in X$ such that

$$d(x_0, M) > f(x_0) - m. \tag{4.60}$$

Then, $x_0 \notin M$ and this is true for all $z \in S(x_0)$.

Indeed, if there were $z \in S(x_0)$ with $f(z) = m$, then we have

$$d(x_0, M) \leq d(x_0, z) \leq f(x_0) - m,$$

which contradicts (4.60).

We also note that (4.60) holds for all $z \in S(x_0)$. Indeed, take $z \in S(x_0)$, $y \in M$, so that

$$d(x_0, y) \leq d(x_0, z) + d(z, y) \leq f(x_0) - f(z) + d(z, y),$$

and this yields

$$d(x_0, M) \leq f(x_0) - f(z) + d(z, M).$$

But $d(x_0, M) > f(x_0) - m$. Then, from (4.60), we obtain

$$f(x_0) - m < f(x_0) - f(z) + d(z, M),$$

that is, $f(z) - m < d(z, M)$. This gives (4.60) with y is in place of x_0.

Since $f(x_0) > m$, by hypothesis there is $x_1 \in S(x_0)$ with $x_1 \neq x_0$. Since $f(x_1) - m < d(x_1, M)$, again it is clear that $x_1 \notin M, f(x_1) < f(x_0)$ and we can again show as above that

$$f(z) - m < d(z, M), \quad \text{for all } z \in S(x_1) \quad \text{and} \quad S(x_1) \cap M = \emptyset.$$

In addition, we select x_1 such that $f(x_1) = \inf\{f(z) : z \in S(x_1)\}$. This is possible since X is complete, f is lower semicontinuous, and $S(x_1)$ is nonempty and closed. Continuing in this way, we generate a sequence $\{x_n\}$ with the above properties. Namely, if $x_0, x_1, \ldots, x_n$ have been chosen so that at $x_i \in S(x_{i-1}), f(x_i) < f(x_{i-1}), f(x_i) = \inf\{f(z) : z \in S(x_i)\}$, $S(x_{i-1}) \cap M = \emptyset$, for $i = 1, 2, \ldots, n$, and $f(z) - m < d(z, M)$ for all $z \in \bigcup_{i=1}^{n} S(x_{i-1})$, then, since $x_n \notin M$, we can choose $x_{n+1} \in S(x_n)$, $x_{n+1} \neq x_n$ with $f(x_{n+1}) = \inf\{f(z) : z \in S(x_n)\}$, and as above we will have

$$f(x_{n+1}) > m, \ f(x_{n+1}) < f(x_n) \text{ and } f(z) - m < d(z, M), \text{ for all } z \in S(x_{n+1}).$$

To see the latter, we write again, just as above, $f(x_n) - m < d(x_n, M) \leq f(x_n) - f(x_{n+1}) + d(x_{n+1}, M)$, giving $f(x_{n+1}) - m < d(x_{n+1}, M)$. Hence, $A(x_{n+1}) \cap M = \emptyset$.

We now have our sequence $\{x_n\}$ consisting of all different elements and $f(x_{n+1}) < f(x_n)$. Since

$$d(x_{n+k}, x_n) \leq \sum_{i=1}^{k} d(x_{n+i}, x_{n+i-1}) \leq \sum_{i=1}^{k} (f(x_{n+i-1}) - f(x_{n+i})) = f(x_n) - f(x_{n+k}),$$

and noting that $f(x_n)$ monotonically decreases to some c, $\{x_n\}$ must be Cauchy. Since X is complete, we can assume that x_n converges to $x \in X$. We now show that $x \in \bigcap_{i=0}^{\infty} S(x_i)$. We first show that, for every n, $x_n \in \bigcap_{i=0}^{n-1} S(x_i)$. This follows from the following:

$$\begin{aligned} d(x_{n-k}, x_n) &\leq \sum_{j=0}^{k-1} d(x_{n-k+j}, x_{n-k+j+1}) \\ &\leq \sum_{j=0}^{k-1} \big(f(x_{n-k+j}) - f(x_{n-k+j+1})\big) \\ &= f(x_{n-k}) - f(x_n), \end{aligned}$$

which shows (recall that all the x_i are outside M) that $x_n \in S(x_{n-k})$, $k = 1, 2, \ldots, n$; hence $x_n \in \bigcap_{i=0}^{n-1} S(x_i)$. It follows immediately from this that $x_k \in \bigcap_{i=0}^{n-1} S(x_i)$ for all $k \geq n$. Since $\bigcap_{i=0}^{n-1} S(x_i)$ is a closed set, $x \in \bigcap_{i=0}^{\infty} S(x_i)$. Thus, $x \in S(x_n)$, and $x \neq x_n$; hence $f(x) < f(x_n)$, and this is a contradiction, since $f(z) \geq f(x_n)$ for all $z \in S(x_n)$. ■

Chapter 5

Equilibrium Problems and Extended Ekeland's Variational Principle

5.1 Equilibrium Problems

The mathematical formulation of the equilibrium problem is to find an element $\bar{x}$ of a set K such that

$$F(\bar{x}, y) \geq 0, \quad \text{for all } y \in K, \tag{5.1}$$

where $F : K \times K \to \mathbb{R}$ is a bifunction such that $F(x, x) = 0$ for all $x \in K$. It is an unified model of several fundamental mathematical problems, namely, optimization problems, saddle point problems, fixed point problems, minimax inequality problems, Nash equilibrium problem, complementarity problems, variational inequality problems, etc. In 1955, Nikaido and Isoda [134] first considered equilibrium problem (5.1) as an auxiliary problem to establish the existence results for Nash's equilibrium points in noncooperative games. In the theory of equilibrium problems, the key contribution was made by Ky Fan [79], whose new existence results contained the original techniques which became a basis for most further existence theorems in the setting of topological vector spaces. That is why equilibrium problem (5.1) is also known as Ky Fan type inequality. Within the context of calculus of variations, motivated mainly by the works of Stampacchia [160], there arises the work of Brézis, Niremberg, and Stampacchia [45] establishing a more general result than that in [79]. In the last three decades, the theory of equilibrium problems emerges as a new direction of research in nonlinear analysis, optimization, optimal control, game theory, mathematical economics, etc. Most of the results on the existence of solutions for equilibrium problems are studied in the setting of topological vector spaces by using some kind of fixed point (Fan-Browder type fixed point) theorem or KKM type theorem. The term "equilibrium problem" was first used by Muu and Oettli [130] and later adopted by Blum and Oettli [38]. For further details, we refer to [3–5, 7–10, 12, 13, 32–36, 38, 54–57, 62–64, 83, 84, 96, 97, 106, 107, 125, 130, 135] and the references therein. In most of the existence results for a solution of equilibrium problems, the convexity of the underlying set K and the bifunction F is assumed; see, for example, [10, 13, 35, 36, 45, 56, 57, 83, 84, 107] and the references therein. Inspired by the work of Blum and Oettli [38] and Oettli and Théra [135], the existence theory for solutions of equilibrium problems has been developed by many researchers in the setting of metric spaces and without any convexity assumption on the underlying set K and bifunction F; see, for example, [4, 7–9, 12, 32, 38, 54, 55, 63, 107, 106, 111, 135] and the references therein.

Ekeland's variational principle for a bifunction F is known as the equilibrium version of Ekeland's variational principle or the extended form of Ekeland's variational principle. In most of the papers that appeared in the literature on the equilibrium version of Ekeland's variational principle, the triangle inequality $F(x,z) \leq F(x,y) + F(y,z)$ for all $x,y,z \in K$ is used; see, for example, [4, 7–9, 12, 32, 38, 55, 117, 119, 135, 139] and the references therein. Castellani et al. [55] considered the class of bifunctions satisfying the triangle inequality and presented several properties of such class of bifunctions. Castellani and Giuli [54] have pointed out the fact that the triangle inequality is rather demanding and therefore restricts the applicability of the related results in the important field of variational inequalities. Therefore, they established, for the first time, the equilibrium version of Ekeland's variational principle without using the triangle inequality. Recently, Cotrina and Svensson [63] discussed the existence of solutions of equilibrium problems by using the finite intersection property. Khanh and Quan [111] studied the existence of solutions of equilibrium problems under the cyclic anti-quasimonotonicity of the bifunction and gave several applications to fixed point theory, variational inequality problems, and noncooperative games. Recently, Al-Homidan et al. [3] introduced the concept of weak sharp solutions for equilibrium problems and gave its characterization by using the equilibrium version of Ekeland's variational principle. It is worth mentioning that the weak sharpness of the solution set provides the finite termination property of the algorithm for solving an optimization problem.

Example 5.1 (a) **Minimization Problem.** Let K be a nonempty set and $f : K \to \mathbb{R}$ be a real-valued function. The *minimization problem* is to find $\bar{x} \in K$ such that

$$f(\bar{x}) \leq f(y), \quad \text{for all } y \in K. \tag{5.2}$$

If we set $F(x,y) = f(y) - f(x)$ for all $x, y \in K$, then the minimization problem (5.2) is equivalent to the equilibrium problem (5.1).

(b) **Saddle Point Problem.** Let K_1 and K_2 be nonempty sets and $L : K_1 \times K_2 \to \mathbb{R}$ be a real-valued bifunction. The *saddle point problem* is to find $(\bar{x}_1, \bar{x}_2) \in K_1 \times K_2$ such that

$$L(\bar{x}_1, y_2) \leq L(\bar{x}_1, \bar{x}_2) \leq L(y_1, \bar{x}_2), \quad \text{for all } (y_1, y_2) \in K_1 \times K_2. \tag{5.3}$$

Set $K := K_1 \times K_2$ and define $F : K \times K \to \mathbb{R}$ by

$$F((x_1, x_2), (y_1, y_2)) = L(y_1, x_2) - L(x_1, y_2),$$

for all $(x_1, x_2), (y_1, y_2) \in K_1 \times K_2$. Then, the saddle point problem (5.3) coincides with the equilibrium problem (5.1).

(c) **Nash Equilibrium Problem.** Let I be a finite set of players. For each $i \in I$, let K_i be the strategy set of the ith player. Let $K := \prod_{i \in I} K_i$. For each player $i \in I$, let $\varphi_i : K \to \mathbb{R}$ be the loss function of the ith player, depending on the strategies of all players. For $x = (x_1, x_2, \ldots, x_n) \in K$, we define $x^i = (x_1, \ldots, x_{i-1}, x_{i+1}, \ldots, x_n)$. The *Nash equilibrium problem* [132, 133] is to find $\bar{x} \in K$ such that for each $i \in I$,

$$\varphi_i(\bar{x}) \leq \varphi_i(\bar{x}^i, y_i), \quad \text{for all } y_i \in K_i. \tag{5.4}$$

Define

$$F(x,y) = \sum_{i=1}^{n} \left(\varphi_i(x^i, y_i) - \varphi_i(x)\right), \quad \text{for all } (x,y) \in K \times K.$$

Then, the Nash equilibrium problem (5.4) is same as the equilibrium problem (5.1).

(d) Fixed Point Problem. Let K be a nonempty subset of $\mathbb{R}^n$ and $f : K \to K$ be a given mapping. The *fixed point problem* is to find $\bar{x} \in K$ such that $f(\bar{x}) = \bar{x}$.

Setting $F(x,y) = \langle x - f(x), y - x \rangle$ for all $x, y \in K$, where $\langle \cdot, \cdot \rangle$ denotes the inner product on $\mathbb{R}^n$. Then, $\bar{x}$ is a solution of the fixed point problem if and only if it is a solution of the equilibrium problem (5.1).

(e) Variational Inequality Problem. Let K be a nonempty subset of $\mathbb{R}^n$ and $\Phi : K \to \mathbb{R}^n$ be a mapping. The *variational inequality problem* is to find $\bar{x} \in K$ such that

$$\langle \Phi(\bar{x}), y - \bar{x} \rangle \geq 0, \quad \text{for all } y \in K, \tag{5.5}$$

where $\langle \cdot, \cdot \rangle$ denotes the inner product on $\mathbb{R}^n$. We set $F(x,y) = \langle \Phi(x), y - x \rangle$ for all $x, y \in K$. Then, the variational inequality problem (5.5) is equivalent to the equilibrium problem (5.1). For further details on variational inequality problems in the setting of finite dimensional spaces, we refer to [11, 77].

5.2 Triangle Inequality and Cyclically Antimonotonicity

Let K be a nonempty subset of a metric space X. We denote by $\mathbb{F}$ the family of bifunctions $F : K \times K \to \mathbb{R}$ which satisfy the following triangle inequality:

$$F(x,y) \leq F(x,z) + F(z,y), \quad \text{for all } x, y, z \in K. \tag{5.6}$$

The Q-functions considered in [4], τ-functions considered in [117], fitting functions considered in [119], W-distance function considered in [105] fall under the class $\mathbb{F}$.

Clearly, for each bifunction $F : K \times K \to \mathbb{R}$ that satisfies the triangle inequality (5.6), we have

$$F(x,y) \leq F(x,x) + F(x,y), \quad \text{that is,} \quad F(x,x) \geq 0, \quad \text{for all } x, y \in K.$$

Also,

$$F(x,y) + F(y,x) \geq F(x,x) \geq 0, \quad \text{for all } x, y \in K. \tag{5.7}$$

A bifunction F which satisfies (5.7) is called *antimonotone*.

Now we present some properties of the class $\mathbb{F}$ which are established in [55].

Proposition 5.1 (a) *If $f_1, f_2 : K \to \mathbb{R}$ are functions such that $f_1(x) + f_2(x) \geq 0$ for all $x \in K$, then the bifunction $F(x,y) = f_1(x) + f_2(y)$ belongs to $\mathbb{F}$.*

(b) *If $F \in \mathbb{F}$ and $\varphi : \mathbb{R} \to \mathbb{R}$ is a nondecreasing and subadditive function ($\varphi(x+y) \leq \varphi(x) + \varphi(y)$), then $\varphi \circ F \in \mathbb{F}$.*

(c) *If $G : K \times K \to \mathbb{R}$ is a bounded bifunction (bounded below as well as bounded above), that is, there exist $L, U \in \mathbb{R}$ such that $L \leq G(x,y) \leq U$ for all $(x,y) \in K \times K$, then the bifunction $F(x,y) = G(x,y) + \alpha$ belongs to $\mathbb{F}$ for all $\alpha > U - 2L$.*

Proof (a) Suppose that $f_1, f_2 : K \to \mathbb{R}$ are functions such that $f_1(x) + f_2(x) \geq 0$ for all $x \in K$. Let $x, y, z \in K$. Then,

$$F(x,y) = f_1(x) + f_2(y),\ F(x,z) = f_1(x) + f_2(z) \text{ and } F(z,y) = f_1(z) + f_2(y).$$

Hence,

$$\begin{aligned} F(x,z) + F(z,y) - F(x,y) &= f_1(x) + f_2(z) + f_1(z) + f_2(y) - f_1(x) - f_2(y) \\ &= f_2(z) + f_1(z) \geq 0. \end{aligned}$$

(b) Suppose that $F \in \mathbb{F}$ and $\varphi : \mathbb{R} \to \mathbb{R}$ is a nondecreasing and subadditive function. Let $x, y, z \in K$. Since $F \in \mathbb{F}$ and φ is nondecreasing, we have

$$(\varphi \circ F)(x,y) = \varphi(F(x,y)) \leq \varphi(F(x,z) + F(z,y)).$$

By the subadditivity of φ, we have

$$(\varphi \circ F)(x,y) \leq \varphi(F(x,z)) + \varphi(F(z,y)) = (\varphi \circ F)(x,z) + (\varphi \circ F)(z,y).$$

Thus, $\varphi \circ F \in \mathbb{F}$.

(c) For any $x, y, z \in K$, we have

$$\begin{aligned} F(x,z) + F(z,y) &= G(x,z) + G(z,y) + 2\alpha \\ &\geq G(x,z) + G(z,y) - 2L + U + \alpha \quad (\text{Since } \alpha > U - 2L) \\ &\geq U + \alpha \geq G(x,y) + \alpha = F(x,y). \end{aligned}$$

This completes the proof. ■

Proposition 5.2 *$F \in -\mathbb{F} \cap \mathbb{F}$ if and only if there exists a function $f : K \to \mathbb{R}$ such that $F(x,y) = f(y) - f(x)$ for all $x, y \in K$.*

Proof If $F(x,y) = f(y) - f(x)$ for all $x, y \in K$, then obviously, $F \in -\mathbb{F} \cap \mathbb{F}$.

Conversely, if $F \in -\mathbb{F} \cap \mathbb{F}$, then

$$F(x,z) = F(x,y) + F(y,z), \quad \text{for all } x, y, z \in K.$$

Hence, for fixed $\bar{z}$ and considering $f \equiv -F(\cdot, \bar{z})$, we have

$$F(x,y) = F(x,\bar{z}) - F(y,\bar{z}) = f(y) - f(x), \quad \text{for all } x, y \in K,$$

which concludes the proof. ■

Remark 5.1 For each $i = 1, 2, \ldots, m$, if $F_i : K \times K \to \mathbb{R}$ belongs to $\mathbb{F}$, then the bifunction $F(x,y) = \max_{i=1}^{m} F_i(x,y)$ also belongs to $\mathbb{F}$.

Proposition 5.3 *Let $F \in \mathbb{F}$ be such that $F(x,x) = 0$ for all $x \in K$. If $F(\cdot, y)$ is upper semicontinuous at y, then $F(x, \cdot)$ is lower semicontinuous at y for every $x \in K$.*

Proof Let $x, y \in K$ and $\varepsilon > 0$ be fixed. By upper semicontinuity of $F(\cdot, y)$ at y, there exists $\delta > 0$ such that

$$F(z, y) \leq F(y, y) + \varepsilon = \varepsilon, \quad \text{for all } z \in S_\delta(y),$$

where $S_\delta(y)$ denotes the open sphere with center at y and radius δ. Since $F \in \mathbb{F}$, we have

$$F(x, y) \leq F(x, z) + F(z, y) \leq F(x, z) + \varepsilon, \quad \text{for all } z \in S_\delta(y),$$

that is, $F(x, \cdot)$ is lower semicontinuous at y for every $x \in K$. ■

Remark 5.2 The assumption "$F(x, x) = 0$ for all $x \in K$" cannot be dropped from Proposition 5.3.
Indeed, consider two nonnegative upper semicontinuous functions $f_1, f_2 : K \to \mathbb{R}$. Then, the bifunction $F(x, y) = f_1(x) + f_2(y)$ belongs to $\mathbb{F}$ and $F(\cdot, y)$ is upper semicontinuous, but $F(x, \cdot)$ is not necessarily lower semicontinuous.

The following example shows that the reverse implication of Proposition 5.3 does not hold.

Example 5.2 Consider a sequence of functions $f_n : \mathbb{R} \to \mathbb{R}$ defined by

$$f_n(x) := \begin{cases} 0, & \text{if } x \leq 0, \\ nx, & \text{if } 0 < x < 1/n, \\ 1, & \text{if } x \geq 1/n. \end{cases}$$

Then, the bifunction $F(x, y) = \sup_{n \in \mathbb{N}} [f_n(x) - f_n(y)]$ belongs to $\mathbb{F}$, $F(x, x) = 0$ for all $x \in K$, and $F(x, \cdot)$ is continuous for any x. Indeed, if $x \leq 0$, then

$$F(x, y) := \begin{cases} 0, & \text{if } y \leq 0, \\ -y, & \text{if } 0 < y < 1, \\ -1, & \text{if } y \geq 1; \end{cases}$$

if $x \in [1/(k+1), 1/k]$ for some $k \in \mathbb{N}$, then

$$F(x, y) := \begin{cases} 1, & \text{if } y \leq 0, \\ 1 - (k+1)y, & \text{if } 0 < y \leq 1 - kx, \\ k(x - y), & \text{if } 1 - kx < y < x, \\ 0, & \text{if } y \geq x; \end{cases}$$

finally, if $x \geq 1$, then

$$F(x, y) := \begin{cases} 1, & \text{if } y \leq 0, \\ 1 - y, & \text{if } 0 < y < 1, \\ 0, & \text{if } y \geq 1. \end{cases}$$

However, since

$$F(x,y)=\begin{cases}0, & \text{if } x\le 0,\\ 1, & \text{if } x>0,\end{cases}$$

it follows that $F(\cdot,0)$ is not upper semicontinuous at 0.

Definition 5.1 Let K be a nonempty subset of a metric space X. A bifunction $F : K\times K\to\mathbb{R}$ is said to be *cyclically monotone* if

$$F(x_0,x_1)+F(x_1,x_2)+\cdots+F(x_m,x_0)\le 0,$$

for all $x_0,x_1,\dots,x_m\in K$ and for any $m\in\mathbb{N}$.

The bifunction F is called *cyclically antimonotone* if the bifunction $-F$ is cyclically monotone, that is, for any $x_0,x_1,\dots,x_m\in K$, we have

$$\sum_{i=0}^{m} F(x_i,x_{i+1})\ge 0, \tag{5.8}$$

where $x_{m+1}=x_0$.

Remark 5.3 If the triangle inequality $F(x,y)\le F(x,z)+F(z,y)$ holds for all $x,y,z\in K$, then F is cyclically antimonotone. Indeed, by (5.7), we have $F(x,y)+F(y,x)\ge 0$, that is, inequality (5.8) holds for $m=2$. By induction, assuming that (5.8) holds for m, then from the triangle inequality, we have

$$\begin{aligned}\sum_{i=0}^{m+1} F(x_i,x_{i+1}) &= \sum_{i=0}^{m-1} F(x_i,x_{i+1})+F(x_m,x_{m+1})+F(x_{m+1},x_0)\\ &\ge \sum_{i=0}^{m-1} F(x_i,x_{i+1})+F(x_m,x_0)\\ &=\sum_{i=0}^{m} F(x_i,x_{i+1})\ge 0.\end{aligned}$$

The following example shows that a cyclically antimonotone bifunction may not satisfy the triangle inequality.

Example 5.3 Let $F : \mathbb{R}\times\mathbb{R}\to\mathbb{R}$ be a bifunction defined by

$$F(x,y)=x^2-xy, \quad \text{for all } x,y\in\mathbb{R}.$$

Then, F is cyclically antimonotone. Indeed, for fix $m\in\mathbb{N}$ and $m+1$ points $x_0,x_1,\dots,x_m\in\mathbb{R}$, we have

$$\begin{aligned}F(x_0,x_1)+F(x_1,x_2)+\cdots+F(x_{m-1},x_m)+F(x_m,x_0) &= x_0^2+x_1^2+\cdots+x_m^2\\ &\quad -x_0x_1-x_1x_2-\cdots-x_{m-1}x_m-x_mx_0.\end{aligned}$$

Let $u = (x_0, x_1, x_2, \dots, x_{m-1}, x_m)$, $v = (x_1, x_2, \dots, x_m, x_0) \in \mathbb{R}^n$. Then,

$$\|u\|^2 = \|v\|^2 = x_0^2 + x_1^2 + \cdots + x_m^2$$

and

$$\langle u, v\rangle = x_1x_2 + \cdots + x_{m-1}x_m + x_mx_0.$$

By Cauchy–Schwarz inequality, we get

$$x_0^2 + x_1^2 + \cdots + x_m^2 - x_0x_1 - x_1x_2 - \cdots - x_{m-1}x_m - x_mx_0 = \|u\|\|v\| - \langle u, v\rangle \geq 0.$$

Thus,

$$F(x_0, x_1) + F(x_1, x_2) + \cdots + F(x_{m-1}, x_m) + F(x_m, x_0) \geq 0,$$

that is, F is cyclically antimonotone.

Now, for $0 \neq u \in \mathbb{R}$, we have

$$F(2u, 0) = 4u^2 > 3u^2 = 4u^2 - 2u^2 + u^2 = F(2u, u) + F(u, 0).$$

Thus, F does not satisfy the triangle inequality.

Proposition 5.4 [54] *Let K be a nonempty subset of a metric space X. A bifunction $F : K \times K \to \mathbb{R}$ is cyclically antimonotone if and only if there exists a function $f : K \to \mathbb{R}$ such that*

$$F(x, y) \geq f(y) - f(x), \quad \text{for all } x, y \in K. \tag{5.9}$$

Proof If F satisfies (5.9), then obviously F is cyclically antimonotone.

Conversely, fix an arbitrary $x_0 \in K$ and define a function $f : K \to \mathbb{R}$ by

$$f(x) := -\inf\{F(x, x_m) + F(x_m, x_{m-1}) + \cdots + F(x_1, x_0)\}, \quad \text{for all } x \in K,$$

where the infimum is taken over all finite sets of points $x_1, x_2, \dots, x_m \in K$ with $m \in \mathbb{N}$. Since F is cyclically antimonotone, for each $m \in \mathbb{N}$ and $x, x_1, x_2, \dots, x_m \in K$, we have

$$F(x, x_m) + F(x_m, x_{m-1}) + \cdots + F(x_1, x_0) \geq -F(x_0, x),$$

which implies that $f(x) \leq F(x_0, x)$ and, in particular, f is real-valued. Now take $x, y \in K$. For each $m \in \mathbb{N}$ and $x_1, x_2, \dots, x_m \in K$, we have

$$F(x, y) + F(y, x_m) + F(x_m, x_{m-1}) + \cdots + F(x_1, x_0) \geq -f(x).$$

Now taking the infimum over all families $x_1, x_2, \dots, x_m \in K$, we deduce

$$F(x, y) - f(y) \geq -f(x),$$

that is, (5.9) holds. ∎

Theorem 5.1 [54] *Let K be a nonempty subset of a metric space X, $F : K \times K \to \mathbb{R}$ be a cyclically antimonotone bifunction and $f : K \to \mathbb{R}$ be a function which satisfies inequality* (5.9). *Consider the following assertions.*

(a) *For some $y \in K$, $x \mapsto F(x, y)$ is bounded above.*
(b) *The function f is bounded below.*
(c) *For some $x \in K$, $y \mapsto F(x, y)$ is bounded below.*

Then, (a) implies (b), and (b) implies (c).

Proof (a) $\Rightarrow$ (b): Assume that the function $x \mapsto F(x, y)$ is bounded above for some $y \in K$. Then, there exists a real number k such that $F(x, y) \leq k$ for all $x \in K$. From inequality (5.9), we obtain

$$f(x) \geq f(y) - F(x, y) \geq f(y) - k, \quad \text{for all } x \in K,$$

and hence f is bounded below.

(b) $\Rightarrow$ (c): Let us fix $x \in K$. From assertion (b), there exists a real number k such that $f(y) \geq k$ for all $y \in K$. Therefore,

$$F(x, y) \geq f(y) - f(x) \geq k - f(x), \quad \text{for all } y \in K,$$

that is, F is bounded below. ∎

Castellani and Giuli [54] showed that the implication (b) implies (a) does not hold.

Example 5.4 Let $K = [-1, \infty)$ and define $F : K \times K \to \mathbb{R}$ by

$$F(x, y) = \begin{cases} x^2 - 5xy + 5y^2, & \text{if } x \neq y, \\ 0, & \text{if } x = y. \end{cases}$$

Then, $F(x, x) = 0$ for all $x \in K$ and $y \mapsto F(\cdot, y)$ is not bounded above for every $y \in K$. However, F satisfies inequality (5.9) for $f(x) = \frac{5}{2}x^2$ which is bounded below.

Theorem 5.2 [54] *Let K be a nonempty subset of a metric space X and $F : K \times K \to \mathbb{R}$ be a cyclically antimonotone bifunction. Consider the following assertions.*

(a) *For each $y \in K$, $x \mapsto F(x, y)$ is upper semicontinuous.*
(b) *There exists a lower semicontinuous function $f : K \to \mathbb{R}$ which satisfies the inequality* (5.9).
(c) *For all $x \in K$ such that $F(x, x) = 0$, the function $y \mapsto F(x, y)$ is upper semicontinuous at x.*

Then, (a) implies (b), and (b) implies (c).

Proof (a) $\Rightarrow$ (b): Assume that the function $x \mapsto F(x, y)$ is upper semicontinuous for all $y \in K$. The function $f : K \to \mathbb{R}$ defined by

$$f(x) := -\inf\{F(x, x_m) + F(x_m, x_{m-1}) + \cdots + F(x_1, x_0)\}, \quad \text{for all } x \in K,$$

where the infimum is taken over all finite sets of points $x_1, x_2, \ldots, x_m \in K$ with $m \in \mathbb{N}$, is lower semicontinuous, and satisfies inequality (5.9) since it is the pointwise supremum of a collection of lower semicontinuous functions.

(b) ⇒ (c): Assume that for every $\varepsilon > 0$, there exists a neighborhood N_x of x such that

$$f(y) \geq f(x) - \varepsilon, \quad \text{for all } y \in N_x.$$

Therefore,

$$F(x,y) \geq f(y) - f(x) \geq -\varepsilon = F(x,x) - \varepsilon, \quad \text{for all } y \in K,$$

that is, F is lower semicontinuous in x. ∎

Definition 5.2 Let K be a nonempty subset of a metric space X. A bifunction $F : K \times K \to \mathbb{R}$ is said to be *quasimonotone* if for all $x, y \in K$,

$$F(x,y) > 0 \quad \Rightarrow \quad F(y,x) \leq 0. \tag{5.10}$$

The bifunction F is called *anti-quasimonotone* if the bifunction $-F$ is quasimonotone, that is, for any $x, y \in K$,

$$F(x,y) < 0 \quad \Rightarrow \quad F(y,x) \geq 0. \tag{5.11}$$

Definition 5.3 Let K be a nonempty subset of a metric space X. A bifunction $F : K \times K \to \mathbb{R}$ is said to be *cyclically quasimonotone* if for all $m \in \mathbb{N}$ and for any $x_1, x_2, \dots, x_m \in K$, there exists $i \in \{1, 2, \dots, m\}$ such that

$$F(x_i, x_{i+1}) \leq 0,$$

where $x_{m+1} := x_1$.

The bifunction F is called *cyclically anti-quasimonotone* if the bifunction $-F$ is cyclically quasimonotone, that is, for all $m \in \mathbb{N}$ and for any $x_1, x_2, \dots, x_m \in K$, there exists $i \in \{1, 2, \dots, m\}$ such that

$$F(x_i, x_{i+1}) \geq 0, \tag{5.12}$$

where $x_{m+1} := x_1$.

Remark 5.4 **(a)** The functions $F(x,y) = f(x) - f(y)$ and $F(x,y) = f(y) - f(x)$, where $f : K \to \mathbb{R}$, are cyclically quasimonotone and cyclically anti-quasimonotone, respectively.

(b) If $F : K \times K \to \mathbb{R}$ is cyclically quasimonotone (respectively, cyclically anti-quasimonotone), then $F(x,x) \leq 0$ (respectively, $F(x,x) \geq 0$) for all $x \in K$.

(c) If $F, G : K \times K \to \mathbb{R}$ are bifunctions such that F is cyclically quasimonotone (respectively, cyclically anti-quasimontone) and $F(x,y) \geq G(x,y)$ (respectively, $G(x,y) \geq F(x,y)$) for all $x, y \in K$, then G is cyclically quasimonotone (respectively, cyclically anti-quasimontone).

cyclic (anti-)monotonicity	⇒	(anti-)monotonicity
⇓		⇓
cyclic (anti-)quasimonotonicity	⇒	(anti-)quasimonotonicity

Figure 5.1 Relations among different kinds of monotonicities

We now give the characterization of cyclically quasimonotone and cyclically anti-quasimonotone bifunctions.

Proposition 5.5 [63] *Let K be a nonempty subset of a metric space X and $F : K \times K \to \mathbb{R}$ be a bifunction. Then, F is cyclically quasimonotone if and only if for any nonempty finite subset $\{x_1, x_2, \dots, x_m\}$ of K, there exists $i_0 \in \{1, 2, \dots, m\}$ such that*

$$\max_{i \in \{1,2,\dots,m\}} F(x_i, x_{i_0}) \leq 0.$$

Proof Let F be cyclically quasimonotone. Assume the contrary that there exists a finite subset $A := \{x_1, x_2, \dots, x_m\}$ of K such that $\left(\bigcap_{i=1}^{m} S(x_i)\right) \cap A = \emptyset$, where $S(x_i) := \{y \in K : F(x_i, y) \leq 0\}$. This is equivalent to

$$\left(\bigcup_{i=1}^{m} (S(x_i))^c\right) \cup A^c = K. \tag{5.13}$$

Set $x_{i(1)} = x_1$. Then by (5.13), there exists x_j with $x_j \neq x_1$ such that $x_1 \in (S(x_j))^c$, that is, $F(x_j, x_1) > 0$.

Set $x_{i(2)} = x_j$. Then apply (5.13) again. Continuing in this way, we define a sequence $\{x_{i(m)}\}_{m \in \mathbb{N}}$ such that

$$f(x_{i(k+1)}, x_{i(k)}) > 0, \quad \text{for all } k \in \mathbb{N}. \tag{5.14}$$

Since the set $\{x_1, x_2, \dots, x_m\}$ is finite, there exist $n, k \in \mathbb{N}$ with $n < k$ such that $x_{i(k+1)} = x_{i(m)}$. We now consider the points

$$\hat{x}_1 = x_{i(m)},\ \hat{x}_2 = x_{i(k)},\ \hat{x}_3 = x_{i(k-1)}, \dots, \hat{x}_{k+1-m} = x_{i(m+1)}$$

which, due to (5.14), satisfy

$$f(\hat{x}_j, \hat{x}_{j+1}) > 0, \quad \text{for all } j = 1, 2, \dots, k+1-m \text{ with } \hat{x}_{k+2-m} = \hat{x}_1.$$

This means that F is not cyclically quasimonotone, and we get a contraction.

Conversely, assume that for given points $\{x_1, x_2, \dots, x_m, x_{m+1}\}$ of K with $x_{m+1} = x_1$, there exists $i_0 \in \{1, 2, \dots, m\}$ such that $F(x_i, x_{i_0}) \leq 0$ for all $i \in \{1, 2, \dots, m\}$. If $i_0 = 1$, then $F(x_n, x_{n+1}) \leq 0$. If $i_0 > 1$, then $F(x_{i_0-1}, x_{i_0}) \leq 0$. ■

Proposition 5.6 [111] *Let K be a nonempty subset of a metric space X and $F : K \times K \to \mathbb{R}$ be a bifunction. Then, F is cyclically anti-quasimonotone if and only if for any nonempty finite subset $\{x_1, x_2, \dots, x_m\}$ $(m \geq 2)$ of K satisfying $F(x_i, x_{i+1}) < 0$ for all $i = 1, 2, \dots, m-1$, we have $F(x_m, x_1) \geq 0$.*

Proof The "only if" part is obvious. For the "if" part, we only have to discuss the case for $m = 1$. For any $x_1 \in K$, suppose that $F(x_1, x_{1+1}) = F(x_1, x_1) < 0$ (where $x_{1+1} := x_1$). Set $x_2 = x_1$ to form a cycle $\{x_1, x_2\}$ of length 2. Then, $x_{2+1} := x_1$ and so $F(x_1, x_1) = F(x_2, x_1) \geq 0$, a contradiction. Thus, $F(x_1, x_{1+1}) \geq 0$. ■

Remark 5.5 We have already seen that if $F : K \times K \to \mathbb{R}$ is a bifunction which satisfies that triangle inequality $F(x, z) \leq F(x, y) + F(y, z)$ for all $x, y, z \in K$, then F is cyclically anti-monotone. Also, every cyclically anti-monotone bifunction is a cyclically anti-quasimonotone.

Proposition 5.7 *A bifunction $F : K \times K \to \mathbb{R}$ is cyclically anti-quasimonotone if*

$$\left.\begin{array}{l} F(x,x) \geq 0 \text{ for all } x \in K, \text{ and for all } x,y,z \in K, \\ F(x,y) < 0 \text{ and } F(y,z) < 0 \text{ imply } F(x,z) < 0. \end{array}\right\} \tag{5.15}$$

Note that the condition (5.15) is equivalent to the following condition:

$$\left.\begin{array}{l} F(x,x) \geq 0 \text{ for all } x \in K, \text{ and whenever } x_1, x_2, \dots, x_m \in K \ (m \geq 2) \\ \text{satisfying } F(x_i, x_{i+1}) < 0 \text{ for all } i = 1, 2, \dots, m-1, \\ \text{we have } F(x_1, x_m) < 0. \end{array}\right\} \tag{5.16}$$

Proof Suppose that there exist $x_1, x_2, \dots, x_m \in K$ ($m \geq 2$) satisfying $F(x_i, x_{i+1}) < 0$ for all $i = 1, 2, \dots, m-1$, but $F(x_m, x_1) < 0$. Then, $F(x_1, x_m) < 0$ and $F(x_m, x_1) < 0$, which imply that $F(x_1, x_1) < 0$ by (5.15), a contradiction. ∎

Example 5.5 [111] Let $K = [a, b]$ where $a \leq -3$ and $b \geq 3$. Consider the bifunctions $F_1, F_2, F_3 : K \times K \to \mathbb{R}$ defined by

$$F_1(x,y) = -x^3 + y^3 + x^2y^2,$$

$$F_2(x,y) = \begin{cases} 1, & \text{if } x \geq y, \\ -1, & \text{if } x < y, \end{cases}$$

$$F_3(x,y) = \begin{cases} y^2, & \text{if } x \geq y, \\ y - 3x^2, & \text{if } x < y. \end{cases}$$

Since $\sum_{i=1}^{m} F_1(x_i, x_{i+1}) = \sum_{i=1}^{m} x_i^2 x_{i+1}^2 \geq 0$ for any $x_1, x_2, \dots, x_m \in K$ with $x_{m+1} := x_1$, F_1 is cyclically antimonotone, and hence, cyclically anti-quasimonotone. However, F_1 neither satisfies the triangle inequality nor the condition (5.15) because $F_1(3, -1) = -19$ and $F_1(-1, -2) = -3$, but $F_1(3, -2) = 1$.

F_2 obviously satisfies the condition (5.15) (and hence cyclically anti-quasimonotone), but it is neither cyclically antimonotone nor satisfies triangle inequality because $F_2(1, 2) + F_2(2, 3) + F_3(3, 1) = -1$.

F_3 is cyclically anti-quasimonotone, but not cyclically antimonotone. Also, F_3 does not satisfy the condition (5.15).

Indeed, take any $x_1, x_2, \dots, x_m \in K$ such that $F_3(x_i, x_{i+1}) < 0$ for $i = 1, 2, \dots, m-1$. From the definition of F_3, we have $x_1 < x_2 < \cdots < x_m$. Also, from the definition of F_3, $F_3(x_m, x_1) = x_1^2 \geq 0$. Thus, F_2 is cyclically anti-quasimonotone.

For $x = -\frac{1}{\sqrt{2}}$, $y = 1$, and $z = 2$, we have $F_3(x, y) = -\frac{1}{2}$, $F_3(y, z) = -1$, $F_3(x, z) = F_3(z, x) = \frac{1}{2}$, and $F_3(x, y) + F_3(y, z) + F_3(z, x) = -1$. Thus, F_3 is not cyclically antimonotone and also does not satisfy the condition (5.15).

5.3 Extended Ekeland's Variational Principle

In view of Proposition 5.4, the triangle inequality

$$F(x,y) \leq F(x,z) + F(z,y), \quad \text{for all } x,y,z \in K, \tag{5.17}$$

can be replaced by the condition

$$\text{there exists a function } f : K \to \mathbb{R} \text{ which satisfies (5.9),}$$

in all the results from now onward.

We present the following equilibrium version of Ekeland's variational principle.

Theorem 5.3 (Extended Ekeland's Variational Principle) *Let K be a nonempty closed subset of a complete metric space (X,d) and $F : K \times K \to \mathbb{R}$ be a bifunction. Assume that $\varepsilon > 0$ and the following conditions are satisfied:*

(i) *For all $x \in K$, $L := \{y \in K : F(x,y) + \varepsilon d(x,y) \leq 0\}$ is closed;*
(ii) *For all $x \in K$, $F(x,x) = 0$;*
(iii) *For all $x,y,z \in K$, $F(x,y) \leq F(x,z) + F(z,y)$.*

If $\inf_{y \in K} F(x_0, y) > -\infty$ for some $x_0 \in K$, then there exists $\bar{x} \in K$ such that

(a) $F(x_0,\bar{x}) + \varepsilon d(x_0,\bar{x}) \leq 0$,

(b) $F(\bar{x},x) + \varepsilon d(\bar{x},x) > 0$ *for all $x \in K$, $x \neq \bar{x}$.*

Proof For the sake of convenience, we set $d_\varepsilon(u,v) = \varepsilon d(u,v)$. Then, d_ε is equivalent to d and (X, d_ε) is complete. For all $x \in K$, define

$$S(x) = \{y \in K : F(x,y) + d_\varepsilon(x,y) \leq 0\}.$$

By condition (i), $S(x)$ is closed for every $x \in K$. From condition (ii), $x \in S(x)$ for all $x \in X$, and therefore $S(x)$ is nonempty for all $x \in X$. So, we can assume that $y \in S(x)$, that is,

$$F(x,y) + d_\varepsilon(x,y) \leq 0, \tag{5.18}$$

and also let $z \in S(y)$. Then,

$$F(y,z) + d_\varepsilon(z,y) \leq 0. \tag{5.19}$$

Adding inequalities (5.18) and (5.18), and using condition (iii), we obtain

$$0 \geq F(x,y) + d_\varepsilon(x,y) + F(y,z) + d_\varepsilon(z,y) \geq F(x,z) + d_\varepsilon(x,z).$$

Therefore, $z \in S(x)$, which implies that $S(y) \subseteq S(x)$.
Define

$$\mathcal{V}(x_0) := \inf_{z \in S(x_0)} F(x_0, z) > -\infty,$$

and construct a sequence in the following manner: There exists $x_1 \in S(x_0)$ such that

$$F(x_0, x_1) < \mathcal{V}(x_0) + \frac{1}{1}.$$

Since $x_1 \in S(x_0)$, we have $S(x_1) \subseteq S(x_0)$. By condition (iii), we have

$$\begin{aligned}\mathcal{V}(x_1) := \inf_{z \in S(x_1)} F(x_1, z) &\geq \inf_{z \in S(x_1)} F(x_0, z) - F(x_0, x_1) \\ &\geq \mathcal{V}(x_0) - F(x_0, x_1) > -\infty.\end{aligned}$$

Then, there exists $x_2 \in S(x_1)$ such that

$$F(x_1, x_2) \leq \mathcal{V}(x_1) + \frac{1}{2}.$$

Again, since $x_2 \in S(x_1)$, we have $S(x_2) \subseteq S(x_1)$. By condition (iii), we have

$$\begin{aligned}\mathcal{V}(x_2) := \inf_{z \in S(x_2)} F(x_2, z) &\geq \inf_{z \in S(x_2)} F(x_1, z) - F(x_1, x_2) \\ &\geq \mathcal{V}(x_1) - F(x_1, x_2) > -\infty.\end{aligned}$$

Continuing in this way, we obtain a sequence $\{x_n\}$ such that

$$x_{n+1} \in S(x_n), \quad F(x_n, x_{n+1}) < \mathcal{V}(x_n) + \frac{1}{n+1},$$

$$\mathcal{V}(x_{n+1}) \geq \mathcal{V}(x_n) - F(x_n, x_{n+1}), \quad \text{for all } n \geq 0,$$

which imply that

$$-\mathcal{V}(x_n) \leq -F(x_n, x_{n+1}) + \frac{1}{n+1} \leq \mathcal{V}(x_{n+1}) - \mathcal{V}(x_n) + \frac{1}{n+1}.$$

Consequently, we obtain

$$\mathcal{V}(x_{n+1}) + \frac{1}{n+1} \geq 0, \quad \text{for all } n \geq 0. \tag{5.20}$$

If $z_1, z_2 \in S(x_n)$, then

$$d(z_1, z_2) \leq d(x_n, z_1) + d(x_n, z_2) \leq -F(x_n, z_1) - F(x_n, z_2) \leq -2\mathcal{V}(x_n).$$

Hence, the diameter of $S(x_n)$, $\operatorname{diam}(S(x_n)) \leq -2\mathcal{V}(x_n)$. Thus, by inequality (5.20), $\operatorname{diam}(S(x_n)) \to 0$ as $n \to \infty$. Since $\{S(x_n)\}$ is a family of closed sets such that $S(x_{n+1}) \subseteq S(x_n)$ for every $n \geq 0$ and $\operatorname{diam}(S(x_n)) \to 0$ as $n \to \infty$, by Cantor's Intersection Theorem 1.4, there exists exactly one point $\bar{x} \in X$ such that

$$\bigcap_{n=0}^{\infty} S(x_n) = \{\bar{x}\}.$$

This implies that $\bar{x} \in S(x_0)$, that is,

$$F(x_0, \bar{x}) + d_\varepsilon(x_0, \bar{x}) \leq 0,$$

and so (a) holds.

Moreover, $\bar{x}$ also belongs to all $S(x_n)$ and, since $S(\bar{x}) \subseteq S(x_n)$ for all n, we have $S(\bar{x}) = \{\bar{x}\}$. It follows that $x \notin S(\bar{x})$ whenever $x \neq \bar{x}$, implying that

$$F(\bar{x}, x) + d_e(\bar{x}, x) > 0,$$

that is, (b) holds. ∎

Remark 5.6 **(a)** For each fixed $x \in K$, if the function $F(x, \cdot) : K \to \mathbb{R}$ is lower semicontinuous, then the condition (i) of Theorem 5.3 holds for every $\varepsilon > 0$.

(b) If for some $x_0 \in K$, the function $F(x_0, \cdot) : K \to \mathbb{R}$ is bounded below, then the assumption "$\inf_{y \in K} F(x_0, y) > -\infty$ for some $x_0 \in K$" in Theorem 5.3 holds.

The following example shows that the condition (i) of Theorem 5.3 holds but $F(x, \cdot) : K \to \mathbb{R}$ is not lower semicontinuous.

Example 5.6 Let $K = [0, \infty)$ and $F : K \times K \to \mathbb{R}$ be defined by

$$F(x, y) = \begin{cases} 0, & \text{if } (x, y) \in \{0\} \times K \text{ or } x = y, \\ 1, & \text{otherwise.} \end{cases}$$

It is easy to see that the condition (i) of Theorem 5.3 holds for $\varepsilon > 0$, but $F(x, \cdot)$ is not lower semicontinuous at 0. Moreover, F also satisfies conditions (ii) and (iii) of Theorem 5.3.

In view of Remark 5.6, we have following corollary.

Corollary 5.1 (Extended Ekeland's Variational Principle) *Let K be a nonempty closed subset of a complete metric space (X, d) and $F : K \times K \to \mathbb{R}$ be a bifunction. Assume that $\varepsilon > 0$ and the following conditions are satisfied:*

(i) *F is bounded below and lower semicontinuous in the second argument;*
(ii) *For all $x \in K$, $F(x, x) = 0$;*
(iii) *For all $x, y, z \in K$, $F(x, y) \leq F(x, z) + F(z, y)$.*

Then, for all $x_0 \in K$, there exists $\bar{x} \in K$ such that

(a) $F(x_0, \bar{x}) + \varepsilon d(x_0, \bar{x}) \leq 0$,

(b) $F(\bar{x}, x) + \varepsilon d(\bar{x}, x) > 0$, *for all $x \in K$, $x \neq \bar{x}$.*

The following example shows that triangle inequality $F(x, y) \leq F(x, z) + F(z, y)$ in Theorem 5.3 and Corollary 5.1 cannot be dropped.

Example 5.7 Let $K = [0, 1]$ and $F : K \times K \to \mathbb{R}$ be defined by $F(x, y) = -\frac{1}{3}\sqrt{|x - y|}$. If $\varepsilon = \frac{1}{2}$ in Theorem 5.3, then F satisfies all the assumptions of Theorem 5.3 except the triangle inequality, that is, condition (iii). However, the conclusion of Theorem 5.3 does not hold.

Remark 5.7 Let K be a nonempty closed subset of a complete metric space (X, d) and $f : K \to \mathbb{R}$ be a bounded below and lower semicontinuous function. Then, any bifunction $F(x, y) = f(y) - f(x)$

trivially satisfies the condition (ii) of Corollary 5.1. There are other functions, not of this form, that fall into the framework of Corollary 5.1. For instance, consider the function

$$F(x,y) = \begin{cases} e^{-d(x,y)} + 1 + f(y) - f(x), & \text{if } x \neq y, \\ 0, & \text{if } x = y. \end{cases}$$

Remark 5.8 For the nonempty closed subset K of a complete metric space X and $f : K \to \mathbb{R}$, Theorem 4.2 and Corollary 5.1 are equivalent in the sense that each one can be derived by using the other.

Proof Let $f : K \to \mathbb{R}$ be a function satisfying the conditions of Theorem 4.2. Taking $F(x,y) = f(y) - f(x)$ for all $x, y \in K$, then F satisfies all the conditions of Corollary 5.1, and hence, we get the conclusion of Theorem 4.2.

Conversely, assume that the hypothesis of Corollary 5.1 and the conclusion of Theorem 4.2 holds. Then for each $\varepsilon > 0$ and $x_0 \in K$, apply Theorem 4.2 to the function $F(x_0, \cdot)$ gives the existence of $\bar{x} \in K$ such that

$$F(x_0, \bar{x}) \leq F(x_0, x_0) - \varepsilon d(x_0, \bar{x}) \tag{5.21}$$

$$F(\bar{x}, x) > F(x_0, \bar{x}) - \varepsilon d(\bar{x}, x), \quad \text{for all } x \in K, x \neq \bar{x}. \tag{5.22}$$

Since $F(x,x) = 0$ for all $x \in K$, inequality (5.21) reduces to

$$F(x_0, \bar{x}) + \varepsilon d(x_0, \bar{x}) \leq 0.$$

On the other hand, by the triangle inequality, we have

$$F(x_0, x) \leq F(x_0, \bar{x}) + F(\bar{x}, x).$$

Thus, inequality (5.22) reduces to

$$F(\bar{x}, x) + \varepsilon d(\bar{x}, x) > 0, \quad \text{for all } x \in K, x \neq \bar{x}.$$

This completes the proof. ∎

As a direct consequence of Corollary 5.1, we can derive the following corollary.

Corollary 5.2 [64] *Let K be a nonempty closed subset of a complete metric space (X,d) and $F : K \times K \to \mathbb{R}$ be a bifunction. Assume that there exists a bifunction $G : K \times K \to \mathbb{R}$ such that $F(x,y) \geq G(x,y)$ for all $(x,y) \in K \times K$. Assume that $\varepsilon > 0$ and the following conditions are satisfied:*

(i) *G is bounded below and lower semicontinuous in the second argument;*
(ii) *For all $x \in K$, $G(x,x) = 0$;*
(iii) *For all $x, y, z \in K$, $G(x,y) \leq G(x,z) + G(z,y)$.*

Then, for all $x_0 \in K$, there exists $\bar{x} \in K$ such that

(a) $G(x_0, \bar{x}) + \varepsilon d(x_0, \bar{x}) \leq 0,$

(b) $F(\bar{x}, x) + \varepsilon d(\bar{x}, x) > 0, \quad$ *for all $x \in K, x \neq \bar{x}$.*

Proof The bifunction G satisfies all the assumptions of Corollary 5.1. Therefore, there exists $\bar{x} \in K$ such that (a) is verified.

Moreover, from the assumption $F(x,y) \geq G(x,y)$ for all $(x,y) \in K \times K$, and the conclusion (b) of Corollary 5.1, we obtain

$$F(\bar{x},x) \geq G(\bar{x},y) > -\varepsilon d(\bar{x},x) > 0, \quad \text{for all } x \in K,\ x \neq \bar{x},$$

that is, (b) holds. ∎

If $f : K \to \mathbb{R}$ is a function and $G(x,y) = f(y) - f(x)$ for all $x, y \in K$, then we get the following result which is established in [54].

Corollary 5.3 *Let K be a nonempty closed subset of a complete metric space (X,d) and $F : K \times K \to \mathbb{R}$ be a bifunction. Assume that the following conditions are satisfied:*

(i) *There exists a function $f : K \to \mathbb{R}$ such that $F(x,y) \geq f(y) - f(x)$ for all $x, y \in K$;*

(ii) *f is bounded below and lower semicontinuous.*

Then, for every $\varepsilon > 0$ and for all $x_0 \in K$, there exists $\bar{x} \in K$ such that

(a) $f(\bar{x}) \leq f(x_0) - \varepsilon d(x_0,\bar{x})$,

(b) $F(\bar{x},x) + \varepsilon d(\bar{x},x) > 0$ *for all* $x \in K, x \neq \bar{x}$.

Castellani and Giuli [54] developed the following example to show that there are many cases where Corollary 5.3 can be applied but not all the assumptions of Corollary 5.1 are satisfied.

Example 5.8 Let $X = \mathbb{R}^n$ be the metric space with the usual metric $d(x,y) = ||x-y||$ and $K = \mathbb{R}^n_+$ be the first orthant. Define $F : K \times K \to \mathbb{R}$ by

$$F(x,y) = ||x-y||^2 e^{-||x-y||/2} + f(y) - f(x), \quad \text{for all } x, y \in K,$$

where $f : K \to \mathbb{R}$ is a lower semicontinuous and bounded below function. Then, the bifunction F satisfies all the assumptions of Corollary 5.1 except the triangle inequality.

Indeed, for a fixed $x \in K$ with $d(x,0) = ||x|| = 1$, we have

$$F(0,2x) = 4e^{-1} + f(2x) - f(0) > 2e^{-1/2} + f(2x) - f(0) = F(0,x) + F(x,2x).$$

However, condition (i) in Corollary 5.3 is satisfied since $||x-y||^2 e^{-||x-y||/2}$ is nonnegative.

It is worth to mention that if f has no minimum, then the equilibrium problem has no solution.

We present the following two examples from [54] which show that the triangle inequality holds but Corollary 5.1 cannot be applied since either $F(x,\cdot)$ is not lower semicontinuous or $F(x,x) > 0$ for all x.

Example 5.9 Let $\mathbb{Q}^c$ denote the set of all irrational numbers and $\delta_{\mathbb{Q}^c}$ denote the indicator function of $\mathbb{Q}^c$ having the value 1 for all irrational numbers and the value 0 for all rational numbers. Let $K = [1, \infty)$ and define the bifunction $F : K \times K \to \mathbb{R}$ by

$$F(x,y) = \frac{1}{y} - \frac{1}{x} + \delta_{\mathbb{Q}^c}(y-x), \quad \text{for all } x, y \in K.$$

Then, F satisfies all the conditions of Corollary 5.1 including the triangle inequality except the lower semicontinuity of $F(x, \cdot)$.

However, there exists a lower semicontinuous function $f(x) = 1/x$ for all $x \in K$ such that $F(x,y) \leq f(y) - f(x)$ for all $(x,y) \in K \times K$.

Example 5.10 Let $K = [1, \infty)$ and $F : K \times K \to \mathbb{R}$ be a bifunction defined by

$$F(x,y) = \frac{2}{y} - \frac{1}{x} + \frac{1}{2x^2}, \quad \text{for all } (x,y) \in K \times K.$$

Then, F satisfies all the conditions of Corollary 5.1 except $F(x,x) = 0$ for all $x \in K$.

Again, as in the above example, there exists a lower semicontinuous function $f(x) = 1/x$ for all $x \in K$ such that $F(x,y) \leq f(y) - f(x)$ for all $(x,y) \in K \times K$.

5.4 Approximate Solutions for Equilibrium Problems

Definition 5.4 Let K be a nonempty set, $F : K \times K \to \mathbb{R}$ be a bifunction, and $\varepsilon > 0$ be given. The element $\bar{x} \in K$ is said to be an *ε-solution* of the equilibrium problem (5.1) if

$$F(\bar{x}, y) \geq -\varepsilon d(\bar{x}, y), \quad \text{for all } y \in K. \tag{5.23}$$

It is called *strictly ε-solution* of the equilibrium problem (5.1) if the inequality (5.23) is strict for all $\bar{x} \neq y$.

Example 5.11 **(a)** The equilibrium problem corresponding to the bifunction defined in Example 5.9 has strict ε-solutions $\bar{x}$ with $\bar{x} > 1/\sqrt{\varepsilon}$.

(b) The equilibrium problem corresponding to the bifunction defined in Example 5.10 has strict ε-solutions $\bar{x}$ with $\bar{x} > 1/\sqrt{\varepsilon}$.

Remark 5.9 **(a)** Note that the conclusion (b) of the extended Ekeland's variational principle (Theorem 5.3, Corollaries 5.1 and 5.3) gives the existence of a strict ε-solution of the equilibrium problem (5.1) for every $\varepsilon > 0$. Moreover, by conditions (ii) and (iii) of Theorem 5.3, we have $F(\bar{x}, \hat{x}) + F(\hat{x}, \bar{x}) \geq F(\bar{x}, \bar{x}) = 0$, and so, $F(\bar{x}, \hat{x}) \geq -F(\hat{x}, \bar{x})$. It follows from conclusion (a) of Theorem 5.3 that

$$F(\bar{x}, \hat{x}) \geq \varepsilon d(\bar{x}, \hat{x}),$$

"localizing," in certain sense, the position of the $\bar{x}$.

(b) The equilibrium problems corresponding to the bifunctions defined in Example 5.11 do not have any solution. However, since all the assumptions of Corollary 5.3 hold, they admit a strict ε-solution for every $\varepsilon > 0$.

(c) Let $-\varepsilon := \inf_{x \in X} F(\hat{x}, x)$. By replacing ε by $1/n$, we obtain from Corollary 5.1 for arbitrary $n \in \mathbb{N}$, the existence of $\bar{x} \in X$ such that

$$F(\hat{x}, \bar{x}) + \frac{1}{n} d(\hat{x}, \bar{x}) \leq 0$$

and

$$F(\bar{x}, x) + \frac{1}{n} d(\bar{x}, x) \geq 0. \tag{5.24}$$

The first inequality implies in particular that

$$F(\hat{x}, \bar{x}) \leq 0, \tag{5.25}$$

and, since $-F(\hat{x}, \bar{x}) \leq \varepsilon$, we have

$$d(\hat{x}, \bar{x}) \leq n\varepsilon. \tag{5.26}$$

In view of Remark 5.9 (a) and Theorem 5.1 and 5.2, we have the following existence result for a strict ε-solution of the equilibrium problem (5.1).

Theorem 5.4 *Let K be a nonempty closed subset of a complete metric space X and $F : K \times K \to \mathbb{R}$ be a cyclically antimonotone bifunction such that for each $y \in K$, the function $x \mapsto F(x, y)$ is upper semicontinuous and bounded above. Then, the equilibrium problem* (5.1) *has a strict ε-solution, for every $\varepsilon > 0$.*

5.5 Existence Results for Solutions of Equilibrium Problems

By using Weierstrass's Theorem 1.19 for lower semicontinuous function from above, Castellani et al. [55] established the following existence result for a solution of the equilibrium problem (5.1).

Theorem 5.5 *Let K be a nonempty compact subset of a metric space X and $F : K \times K \to \mathbb{R}$ be a function such that $F(x, y) \leq F(x, z) + F(z, y)$ for all $x, y, z \in K$. If for all $\bar{z} \in K$, the function $F(\bar{z}, \cdot)$ is lower semicontinuous from above, then the equilibrium problem* (5.1) *has a solution.*

Proof For fixed $\bar{z} \in K$, consider the following minimization problem:

$$\min_{y \in K} F(\bar{z}, y). \tag{5.27}$$

Then by Theorem 1.19, there exists a solution $\bar{x} \in K$ of the minimization problem (5.27). Since F satisfies the triangle inequality and $\bar{x} \in K$ is a solution of the minimization problem (5.27), we have

$$F(\bar{x}, y) \geq F(\bar{z}, y) - F(\bar{z}, \bar{x}) \geq 0, \quad \text{for all } y \in K,$$

that is, $\bar{x} \in K$ is a solution of the equilibrium problem (5.1). ■

Remark 5.10 **(a)** In addition to the assumption of Theorem 5.5, if we further assume that the function $F(\cdot, y)$ is upper semicontinuous for each fixed $y \in K$, then the solution set of the equilibrium problem (5.1) is a closed subset of the compact set K, and hence compact.

(b) By using Weierstrass's Theorem 1.19 for upper semicontinuous function from below and by considering the maximization problem

$$\max_{x \in K} F(x, \bar{z}), \tag{5.28}$$

one can easily replace the assumption "If for all $\bar{z} \in K$, the function $F(\bar{z}, \cdot)$ is lower semicontinuous from above" in Theorem 5.5 by the assumption "If for all $\bar{z} \in K$, the function $F(\cdot, \bar{z})$ is upper semicontinuous from below."

In view of Proposition 5.4, the following result is equivalent to Theorem 5.5.

Theorem 5.6 *Let K be a nonempty compact subset of a metric space X and $F : K \times K \to \mathbb{R}$ be a function such that there exists a function $f : K \to \mathbb{R}$ which satisfies (5.9). If for all $\bar{z} \in K$, the function $F(\bar{z}, \cdot)$ is lower semicontinuous from above, then the equilibrium problem (5.1) has a solution.*

As an application of Theorem 5.3, we derive the following existence results for a solution of the equilibrium problem (5.1).

Theorem 5.7 (Extended Takahashi's Minimization Theorem) *Let K be a nonempty closed subset of a complete metric space (X, d) and $F : K \times K \to \mathbb{R}$ be lower semicontinuous in the second argument. Assume that the following conditions hold:*

(i) *$F(x,x) = 0$ for all $x \in K$;*

(ii) *$F(x,y) \leq F(x,z) + F(z,y)$ for all $x, y, z \in K$;*

(iii) *There exists $\hat{x} \in K$ such that $\inf_{y \in K} F(\hat{x}, y) > -\infty$.*

Further, assume that the following extended Takahashi's condition holds:

$$\begin{cases} \text{Assume that for every } x \in K \text{ with } \inf_{y \in K} F(x,y) < 0, \\ \text{there exists } y \in K,\ y \neq x \text{ such that } F(x,y) + d(x,y) \leq 0. \end{cases} \tag{5.29}$$

Then, there exists a solution $\bar{x} \in K$ of the equilibrium problem (5.1).

Proof By Theorem 5.3, there exists $\bar{x} \in K$ such that

$$F(\bar{x}, y) + \varepsilon d(\bar{x}, y) > 0, \quad \text{for all } y \in K,\ y \neq \bar{x}. \tag{5.30}$$

We claim that $\bar{x}$ is a solution of the equilibrium problem (5.1). Otherwise, there exists $y \in K$ such that $F(\bar{x}, y) < 0$. From assumption (5.29), we obtain $z \in K$ with $z \neq \bar{x}$ and $F(\bar{x}, z) + \varepsilon d(\bar{x}, z) \leq 0$ which contradicts inequality (5.30). ∎

Remark 5.11 Theorem 5.7 guarantees the existence of a solution of the equilibrium problem (5.1) without any compactness or convexity assumption on the underlying set.

The following result guarantees the existence of a solution of the minimization problem for a bounded below function on a closed set.

Theorem 5.8 *Let K be a nonempty closed subset of a complete metric space (X, d) and $f : K \to \mathbb{R}$ be a bounded below function. If for every $x \in K$ with $\inf_{y \in K} f(y) < f(x)$, there exists $z \in K$ such that $z \neq x$ and $f(z) + \varepsilon d(x,z) \leq f(x)$, then there exists $\bar{x} \in K$ such that $f(\bar{x}) \leq f(y)$ for all $y \in K$.*

Proof Define $F : K \times K \to \mathbb{R}$ by

$$F(x,y) = f(y) - f(x), \quad \text{for all } x, y \in K.$$

Then, F satisfies all of the assumptions of Theorem 5.7. So, there exists $\bar{x} \in K$ such that $F(\bar{x}, y) \geq 0$ for all $y \in K$. ∎

Let $\mathbb{S} := \{x \in K : F(x, y) \geq 0 \text{ for all } y \in K\}$, that is, $\mathbb{S}$ is the set of solutions of the equilibrium problem (5.1). Then, the extended Takahashi's condition (5.29) reads as

$$\text{for all } x \in K \setminus \mathbb{S}, \text{ there exists } y \in K,\ y \neq x \text{ such that } F(x, y) + d(x, y) \leq 0.$$

The extended Ekeland's variational principle states for the same class of functions as

$$\text{there exists } x \in K \text{ such that } F(x, y) + d(x, y) > 0, \quad \text{for all } y \in K,\ y \neq x.$$

If we define

$$S(x) = \{y \in K : F(x, y) + d(x, y) \leq 0\},$$

then the extended Takahashi's condition (5.29) can be reformulated as

$$\text{for all } x \in K \setminus \mathbb{S},\ S(x) \neq \{x\}.$$

Remark 5.12 Since extended Ekeland's variational principle and extended Takahashi's minimization theorem are equivalent (see Theorem 5.16), we can say that the only points which satisfy the assertions of extended Ekeland's variational principle are the solutions of equilibrium problem (5.1).

We mention the converse of the extended Takahashi's Minimization Theorem 5.7.

Theorem 5.9 *Let K be a nonempty closed subset of a complete metric space (X, d), $F : K \times K \to \mathbb{R}$ be lower semicontinuous in the second argument and satisfy conditions (i)–(iii) in Theorem 5.7. If there exists a solution $\bar{x} \in K$ of equilibrium problem* (5.1) *such that $F(y, \bar{x}) + d(y, \bar{x}) \leq 0$ for all $y \in K$, then F satisfies extended Takahashi's condition* (5.29).

Proof Assume that every $x \in K$ satisfies

$$\inf_{y \in K} F(x, y) < 0. \tag{5.31}$$

By hypothesis, there exists $\bar{x} \in K$ such that

$$F(\bar{x}, y) \geq 0, \text{ for all } y \in K, \tag{5.32}$$
$$\text{and } F(y, \bar{x}) + d(y, \bar{x}) \leq 0, \text{ for all } y \in K. \tag{5.33}$$

In view of (5.31), the inequality (5.33) holds only for all $y \in K$, $y \neq \bar{x}$. Hence, we get the conclusion. ∎

Lemma 5.1 *Let K be a nonempty closed subset of a complete metric space (X, d), $F : K \times K \to \mathbb{R}$ be lower semicontinuous in the second argument and satisfy conditions (i)–(iii) of Theorem 5.7. Assume that the following alternative form of extended Takahashi's condition holds:*

$$S(x) = \{y \in K : F(x, y) + d(x, y) \leq 0\} \neq \{x\}, \quad \textit{for all } x \in K \setminus \mathbb{S}. \tag{5.34}$$

Then, $S(x) \cap \mathbb{S} \neq \varnothing$ whenever $x \notin \mathbb{S}$.

Proof For each $x \in K$, consider the restriction F_x of F on $S(x) \times S(x)$. Then, F_x is lower semicontinuous in the second argument and $\inf_{y \in S(x)} F(x, y) > -\infty$ for some $x \in S(x)$ because $S(x)$ is nonempty and closed for each $x \in K$. Thus, F_x satisfies all the conditions of extended Ekeland's variational principle (in short, EEVP). By applying Corollary 5.1 for F_x, there exists $\bar{x} \in S(x)$ such that

$$F_x(\bar{x}, y) + d(\bar{x}, y) > 0, \quad \text{for all } y \in S(x),\ y \neq \bar{x}. \tag{5.35}$$

We need to prove that

$$F(\bar{x}, y) + d(\bar{x}, y) > 0, \quad \text{for all } y \in K,\ y \neq \bar{x}. \tag{5.36}$$

Assume that (5.36) does not hold. Then, there exists $u \in K$ such that

$$F(\bar{x}, u) + d(\bar{x}, u) \leq 0. \tag{5.37}$$

Since $\bar{x} \in S(x)$, we have

$$F(x, \bar{x}) + d(x, \bar{x}) \leq 0. \tag{5.38}$$

Combining (5.37) and (5.38), we get

$$F(\bar{x}, u) + d(\bar{x}, u) + F(x, \bar{x}) + d(x, \bar{x}) \leq 0.$$

By utilizing the triangle inequality and condition (iii), we have $d(x, u) + F(x, u) \leq 0$, and hence, $u \in S(x)$. This is a contradiction because the inequality (5.35) for $y = u$ and the inequality (5.37) cannot hold simultaneously. Therefore, inequality (5.36) holds, and in view of Remark 5.12, $\bar{x}$ is a solution of the equilibrium problem (5.1). This is true for all $x \in K \setminus \mathbb{S}$, which completes the proof. ∎

Remark 5.13 Let $\{t_n\}$ be a sequence of positive real numbers which is bounded below by some positive number. If conditions of Theorem 5.3 hold for $\varepsilon = t_n$, for all n, then the sequence of approximate solutions $\{x_n\}$ obtained by Theorem 5.3 corresponding to t_n is bounded. Indeed, putting $\alpha = \inf_{y \in K} F(x_0, y)$ and $\beta = \inf_{n \in \mathbb{N}} t_n$ and using Theorem 5.3 with $\varepsilon = t_n$ for every n, we have

$$d(x_0, x_n) \leq \frac{1}{t_n} - F(x_0, x_n) \leq \frac{1}{t_n} - \inf_{y \in K} F(x_0, y) \leq \frac{1}{\beta} - \alpha.$$

Theorem 5.10 *Let K be a nonempty compact subset of a metric space (X, d), $F : K \times K \to \mathbb{R}$ be a bifunction and $\{t_n\}$ be a decreasing sequence of positive real numbers such that $t_n \to 0$. Assume that*

(i) $L := \{y \in K : F(x, y) + t_n d(x, y) \leq 0\}$ *is closed for every $x \in K$ and for all $n \in \mathbb{N}$,*
(ii) $F(x, x) = 0$, *for all $x \in K$,*
(iii) $F(x, y) \leq F(x, z) + F(z, y)$, *for all $x, y, z \in K$,*
(iv) $U := \{y \in K : F(y, x) + t_n d(y, x)\}$ *is closed for every $x \in K$ and for all $n \in \mathbb{N}$.*

If $\inf_{y \in K} F(x_0, y) > -\infty$ for some $x_0 \in K$, then the set of solutions of the equilibrium problem (5.1) *is nonempty.*

Proof By Theorem 5.3, for each $n \in \mathbb{N}$, there exists t_n-solution of the equilibrium problem (5.1), that is, there exists $x_n \in K$ such that

$$F(x_n, y) \geq -t_n d(x_n, y), \quad \text{for all } y \in K. \tag{5.39}$$

Since K is compact, we can choose a subsequence $\{x_{n_k}\}$ of $\{x_n\}$ such that $x_{n_k} \to \bar{x}$ as $k \to \infty$. By using inequality (5.39) and since $\{t_n\}$ is decreasing sequence, for every fixed positive integer k_0, we have

$$x_{n_k} \in \{y \in K : F(x, y) + t_{n_{k_0}} d(x, y)\}, \quad \text{for all } k \geq k_0.$$

By condition (iv) and $x_{n_k} \to \bar{x}$, we deduce that

$$F(\bar{x}, y) + t_{n_{k_0}} d(\bar{x}, y) \geq 0, \quad \text{for all } y \in K. \tag{5.40}$$

Since n_{k_0} is arbitrary and t_{n_k} approaches zero as $k \to +\infty$, inequality (5.40) implies that $\bar{x}$ is a solution of the equilibrium problem (5.1). ■

We now consider the noncompact case. For the rest of this section, we assume that (X, d) is a metric space with Heine–Borel property, that is, each closed bounded subset of X is compact. Let K be a closed subset of X and $F : K \times K \to \mathbb{R}$ be a given bifunction.

Consider the following coercivity condition:

$$\exists\, S_r[c] : \forall x \in K \backslash K_r,\ \exists\, y \in K \text{ satisfying } d(y, c) < d(x, c) \text{ and } F(x, y) \leq 0, \tag{5.41}$$

where $K_r = K \cap S_r[c]$ and $S_r[c] = \{y \in X : d(c, y) \leq r\}$.

Theorem 5.11 *Let K be a nonempty closed subset of (X, d) and $\{t_n\}$ be a decreasing sequence of positive real numbers such that $t_n \to 0$. Suppose that $F : K \times K \to \mathbb{R}$ satisfies conditions (i)–(iv) of Theorem 5.10. If $\inf_{y \in K} f(x_0, y) > -\infty$ for some $x_0 \in K$, and the coercivity condition* (5.41) *holds, then the set of solutions of the equilibrium problem* (5.1) *is nonempty.*

Proof For each $x \in K$, consider the nonempty set

$$S(x) = \{y \in K : d(y, c) \leq d(x, c) \text{ and } F(x, y) \leq 0\}.$$

Observe that for every $x, y \in K_r$, $y \in S(x)$ implies that $S(y) \subseteq S(x)$. Indeed, for $z \in S(y)$ we have $d(z, c) \leq d(y, c) \leq d(x, c)$ and by condition (iii) of Theorem 5.10 $F(x, z) \leq F(x, y) + F(y, z) \leq 0$. Since K_r is nonempty and compact, by Theorem 5.10, there exists $x_r \in K_r$ such that

$$F(x_r, y) \geq 0, \quad \text{for all } y \in K_r. \tag{5.42}$$

Suppose that there exists $x \in K$ with $F(x_r, x) < 0$ and put

$$a = \min_{y \in S(x)} d(y, c).$$

We distinguish two cases.

CASE I: $a \le r$. Let $y_0 \in S(x)$ such that $d(y_0, c) = a \le r$. Then, we have $F(x, y_0) \le 0$. Since $F(x_r, x) < 0$, it follows by condition (iii) of Theorem 5.10 that

$$F(x_r, y_0) \le F(x_r, x) + F(x, y_0) < 0,$$

which contradicts (5.42).

CASE II: $a > r$. Let again $y_0 \in S(x)$ such that $d(y_0, c) = a > r$. Then by condition (5.41), we can choose an element $y_1 \in K$ with $d(y_1, c) < d(y_0, c) = a$ such that $F(y_0, y_1) \le 0$. Thus, $y_1 \in S(y_0) \subseteq S(x)$. Hence,

$$d(y_1, c) < a = \min_{y \in S(x)} d(y, c),$$

which is a contradiction. Therefore, there is no $x \in K$ such that $F(x_r, x) < 0$, that is, x_r is a solution of the equilibrium problem (5.1) on K. ■

We now present an existence result for solutions of the equilibrium problem (5.1) under the cyclically anti-quasimonotonicity of the involved bifunction. It is established and referred to as the Weierstrass theorem for bifunctions by Khanh and Quan [111].

Theorem 5.12 [111] *Let K be a nonempty compact subset of a metric space (X, d) and $F : K \times K \to \mathbb{R}$ be cyclically anti-quasimonotone such that the level set $L_{\geq} F(\cdot, y) := \{x \in K : F(x, y) \ge 0\}$ is closed for each $y \in K$. Then, there exists $\bar{x} \in K$ such that $\inf_{y \in K} F(\bar{x}, y) \ge 0$. If, in addition, $F(x, x) = 0$ for all $x \in K$, then $\bar{x} \in K$ is a solution of the equilibrium problem* (5.1).

Proof Suppose the contrary that for each $x \in K$, $\inf_{y \in K} F(x, y) < 0$. Then for each $x \in K$, there exists $y_x \in K$ such that $F(x, y_x) < 0$. Since the level set $L_{\geq} F(\cdot, y_x)$ is closed, the set $V(y_x) := \{z \in K : F(z, y_x) < 0\}$ is an open neighborhood of x in K, and therefore, the family $\{V(y_x) : x \in K\}$ is an open cover of K. By compactness of K, there exist $x_1, x_2, \dots, x_m \in K$ such that $K = \bigcup_{i=1}^{m} V(y_{x_i})$. By anti-quasimonotonicity of F, we have $F(y_{x_i}, y_{x_i}) \ge 0$, that is, $y_{x_i} \notin V(y_{x_i})$ for all $i = 1, 2, \dots, m$. When $y_{x_1} \notin V(y_{x_1})$, without loss of generality, we can assume that $y_{x_1} \in V(y_{x_2})$. Then, $F(y_{x_1}, y_{x_2}) < 0$, and by the cyclic anti-quasimonotonicity of F, $F(y_{x_2}, y_{x_1}) \ge 0$, that is, $y_{x_2} \notin V(y_{x_1})$. Thus, $y_{x_2} \notin V(y_{x_1}) \cap V(y_{x_2})$. We can assume further that $y_{x_2} \in V(y_{x_2})$. Then, $F(y_{x_1}, y_{x_2}) < 0$ and $F(y_{x_2}, y_{x_3}) < 0$, and the cyclic anti-quasimonotonicity of F implies that $F(y_{x_3}, y_{x_1}) \ge 0$ and $F(y_{x_3}, y_{x_2}) \ge 0$, that is, $y_{x_3} \notin V(y_{x_1})$ and $y_{x_3} \notin V(y_{x_2})$. Hence, $y_{x_3} \notin V(y_{x_1}) \cup V(y_{x_2}) \cup V(y_{x_3})$. Continuing in this way, we obtain $y_{x_i} \notin \bigcup_{j=1}^{i} V(y_{x_j})$ for all $i = 1, 2, \dots, m$. In particular, we have $y_{x_m} \notin \bigcup_{j=1}^{m} V(y_{x_j}) = K$, a contradiction. Thus, there must exist $\bar{x} \in K$ such that $\inf_{y \in K} F(\bar{x}, y) \ge 0$,

When $F(\bar{x}, \bar{x}) = 0$, $\inf_{y \in K} F(\bar{x}, y) \ge 0$ implies that $\bar{x}$ is a solution of the equilibrium problem (5.1). ■

Remark 5.14 We know that the bifunction $F : K \times K \to \mathbb{R}$ defined by $F(x, y) = f(y) - f(x)$ for all $x, y \in K$ is cyclically anti-quasimonotone. Therefore, we can obtain Weierstrass's Extreme Value Theorem A.7 by Theorem 5.12.

As pointed out by Khanh and Quan [111], Theorem 5.12 is really more general than Weierstrass's Theorem A.6.

Indeed, let us consider the function $f : [a, b] \to \mathbb{R}$ defined by $f(x) = x + i[x]$, where $b - a > 1$ and $i[x]$ is the integer part of x. Then, it can be easily seen that f is not lower semicontinuous at each integer $x \in (a, b]$, but $F(x, y) = f(y) - f(x) = y - x + i[x] - i[y]$ satisfies the assumptions of Theorem 5.12.

Example 5.12 [111] Let

$$K = \left\{(u,v) \in \mathbb{R}^2 : (u+v)^2 + 4(u-v)^2 \leq 32 \leq 4(u+v)^2 + 16(u-v)^2\right\}$$

and $F : K \times K \to \mathbb{R}$ be defined by

$$F(x,y) = \min\left\{\left(u_1 - \frac{1}{2}u_2\right)(u_1 - u_2), |v_1 - v_2|\right\},$$

for all $x = (u_1, u_2)$, $y = (v_1, v_2) \in K$. Then, it can be easily seen that K is closed and bounded, and hence, compact. Now we show that F is cyclically anti-quasimonotone. For that, let us consider $x_1, x_2, \dots, x_m \in K$ $(m \geq 2)$, where $x_i = (u_i, v_i)$, satisfying $F(x_i, x_{i+1}) < 0$ for all $i = 1, 2, \dots, m-1$. Then, we have $\left(u_i - \frac{1}{2}u_{i+1}\right)(u_i - u_{i+1}) < 0$, which is equivalent to $\min\left\{u_{i+1}, \frac{1}{2}u_{i+1}\right\} < u_i < \max\left\{u_{i+1}, \frac{1}{2}u_{i+1}\right\}$, for all $i = 1, 2, \dots, m-1$. It follows that $0 < u_1 < u_2 < \cdots < u_m$ if $u_m \geq 0$, and $u_m < u_{m-1} < \cdots < u_1 < 0$ if $u_m < 0$. This implies that

$$F(x_m, x_1) = \min\left\{\left(u_m - \frac{1}{2}u_1\right)(u_m - u_1), |v_m - v_1|\right\} \geq 0.$$

Thus, F is cyclically anti-quasimonotone.

For each $y = (u_y, v_y) \in K$, we can see that the level set

$$L_{\geq}F(\cdot, y) = \left\{(u,v) \in K : u \leq \min\left\{\frac{1}{2}u_y, u_y\right\}\right\} \cup \left\{(u,v) \in K : u \geq \max\left\{\frac{1}{2}u_y, u_y\right\}\right\}$$

is closed. Moreover, $F(x,x) = 0$ for all $x \in K$. Thus, by Theorem 5.12, there exists $\bar{x} \in K$ such that $F(\bar{x}, y) \geq 0$ for all $y \in K$.

Note that this bifunction F is not cyclically antimonotone. Indeed, consider two points $x = \left(\frac{3}{2}, 0\right)$ and $y = \left(2, \frac{1}{8}\right)$ in K. Then, $F(x,y) + F(y,x) = -\frac{1}{8} < 0$. Also, F does not satisfy triangle inequality.

Corollary 5.4 [111] *Let K be a nonempty compact subset of a metric space (X, d) and $F : K \times K \to \mathbb{R}$ be cyclically quasimonotone such that the level sets $L_{\leq}(x, \cdot) := \{y \in K : F(x,y) \leq 0\}$ and $L_{\geq}F(\cdot, x) := \{y \in K : F(y,x) \geq 0\}$ are closed for each $y \in K$. Assume that the following condition holds.*

(A) *If $F(y,x) \leq 0$ and $U(x)$ is an open neighborhood of x, then there exist $z \in U(x)$ and $\alpha, \beta \in (0, \infty)$ such that $\alpha F(z,y) + \beta F(z,x) \geq 0$.*

Then, there exists a solution $\bar{x} \in K$ of the equilibrium problem (5.1).

Proof Let $G : K \times K \to \mathbb{R}$ be defined by $G(x,y) = -F(y,x)$ for all $x, y \in K$. Then, G is cyclically anti-quasimonotone, and

$$L_{\geq}F(\cdot, y) = \{x \in K : G(x,y) \geq 0\} = \{x \in K : F(y,x) \leq 0\} = L_{\leq}F(y, \cdot)$$

is closed for all $y \in K$. By Theorem 5.12, there exists $\bar{x} \in K$ such that

$$\inf_{y \in K} G(\bar{x}, y) \geq 0, \text{ that is, } \sup_{y \in K} F(y, \bar{x}) \leq 0.$$

Now suppose that $\inf_{y\in K} F(\bar{x}, y) < 0$. Then, there exists $\bar{y} \in K$ such that $F(\bar{x}, \bar{y}) < 0$, that is, $\bar{x} \in L_{<}F(\cdot, \bar{y}) = \{x \in K : F(x, \bar{y}) < 0\} = K \setminus L_{\geq}F(\cdot, \bar{y})$, an open set. Since $F(\bar{y}, \bar{x}) \leq 0$, by condition A, there exist $z \in L_{<}F(\cdot, \bar{y})$ and $\alpha, \beta \in (0, \infty)$ such that $\alpha F(z, \bar{y}) + \beta F(z, \bar{x}) \geq 0$. This is impossible because $F(z, \bar{y}) < 0$ and $F(z, \bar{x}) \leq 0$. Thus, $\inf_{y\in K} F(\bar{x}, y) \geq 0$. For $y = \bar{x}$, we have $F(\bar{x}, \bar{x}) \leq 0$ and $F(\bar{x}, \bar{x}) \geq 0$, that is, $F(\bar{x}, \bar{x}) = 0$. Hence, $\bar{x} \in K$ is a solution of the equilibrium problem (5.1). ∎

Although we have derived the Corollary 5.4 by using Theorem 5.12, but they are different as the following example shows.

Example 5.13 [111] Let $K = \{(u, v) \in \mathbb{R}^2 : 0 \leq uv, u^2 + v^2 \leq 25\}$ and $F : K \times K \to \mathbb{R}$ be defined as follows: for all $x = (u_1, v_1), y = (u_2, v_2) \in K$,

$$F(x, y) = G(x, y)(u_2 - u_1) + H(x, y)(v_2 - v_1),$$

where $G(x, y) = 5u_2^2 + 5v_2^2 + u_1 - u_2 + 10$ and $H(x, y) = 5u_2^2 + 5v_2^2 + v_1 - v_2 + 10$. It can be easily seen that K is compact.

For $x = (0, 0)$, $y = (0, 1)$, and $z = (1, 0)$, we have

$$F(y, z) = F(z, y) = -2 \quad \text{and} \quad F(x, y) + F(y, z) + F(z, x) = 1.$$

Hence, F is neither cyclically anti-quasimonotone nor cyclically monotone. Since

$$F(x, y) = (5u_2^2 + 5v_2^2 + 10)((u_2 + v_2) - (u_1 + v_1)) - (u_2 - u_1)^2 - (v_2 - v_1)^2,$$

it can be easily checked that F is cyclically quasimonotone and the level sets $L_{\leq}F(x, \cdot)$ and $L_{\geq}F(\cdot, y)$ are closed.

Now we verify the condition A. If $U(x)$ is an open neighborhood of x, then there exists $\delta > 0$ such that $S_\delta(x) \cap K \subseteq U(x)$. For $x, y \in K$, $u_1, u_2, v_1, v_2 \in [-5, 5]$, we have $G(x, y) = 4u_2^2 + 5v_2^2 + \left(u_2 - \frac{1}{2}\right)^2 + u_1 + \frac{39}{4} > 0$ and, similarly, $G(y, x) > 0$, $H(x, y) > 0$, $H(y, x) > 0$ for all $x, y \in K$. So we have three possibilities for $F(y, x) = G(y, x)(u_1 - u_2) + H(y, x)(v_1 - v_2)$ to be nonpositive.

CASE I: ($u_1 \leq u_2$ and $v_1 \leq v_2$). Choose $z = x$. Then $F(z, x) = 0$ and $F(z, y) > 0$.

CASE II: ($u_1 < u_2$ and $v_1 > v_2$). For $\varepsilon = \min\left\{\frac{\delta}{2}, \frac{u_1 - u_2}{2}\right\}$ and $z = (u_1, v_1 - \varepsilon)$, $z \in S_\delta(x) \cap K$ and $F(z, x) > 0$.

CASE III: ($u_1 > u_2$ and $v_1 < v_2$). Take $\varepsilon = \min\left\{\frac{\delta}{2}, \frac{u_1 - u_2}{2}\right\}$ and $z = (u_1 - \varepsilon, v_1)$; then $z \in S_\delta(x) \cap K$ and $F(z, x) > 0$.

In all three cases, we can find $\alpha, \beta \in (0, \infty)$ such that $\alpha F(z, y) + \beta F(z, x) \geq 0$. Thus, by Corollary 5.4, there exists $\bar{x} \in K$ such that $F(\bar{x}, y) \geq 0$ for all $y \in K$.

5.6 Some Equivalent Results to Existence Results for Solutions of Equilibrium Problems

To find an element in the intersection of a family of sets is known as *nonempty-intersection problem*. If the sets of this family are convex, it is called the *convex feasibility problem*. Many mathematical problems can be written in the form of nonempty-intersection problem. Namely, consider a set-valued map $T : X \rightrightarrows X$ defined by $T(y) = \{x \in X : F(x, y) \geq 0\}$ for all $y \in X$. Then, $\bar{x} \in X$ is a solution of the equilibrium problem (5.1) if and only if $\bar{x} \in \bigcap_{y \in X} T(y)$. In this section, we discuss some results which provide the nonemptiness of $\bigcap_{y \in X} T(y)$ and some results which are equivalent to the existence results for solutions of the equilibrium problem (5.1). Most of the results of this section are derived by Khanh and Quan [111].

Definition 5.5 Let X be a set and K be a nonempty subset of X. A set-valued map $T : K \rightrightarrows X$ is said to be *anti-cyclic* if for any $x_1, x_2, \dots, x_m \in K$, there exists an $i \in \{1, 2, \dots m\}$ such that $x_i \in T(x_{i+1})$, where $x_{m+1} := x_1$.

Remark 5.15 (a) A set-valued map $T : K \rightrightarrows X$ is anti-cyclic if and only if whenever $x_1, x_2, \dots, x_m \in K$ $(m \geq 2)$ satisfy $x_i \notin T(x_{i+1})$ for all $i = 1, 2, \dots, m-1$, we have $x_m \in T(x_1)$.

(b) If a set-valued map $T : K \rightrightarrows X$ is anti-cyclic, then for all $x, y \in K$, $x \in T(x)$, $x \in T(y)$ or $y \in T(x)$.

(c) If $x \in T(x)$ for all $x \in K$ and, for $x, y, z \in K$ with $x \notin T(y)$ and $y \notin T(z)$, one has $x \notin T(z)$, then T is anti-cyclic.
Indeed, by assumption, $x_1 \notin T(x_m)$ whenever $x_1, x_2, \dots, x_m \in K$ satisfy $x_i \notin T(x_{i+1})$ for all $i = 1, 2, \dots, m-1$. Hence, if $x_m \notin T(x_1)$, then we simultaneously have $x_1 \notin T(x_m)$ and $x_m \notin T(x_1)$, and so $x_1 \notin T(x_1)$ (here $x = z = x_1, y = x_m$). This contradicts the assumption that $x \in T(x)$ for all $x \in K$.

(d) If $T : K \rightrightarrows X$ is anti-cyclic, then the restriction map $T|_D : D \rightrightarrows D$ defined by $T|_D(x) = T(x) \cap D$, for all $x \in D$, is also anti-cyclic for any subset D of K.

(e) If $T : K \rightrightarrows X$ is anti-cyclic, then $\overline{T} : K \rightrightarrows X$ defined by $\overline{T}(x) = \overline{T(x)}$, for all $x \in K$, is also anti-cyclic.
Indeed, for any $x_1, x_2, \dots, x_m \in K$ with $x_i \in \overline{T}(x_{i+1}) = \overline{T(x_{i+1})}$ for $i = 1, 2, \dots, m-1$, one has $x_i \notin T(x_{i+1})$ for all $i = 1, 2, \dots, m-1$. By anti-cyclicity of T, we have $x_m \in T(x_1) \subseteq \overline{T(x_1)} = \overline{T}(x_1)$.

Example 5.14 Let (X, d) be a metric space, K be a nonempty subset of X, and $f : K \to X$ be a single-valued function. Then, the set-valued map $T : K \rightrightarrows X$ defined by

$$T(x) = \{y \in X : d(x, f(x)) \geq d(y, f(y))\}, \quad \text{for all } x \in K,$$

is anti-cyclic.

Example 5.15 Let (X, d) be a metric space, K be a nonempty subset of X, and $\{S_{r_n}[x_n]\}_{n \in \mathbb{N}}$ be a sequence of closed spheres such that $x_n \in K$, $r_n > 0$ and $S_{r_1}[x_1] \supseteq S_{r_2}[x_2] \supseteq \cdots \supseteq S_{r_n}[x_n] \supseteq \cdots$. Then, the set-valued map $T : K \rightrightarrows X$ defined by

$$T(x) = \begin{cases} S_{r_n}[x_n], & \text{if } x = x_n,\ n = 1, 2, \dots, \\ K, & \text{otherwise,} \end{cases}$$

is anti-cyclic.

The following theorem on nonemptiness of an intersection of anti-cyclic maps is equivalent to Theorem 5.12 in the sense that one can be derived by using the other.

Theorem 5.13 [111] *Let K be a nonempty compact subset of a metric space (X, d) and $T : K \rightrightarrows X$ be a closed valued and anti-cyclic set-valued map. Then,*

$$\bigcap_{x \in K} T(x) \neq \emptyset.$$

Proof Theorem 5.12 ⇒ Theorem 5.13: Let $F : K \times K \to \mathbb{R}$ be defined by

$$F(x,y) = \begin{cases} 0, & \text{if } x \in T(y), \\ -1, & \text{otherwise.} \end{cases} \tag{5.43}$$

Take any $x_1, x_2, \dots, x_m \in K$ ($m \geq 2$) such that $F(x_i, x_{i+1}) < 0$ for all $i = 1, 2, \dots, m-1$. Then by (5.43), $x_i \notin T(x_{i+1})$ for all $i = 1, 2, \dots, m-1$. Since T is anti-cyclic, $x_m \in T(x_1)$. Again by (5.43), we have $F(x_m, x_1) \geq 0$. Thus, F is cyclically anti-quasimonotone. Moreover, $L_{\geq}F(\cdot, y) = T(y)$ is closed. Then by Theorem 5.12, there exists $\bar{x} \in K$ such that $\inf_{y \in K} F(\bar{x}, y) \geq 0$ which is equivalent to $\bar{x} \in \bigcap_{y \in K} T(y)$.

Theorem 5.13 ⇒ Theorem 5.12: Assume that $F : K \times K \to \mathbb{R}$ satisfies the conditions of Theorem 5.12. Define $T : K \rightrightarrows K$ by

$$T(y) = L_{\geq}F(\cdot, y) = \{x \in K : F(x,y) \geq 0\}, \quad \text{for all } y \in K.$$

Then, T is closed valued. The cyclic anti-quasi-monotonicity of F is equivalent to T being anti-cyclic. Thus, by Theorem 5.13, we have

$$\bigcap_{y \in K} T(y) = \bigcap_{y \in K} \{x \in K : F(x,y) \geq 0\} \neq \emptyset,$$

which means that there exists $\bar{x} \in K$ such that $\inf_{y \in K} F(\bar{x}, y) \geq 0$. ∎

Definition 5.6 [120] Let K be a nonempty subset of a metric space X. A set-valued map $T : K \rightrightarrows X$ is said to be *intersectionally closed*

$$\bigcap_{x \in K} \overline{T(x)} = \overline{\left(\bigcap_{x \in K} T(x) \right)}.$$

Clearly, if T is intersectionally closed and $\bigcap_{x \in K} \overline{T(x)} \neq \emptyset$, then $\bigcap_{x \in K} T(x) \neq \emptyset$.

By applying Theorem 5.13 for the set-valued map $\overline{T}$, we obtain the following result.

Corollary 5.5 *Let K be a nonempty compact subset of a metric space (X, d) and $T : K \rightrightarrows X$ be a insersectionally closed and anti-cyclic set-valued map. Then,*

$$\bigcap_{x \in K} T(x) \neq \emptyset.$$

Definition 5.7 Let K be a nonempty subset of a metric space (X, d) and $T : X \rightrightarrows K$ be a set-valued map. A point $\bar{x} \in X$ is called a *maximal element* of T if $T(\bar{x}) = \emptyset$.

Remark 5.16 Let $T : X \rightrightarrows X$ be a set-valued map. Define a binary relation $\prec_T$ on X by

$$x \prec_T u \quad \text{if and only if} \quad u \in T(x).$$

Then, $\bar{x} \in X$ is a maximal element with respect to $\prec_T$, that is, no $u \in X$ satisfies $\bar{x} \prec_T u$ if $T(\bar{x}) = \emptyset$, or, equivalently, $\bar{x} \in \bigcap_{x \in X} \left(X \setminus T^{-1}(x) \right)$.

By restricting Theorem 5.13 for the set-valued map $X \setminus T^{-1}$, we obtain the following maximal element theorem.

Theorem 5.14 [111] *Let K be a nonempty compact subset of a metric space (X, d) and $T : X \rightrightarrows K$ be a set-valued map such that the set $T^{-1}(x) = \{y \in K : y \in T(x)\}$ is open in X. If the set-valued map $X \setminus T^{-1}$ is anti-cyclic, then T has maximal element.*

Let K be a nonempty subset of a metric space X and $T : K \rightrightarrows X$ be a set-valued map. Then for all $i = 1, 2, \ldots m$, define set-valued map $T^i : K \rightrightarrows X$ by, for all $x \in K$,

$$\begin{aligned} T^1(x) &= T(x) \\ T^2(x) &= T(T(x) \cap K) \\ T^3(x) &= T(T^2(x) \cap K) \\ &\vdots \\ T^m(x) &= T(T^{m-1}(x) \cap K). \end{aligned}$$

Definition 5.8 Let K be a nonempty subset of a metric space X and $T : K \rightrightarrows X$ be a set-valued map. A element $\bar{x} \in K$ is said to be an *m-periodic point* of T if $\bar{x} \in T^m(\bar{x})$.

The smallest number m for which $\bar{x} \in T^m(\bar{x})$ is called the *period* of $\bar{x}$.

Clearly, a fixed point is a 1-periodic point, and an m-periodic point of T is a fixed point of T^m. By using Theorem 5.13, we have the following result.

Theorem 5.15 [111] *Let K be a nonempty compact subset of a metric space (X, d) and $T : K \rightrightarrows X$ be a set-valued map.*

(a) *If $K = \bigcup_{x \in K} \mathrm{int}_K T^{-1}(x)$, then T has an m-periodic point for some $m \in \mathbb{N}$.*

(b) *If $K = \bigcup_{x \in K} \mathrm{int}_K T^{-1}(x)$ and $T(y) \subseteq T(x)$ for all $x \in K$ and $y \in T(x) \cap K$, then T has a fixed point.*

Proof Define a set-valued map $\Phi : K \rightrightarrows X$ by

$$\Phi(x) = K \setminus \mathrm{int}_K T^{-1}(x), \quad \text{for all } x \in K.$$

Then, Φ is closed valued. Since $K = \bigcup_{x \in K} \mathrm{int}_K T^{-1}(x)$, we have

$$\bigcap_{x \in K} \Phi(x) = X \setminus \left(\bigcup_{x \in K} \mathrm{int}_K T^{-1}(x) \right) = \emptyset.$$

By Theorem 5.13, we conclude that Φ is not anti-cyclic. Thus, there exist $x_1, x_2, \ldots, x_m \in K$ such that $x_1 \notin \Phi(x_2)$, $x_2 \notin \Phi(x_3)$, ..., $x_{m-1} \notin \Phi(x_m)$, and $x_m \notin \Phi(x_1)$. It follows that $x_1 \notin \mathrm{int}_K \Phi(x_2) \subseteq \Phi(x_2)$, $x_2 \notin \mathrm{int}_K \Phi(x_3) \subseteq \Phi(x_3)$, ..., $x_{m-1} \notin \mathrm{int}_K \Phi(x_m) \subseteq \Phi(x_m)$, and $x_m \notin \mathrm{int}_K \Phi(x_1) \subseteq \Phi(x_1)$. Hence,

$$x_2 \in T(x_1),\ x_3 \in T(x_2),\ \ldots,\ x_m \in T(x_{m-1}),\ \text{and } x_1 \in T(x_m), \tag{5.44}$$

that is,

$$(x_i, x_{i+1}) \in \mathrm{Graph}(T), \quad \text{for all } i = 1, 2, \ldots, m, \text{ where } x_{m+1} := x_1. \tag{5.45}$$

(a) From (5.44), we deduce that $x_i \in T^m(x_i)$ for all $i = 1, 2, \ldots, m$.

(b) From (5.44) and the hypothesis, we have

$$T(x_1) \subseteq T(x_m) \subseteq T(x_{m-1}) \subseteq \cdots \subseteq T(x_2) \subseteq T(x_1).$$

Hence, $T(x_1) = T(x_2) = \cdots = T(x_m)$, and so each x_i, $i = 1, 2, \ldots, m$, is a fixed point of T. ■

Remark 5.17 The condition "$K = \bigcup_{x\in K} \text{int}_K T^{-1}(x)$" is satisfied if for each $x \in K$, $T(x) \cap K \neq \emptyset$ and $T^{-1}(x)$ is open in K.

Example 5.16 [111] Let $K = [0, 1]$ and $T : [0, 1] \rightrightarrows \mathbb{R}$ be a set-valued map defined by $T(x) = \mathbb{Q}$ for all $x \in [0, 1]$, where $\mathbb{Q}$ denotes the set of all rational numbers. Then,

$$T^{-1}(x) = \begin{cases} [0, 1], & \text{if } x \in \mathbb{Q} \cap [0, 1], \\ \emptyset, & \text{otherwise,} \end{cases}$$

and

$$\bigcup_{x\in K} \text{int}_K T^{-1}(x) \supseteq \text{int}_K T^{-1}(1/2) = [0, 1] = K.$$

Let $y \in T(x) = \mathbb{Q}$, then $T(y) = T(x)$. Hence, by Theorem 5.15 (b), T has a fixed point. In fact, all points $x \in \mathbb{Q} \cap [0, 1]$ are fixed points of T.

Since $\bar{x}$ is a fixed point of T if and only if $\bar{x}$ is also a fixed point of T^{-1}, from Theorem 5.15 (b), we obtain the following result.

Corollary 5.6 *Let K be a nonempty compact subset of a metric space (X, d) and $T : X \rightrightarrows K$ be a set-valued map. Assume that*

(i) $K = \bigcup_{x\in K} \text{int}_K T(x)$;

(ii) *For all $x \in K$ and $y \in T(x) \cap K$, $T^{-1}(x) \subseteq T^{-1}(y)$.*

Then, T has a fixed point.

Proof Define a set-valued map $S : K \rightrightarrows K$ by $S(x) = T^{-1}(x)$ for all $x \in K$. Then, S satisfies all the conditions of Theorem 5.15 (b). Hence, there exists $\bar{x} \in K$ such that $\bar{x} \in S(\bar{x}) = T^{-1}(\bar{x})$, that is, $\bar{x} \in T(\bar{x})$. ■

5.7 Some Equivalent Results to Extended Ekeland's Variational Principle

The following result provides the equivalence among extended Ekelend's variational principle, extended Takahashi's minimization theorem, Caristi–Kirk fixed point theorem for set-valued maps, and Oettli–Théra theorem.

Theorem 5.16 *Let (X, d) be a complete metric space and K be a nonempty closed subset of X. Let $F : K \times K \to \mathbb{R}$ be a bifunction such that it is lower semicontinuous in the second argument and satisfies the following conditions:*

(i) *For all $x \in K$, $F(x,x) = 0$;*
(ii) *For all $x, y, z \in K$, $F(x,y) \leq F(x,z) + F(z,y)$.*

Assume that there exists $\hat{x} \in K$ such that $\inf_{x \in K} F(\hat{x}, x) > -\infty$. Let

$$\hat{S} := \{x \in K : F(\hat{x}, x) + d(\hat{x}, x) \leq 0\} \tag{5.46}$$

(From (i) it follows that $\hat{x} \in \hat{S} \neq \emptyset$). Then, the following statements are equivalent:

(a) (Extended Ekeland's Variational Principle). *There exists $\bar{x} \in \hat{S}$ such that*

$$F(\bar{x}, x) + d(\bar{x}, x) > 0, \quad \text{for all } x \neq \bar{x}. \tag{5.47}$$

(b) (Extended Takahashi's Minimization Theorem). *Assume that*

$$\begin{cases} \text{for every } \hat{x} \in \hat{S} \text{ with } \inf_{x \in K} F(\hat{x}, x) < 0, \text{ there exists} \\ x \in K \text{ such that } F(\hat{x}, x) + d(\hat{x}, x) \leq 0, \quad \text{for all } x \neq \hat{x}. \end{cases} \tag{5.48}$$

Then, the equilibrium problem (5.1) *has a solution.*

(c) (Caristi–Kirk Fixed Point Theorem). *Let $T : K \rightrightarrows K$ be a set-valued map such that*

$$\begin{cases} \text{for every } \hat{x} \in \hat{S}, \text{ there exists} \\ x \in T(\hat{x}) \text{ satisfying } F(\hat{x}, x) + d(\hat{x}, x) \leq 0. \end{cases} \tag{5.49}$$

Then, there exists $\bar{x} \in \hat{S}$ such that $\bar{x} \in T(\bar{x})$.

(d) (Oettli–Théra Theorem). *Let $D \subset X$ have the property that*

$$\begin{cases} \text{for every } \hat{x} \in \hat{S} \setminus D, \quad \text{there exists } x \in K \text{ such that} \\ F(\hat{x}, x) + d(\hat{x}, x) \leq 0, \quad \text{for all } x \neq \hat{x}. \end{cases} \tag{5.50}$$

Then, there exists $\bar{x} \in \hat{S} \cap D$.

Proof (a) $\Rightarrow$ (d): Let (a) and the hypothesis of (d) hold. Then, (a) gives $\bar{x} \in \hat{S}$ such that $F(\bar{x}, x) + d(\bar{x}, x) > 0$, for all $x \neq \bar{x}$. From (5.50), we have $\bar{x} \in D$. Hence, $\bar{x} \in \hat{S} \cap D$ and (d) holds.
(d) $\Rightarrow$ (a): Let (d) hold. For all $\hat{x} \in K$, define

$$\Gamma(\hat{x}) = \{x \in K : F(\hat{x}, x) + d(\hat{x}, x) \leq 0,\ x \neq \hat{x}\}.$$

Choose $D := \{\hat{x} \in K : \Gamma(\hat{x}) = \emptyset\}$. If $\hat{x} \notin D$, then from the definition of D, there exists $x \in \Gamma(\hat{x})$. Hence, (5.50) is satisfied, and by (d), there exists $\bar{x} \in \hat{S} \cap D$. Then, $\Gamma(\bar{x}) = \emptyset$, that is, $F(\hat{x}, x) + d(\hat{x}, x) > 0$ for all $x \neq \bar{x}$. Hence, (a) holds.
(b) $\Rightarrow$ (d): Suppose that both (b) and the hypothesis of (d) hold. Assume the contrary that $\hat{x} \notin D$ for all $\hat{x} \in \hat{S}$. Then, by (5.50), for all $\hat{x} \in \hat{S}$

$$\text{there exists } x \neq \hat{x} \text{ with } F(\hat{x}, x) + d(\hat{x}, x) \leq 0. \tag{5.51}$$

Hence, (5.49) is satisfied. By (b), there exists $\bar{x} \in \hat{S}$ such that $F(\bar{x}, x) \geq 0$, for all $x \in K$. This implies that $F(\bar{x}, x) + d(\bar{x}, x) > 0$, for all $x \in K$, $x \neq \bar{x}$, a contradiction with (5.51). Hence, $\hat{x} \in D$ for some $\hat{x} \in \hat{S}$ and (d) holds.

(d) ⇒ (b): Suppose that both (d) and the hypothesis of (b) hold. Choose $D := \{\hat{x} \in K : \inf_{x \in K} F(\hat{x}, x) \geq 0\}$. Then, (5.50) follows from (5.48), and (d) furnishes some $\bar{x} \in \hat{S} \cap D$. It follows from the definition of D that $\inf_{x \in K} F(\bar{x}, x) \geq 0$. Hence, (b) holds.

(c) ⇒ (d): Let (c) and the hypothesis of (d) hold. Define a set-valued map $T : K \rightrightarrows K$ by

$$T(\hat{x}) = \{x \in K : x \neq \hat{x}\}.$$

Assume the contrary that $\hat{x} \notin D$ for all $\hat{x} \in \hat{S}$. Then, (5.49) follows from (5.50), and by (c) there exists $\bar{x} \in T(\bar{x})$. But this is clearly impossible from the definition of T. Hence, $\hat{x} \in D$ for some $\hat{x} \in \hat{S}$ and (d) holds.

(d) ⇒ (c): Suppose that both (d) and the hypothesis of (c) hold. Choose $D := \{\hat{x} \in K : \hat{x} \in T(\hat{x})\}$. Then, (5.50) follows from (5.49), and (d) furnishes some $\bar{x} \in \hat{S} \cap D$ which, from the definition of D, necessarily belongs to $T(\bar{x})$. Hence, (c) holds. ■

5.8 Weak Sharp Solutions for Equilibrium Problems

In 1979, Polyak [143] (see also [144]) introduced the concept of a sharp minimum, also known as a strong isolated minimum or a strong unique local minimum, for real-valued functions under the assumption that an optimization problem has a unique solution. It became an important tool in analysis of the perturbation behavior of certain classes of optimization problems as well as in the convergence analysis of algorithms designed to solve these problems. As a generalization of sharp minima, Ferris [82] (see also [51]) introduced and studied the weak sharp minima for real-valued functions to include the possibility of non-unique solutions. During the last two decades the study of weak sharp minima has drawn much attention motivated by its importance in the treatment of sensitivity analysis, error bounds, and convergence analysis for a wide range of optimization algorithms; see, for example, [51, 48–50, 82] and the references therein. By using Takahashi's minimization theorem [168], Daffer et al. [66] and Hamel [91], separately, studied the weak sharp minima for a class of lower semicontinuous real-valued functions in the setting of metric spaces.

In 1998, Marcotte and Zhu [124] introduced the notion of weak sharp solutions of variational inequality problems. They derived the necessary and sufficient condition for a solution set to be weakly sharp. They also studied the finite convergence of iterative algorithms for solving variational inequalities whose solution set is weakly sharp. Zhou and Wang [184] re-examined the unified treatment of finite termination of a class of iterative algorithms, and showed that some results given by Marcotte and Zhu [124] remain intact even if some conditions are relaxed. Wu and Wu [176] presented several equivalent (and sufficient) conditions for weak sharp solutions of variational inequalities in the setting of Hilbert spaces. They gave a finite convergence result for a class of algorithms for solving variational inequalities. By using the dual gap function, Zhang et al. [180] characterized the directional derivative and subdifferential of the dual gap function. Based on these, they proposed a better understanding of the concepts of a global error bound, weak sharpness, and minimum principle sufficiency property for variational inequalities, where the operator involved is pseudo-monotone. Hu and Song [94] extended the concept of weak sharp solutions for variational inequalities from finite dimensional spaces/Hilbert spaces to reflexive, strictly convex, and smooth

Banach spaces. They presented its equivalent characterizations and established finite convergence of proximal point algorithm for variational inequalities in terms of the weak sharpness of the solution set.

In 2015, Al-Homidan et al. [3] studied the weak sharp solutions for equilibrium problems. By using the extended form of Takahashi's minimization theorem and a gap function [125], they studied the weak sharp solutions for equilibrium problems in the setting of metric spaces. In this section, we present the results established in [3].

Definition 5.9 Let K be a nonempty set. A function $g : K \to \mathbb{R}$ is said to be a *gap function* for equilibrium problem (5.1) if

(a) $g(x) \geq 0$ for all $x \in K$;

(b) $g(x) = 0$ and $x \in K$ if and only if $x \in K$ is a solution of the equilibrium problem (5.1).

Mastroeni [125] studied the gap function for the equilibrium problem (5.1). He observed that the function

$$g(x) := \sup_{y \in K}[-F(x, y)] \tag{5.52}$$

is a gap function for equilibrium problem (5.1).

We say that the equilibrium problem has weak sharp solutions if

$$d(x, \mathbb{S}) \leq g(x), \quad \text{for all } x \in K, \tag{5.53}$$

where $\mathbb{S}$ is the set of solutions of the equilibrium problem (5.1) and $d(x, \mathbb{S}) = \inf_{\bar{x} \in \mathbb{S}} d(x, \bar{x})$.

Theorem 5.17 *Let K be a nonempty closed subset of a complete metric space (X, d), $F : K \times K \to \mathbb{R}$ be lower semicontinuous in the second argument and satisfy the following conditions:*

(i) $F(x, x) = 0$ *for all* $x \in K$;

(ii) $F(x, y) \leq F(x, z) + F(z, y)$ *for all* $x, y, z \in K$;

(iii) *There exists* $\hat{x} \in K$ *such that* $\inf_{y \in K} F(\hat{x}, y) > -\infty$.

Assume that for every $x \in K$ with $\inf_{y \in K} F(x, y) < 0$, there exists $y \in K$, $y \neq x$ such that $F(x, y) + d(x, y) \leq 0$. Then, the equilibrium problem (5.1) *has weak sharp solutions.*

Proof For all $x \in K$, define

$$S(x) = \{y \in K : F(x, y) + d(x, y) \leq 0\}.$$

Then, by lower semicontinuity of F in the second argument, $S(x)$ is closed for all $x \in K$. By Theorem 5.7, $\mathbb{S}$ is nonempty. Clearly, $S(x) \neq \emptyset$ as $x \in S(x)$.

For all $y \in S(x)$, $F(x, y) \leq 0$. Indeed, for all $y \in S(x)$, we have

$$F(x, y) + d(x, y) \leq 0 \Leftrightarrow 0 \leq d(x, y) \leq -F(x, y) \Leftrightarrow F(x, y) \leq 0.$$

Suppose to the contrary that there exists $x_0 \in K$ such that

$$d(x_0, \mathbb{S}) > g(x_0). \tag{5.54}$$

Then, $x_0 \notin \mathbb{S}$. Indeed, if $x_0 \in \mathbb{S}$, then $d(x_0, \mathbb{S}) = \inf_{y \in \mathbb{S}} d(x_0, y) = 0$, and so $g(x_0) < 0$ which contradicts the fact that $g(x) \geq 0$ for all $x \in K$ because g is a gap function.

For all $y \in S(x_0)$, $d(y, \mathbb{S}) > g(y)$. Indeed, take $y \in S(x_0)$ and $z \in \mathbb{S}$, then $d(x_0, y) \leq -F(x_0, y)$. Therefore,

$$d(x_0, z) \leq d(x_0, y) + d(y, z) \leq -F(x_0, y) + d(y, z),$$

that is, $d(x_0, z) \leq d(y, z) - F(x_0, y)$. Taking inf over $\mathbb{S}$ on both the sides, we obtain

$$\inf_{z \in \mathbb{S}} d(x_0, z) \leq \inf_{z \in \mathbb{S}} d(y, z) - F(x_0, y),$$

that is, $d(x_0, \mathbb{S}) \leq d(y, \mathbb{S}) - F(x_0, y)$. By (5.54), we have

$$g(x_0) < d(y, \mathbb{S}) - F(x_0, y). \tag{5.55}$$

By condition (ii), for all $v \in K$, we have

$$F(x_0, v) \leq F(x_0, y) + F(y, v) \quad \Leftrightarrow \quad -F(y, v) \leq -F(x_0, v) + F(x_0, y).$$

Taking sup over K both the sides, we get

$$\sup_{v \in K}[-F(y, v)] \leq \sup_{v \in K}[-F(x_0, v)] + F(x_0, y).$$

This implies that

$$g(y) \leq g(x_0) + F(x_0, y). \tag{5.56}$$

Combining (5.55) and (5.56), we obtain $g(y) < d(y, \mathbb{S})$ for all $y \in S(x_0)$.

Since $x_0 \notin \mathbb{S}$, there exists $y \in K$ such that $F(x_0, y) < 0$, and so $\inf_{y \in K} F(x_0, y) < 0$. By hypothesis, there exists $x_1 \in K$ such that $x_1 \neq x_0$ and $F(x_0, x_1) + d(x_0, x_1) \leq 0$, that is, $x_1 \in S(x_0)$ with $x_1 \neq x_0$. Since $g(x_1) < d(x_1, \mathbb{S})$, then clearly $x_1 \notin \mathbb{S}$ and $F(x_0, x_1) < 0$ because $-F(x_0, x_1) \geq d(x_0, x_1) > 0$ since $x_0 \neq x_1$. We can again show as above that $g(y) < d(y, \mathbb{S})$ for all $y \in S(x_1)$ and $S(x_1) \cap \mathbb{S} = \emptyset$. In addition, we choose x_1 such that

$$F(x_0, x_1) = \inf\{F(x_0, x) : x \in S(x_0)\},$$

where inf exists since K is a closed subset of a complete metric space X, $S(x_0)$ is closed and F is lower semicontinuous in the second argument. Continuing in this way, we generate a sequence $\{x_n\}$ with the following properties:

- there exists $x_i \neq x_{i-1}$ for all $i = 1, 2, \ldots, n$.
- $x_i \in S(x_{i-1})$ for all $i = 1, 2, \ldots, n$.
- $F(x_{i-1}, x_i) < 0$ for all $i = 1, 2, \ldots, n$. Indeed, since $x_i \in S(x_{i-1})$, we have $d(x_{i-1}, x_i) + F(x_{i-1}, x_i) \leq 0$. This implies that $-F(x_{i-1}, x_i) \geq d(x_{i-1}, x_i) > 0$ as $x_{i-1} \neq x_i$. Thus, $F(x_{i-1}, x_i) < 0$.
- $F(x_{i-1}, x_i) = \inf\{F(x_{i-1}, x) : x \in S(x_{i-1})\}$ for all $i = 1, 2, \ldots, n$.
- $S(x_i) \cap \mathbb{S} = \emptyset$ for all $i = 1, 2, \ldots, n$.
- $g(y) < d(y, \mathbb{S})$ for all $y \in \bigcup_{i=1}^{n} S(x_i)$.

Since $x_n \notin \mathbb{S}$, we can choose $x_{n+1} \in S(x_n)$, $x_{n+1} \neq x_n$ with $F(x_n, x_{n+1}) = \inf\{F(x_n, x) : x \in S(x_n)\}$. As above, we also have

$$x_{n+1} \notin \mathbb{S}, \quad F(x_n, x_{n+1}) < 0 \text{ and} \tag{5.57}$$

$$g(y) < d(y, \mathbb{S}), \quad \text{for all } y \in S(x_{n+1}). \tag{5.58}$$

To see this, let $y \in S(x_{n+1})$ and $\bar{x} \in \mathbb{S}$. Then,

$$d(x_{n+1}, \bar{x}) \leq d(x_{n+1}, y) + d(y, \bar{x}) \leq -F(x_{n+1}, y) + d(y, \bar{x}).$$

Taking sup over $\mathbb{S}$, we obtain

$$d(x_{n+1}, \mathbb{S}) \leq -F(x_{n+1}, y) + d(y, \mathbb{S}).$$

Since $-F(x_{n+1}, y) < g(x_{n+1})$, we have $d(y, \mathbb{S}) - F(x_{n+1}, y) > -F(x_{n+1}, x)$, and therefore,

$$\begin{aligned} d(y, \mathbb{S}) &> F(x_{n+1}, y) - F(x_{n+1}, x) \\ &\geq F(x_{n+1}, y) - F(x_{n+1}, y) - F(y, x). \end{aligned}$$

Thus, $d(y, \mathbb{S}) > g(y)$. Since as above

$$\begin{aligned} d(x_{n+1}, \mathbb{S}) &> F(x_n, x_{n+1}) - F(x_n, x) \\ &\geq F(x_n, x_{n+1}) - F(x_n, x_{n+1}) - F(x_n, x), \end{aligned}$$

we have $d(x_{n+1}, \mathbb{S}) > \sup_{x \in K}[-F(x_n, x)] = g(x_{n+1})$, and hence, $S(x_{n+1}) \cap \mathbb{S} = \emptyset$.

So, the sequence $\{x_n\}$ consisting different elements and $F(x_n, x_{n+1}) < 0$. Since

$$d(x_{n+k}, x_n) \leq \sum_{i=1}^{k} d(x_{n+i}, x_{n+i-1}) \leq \sum_{i=1}^{k} -F(x_{n+i}, x_{n+i-1}) \leq F(x_n, x_{n+k}),$$

and $F(x_n, x_{n+k})$ monotonically decreasing to some point, $\{x_n\}$ is a Cauchy sequence in a closed subset K of a complete metric space X, so we can assume that x_n converges to some point $x \in K$. We show that $x \in \bigcap_{i=0}^{\infty} S(x_i)$. To prove it, we show that for every n, $x_n \in \bigcap_{i=0}^{n-1} S(x_i)$.

Since

$$\begin{aligned} d(x_{n-k}, x_n) &\leq \sum_{j=0}^{k-1} d(x_{n-k+j}, x_{n-k+j+1}) \\ &\leq \sum_{j=0}^{k-1} -F(x_{n-k+j+1}, x_{n-k+j}) \\ &= -F(x_{n-k}, x_n), \end{aligned}$$

we have $x_n \in S(x_{n-k})$ for all $k = 1, 2, \ldots n$ (recall that $x_i \notin \mathbb{S}$). Therefore, $x_n \in \bigcap_{i=0}^{n-1} S(x_i)$, and hence, $x_k \in \bigcap_{i=0}^{n-1} S(x_i)$ for all $k \geq n$. Since $\bigcap_{i=0}^{n-1} S(x_i)$ is a closed set, $x \in \bigcap_{i=0}^{\infty} S(x_i)$. Thus, $x \in S(x_n)$, and $x \neq x_n$, and therefore, $F(x_n, x) < -d(x_n, x) < 0$ which contradicts the fact that $F(x_n, y) \geq 0$ for all $y \in S(x_n)$. ■

Inspired by Hamel [91], we have the following the alternative proof of Theorem 5.17.

Alternative Proof of Theorem 5.17 By Lemma 5.1, for each $x \in K \setminus \mathbb{S}$, we find $z \in S(x) \cap \mathbb{S}$ (depending on x). Then, $F(x, z) + d(x, z) \leq 0$. Since $d(x, \mathbb{S}) \leq d(x, z)$, we have

$$F(x, z) + d(x, \mathbb{S}) \leq F(x, z) + d(x, z) \leq 0, \quad \text{for this } z \in S(x) \cap \mathbb{S}.$$

Since for each $x \in K \setminus \mathbb{S}$, we find $z \in S(x) \cap \mathbb{S}$, and therefore, we have

$$F(x, z) + d(x, \mathbb{S}) \leq 0, \quad \text{for all } x \in K \text{ and all } z \in \mathbb{S}.$$

Thus, $d(x, \mathbb{S}) \leq \sup_{z \in \mathbb{S}} \left[-F(x, z)\right] = g(x)$ for all $x \in K$. Hence, the equilibrium problem (5.1) has weak sharp solutions. ■

Remark 5.18 The first proof of Theorem 5.17 is constructive and does not dependent on Lemma 5.1. While the alternative proof of Theorem 5.17 is analytical and based on Lemma 5.1.

Chapter 6

Some Applications of Fixed Point Theory

The fixed point theory has numerous applications within mathematics and also in the diverse fields as biology, chemistry, economics, engineering, game theory, management, social sciences, etc. The Banach contraction principle is one of the most widely applicable fixed point theorems in all of analysis.

In this chapter, we focus on applications of the Banach contraction principle and its variants to the following problems:

- System of linear equations
- Differential equations and second-order two-point boundary-value problems
- Linear and nonlinear Volterra integral equations, Fredholm integral equations, and mixed Volterra–Fredholm integral equations

6.1 Application to System of Linear Equations

In this section, we present an application of the Banach contraction principle to the system of linear equations (2.1).

As we have seen in Example 2.3, the system of linear equations (2.1) can be reformulated as follows:

$$\mathbf{x} = (\mathbf{I} - \mathbf{A})\mathbf{x} + \mathbf{b}, \tag{6.1}$$

where

$$\mathbf{x} = \begin{pmatrix} x_1 \\ x_2 \\ \vdots \\ x_n \end{pmatrix}, \quad \mathbf{b} = \begin{pmatrix} b_1 \\ b_2 \\ \vdots \\ b_n \end{pmatrix}, \quad \mathbf{A} = \begin{pmatrix} a_{11} & a_{12} & \cdots & a_{1n} \\ a_{21} & a_{22} & \cdots & a_{2n} \\ \cdots & \cdots & \cdots & \cdots \\ \cdots & \cdots & \cdots & \cdots \\ \cdots & \cdots & \cdots & \cdots \\ a_{n1} & a_{n2} & \cdots & a_{nn} \end{pmatrix},$$

and $\mathbf{I}$ is the identity matrix.

If $T : \mathbb{R}^n \to \mathbb{R}^n$ is a matrix transformation defined by

$$T(\mathbf{x}) = (\mathbf{I} - \mathbf{A})\mathbf{x} + \mathbf{b}, \tag{6.2}$$

then finding a solution of the system (6.1) is equivalent to find a fixed point of T. So, if T, defined by (6.2), can be proved to be a contraction mapping, then one can use the Banach contraction principle to obtain a unique fixed point of T, and hence, a unique solution of the system (6.1).

The conditions under which T is a contraction mapping depend on the choice of the metric on $X = \mathbb{R}^n$. Here we discuss only one case and have left two others for exercise.

Theorem 6.1 *Let $X = \mathbb{R}^n$ be a metric space with the metric $d_\infty(\boldsymbol{x},\boldsymbol{y}) = \max_{1\le i\le n} |x_i - y_i|$, where $\boldsymbol{x}$ and $\boldsymbol{y}$ are vectors in $\mathbb{R}^n$. If*

$$\max_{1\le i\le n} \sum_{j=1}^{n} |\alpha_{ij}| \le \alpha < 1, \quad \text{where } \alpha_{ij} = -a_{ij} + \delta_{ij}, \text{ and } \delta_{ij} = \begin{cases} 1, & \text{for } i = j, \\ 0, & \text{for } i \ne j, \end{cases} \tag{6.3}$$

then the linear system (6.1) *of n linear equations in n unknowns has a unique solution.*

Proof Since $X = \mathbb{R}^n$ with respect to the metric d_∞ is complete, it is sufficient to prove that the mapping T defined by (6.2) is a contraction. Indeed,

$$\begin{aligned}
d_\infty(T(\mathbf{x}), T(\mathbf{y})) &= \max_{1\le i\le n} \left| \sum_{j=1}^{n} \alpha_{ij}\left(x_j - y_j\right) \right| \\
&\le \max_{1\le i\le n} \sum_{j=1}^{n} |\alpha_{ij}| \; |x_j - y_j| \\
&\le \max_{1\le i\le n} \left(\max_{1\le j\le n} |x_j - y_j| \right) \sum_{j=1}^{n} |\alpha_{ij}| \\
&= \max_{1\le i\le n} \sum_{j=1}^{n} |\alpha_{ij}| \, d_\infty(\mathbf{x}, \mathbf{y}) \\
&\le \alpha d_\infty(\mathbf{x}, \mathbf{y}).
\end{aligned}$$

Thus, T is a contraction mapping. By Banach contraction principle (Theorem 2.1), the linear systems (6.1) has a unique solution. ∎

Example 6.1 Consider the following system of linear equations:

$$\left.\begin{aligned}
\frac{1}{2}x_1 + \frac{1}{4}x_3 &= 1 \\
\frac{1}{2}x_2 + \frac{1}{4}x_4 &= 2 \\
\frac{1}{4}x_1 + \frac{1}{2}x_3 &= 1 \\
\frac{1}{4}x_2 + \frac{1}{2}x_4 &= 1.
\end{aligned}\right\} \tag{6.4}$$

With

$$\mathbf{A} = \begin{pmatrix} 1/2 & 0 & 1/4 & 0 \\ 0 & 1/2 & 0 & 1/4 \\ 1/4 & 0 & 1/2 & 0 \\ 0 & 1/4 & 0 & 1/2 \end{pmatrix}, \quad \mathbf{x} = \begin{pmatrix} x_1 \\ x_2 \\ x_3 \\ x_4 \end{pmatrix} \quad \text{and} \quad \mathbf{b} = \begin{pmatrix} 1 \\ 2 \\ 1 \\ 1 \end{pmatrix},$$

the system of linear equations (6.4) can be written as

$$\mathbf{x} = (\mathbf{I} - \mathbf{A})\mathbf{x} + \mathbf{b}. \tag{6.5}$$

Note that

$$\mathbf{I} - \mathbf{A} = \begin{pmatrix} 1 & 0 & 0 & 0 \\ 0 & 1 & 0 & 0 \\ 0 & 0 & 1 & 0 \\ 0 & 0 & 0 & 1 \end{pmatrix} - \begin{pmatrix} 1/2 & 0 & 1/4 & 0 \\ 0 & 1/2 & 0 & 1/4 \\ 1/4 & 0 & 1/2 & 0 \\ 0 & 1/4 & 0 & 1/2 \end{pmatrix} = \begin{pmatrix} \frac{1}{2} & 0 & -\frac{1}{4} & 0 \\ 0 & \frac{1}{2} & 0 & -\frac{1}{4} \\ -\frac{1}{4} & 0 & \frac{1}{2} & 0 \\ 0 & -\frac{1}{4} & 0 & \frac{1}{2} \end{pmatrix}.$$

Here,

$$\begin{aligned} &\max_{1 \le i \le 4} \sum_{j=1}^{4} |\alpha_{ij}| \\ &= \max\{|a_{11}| + |a_{12}| + |a_{13}| + |a_{14}|, |a_{21}| + |a_{22}| + |a_{23}| + |a_{24}|, \dots, \\ &\quad \dots, |a_{41}| + |a_{42}| + |a_{43}| + |a_{44}|\} \\ &= \max\left\{\frac{1}{2} + \frac{1}{4}, \frac{1}{2} + \frac{1}{4}, \frac{1}{2} + \frac{1}{4}, \frac{1}{2} + \frac{1}{4}\right\} \\ &= \frac{3}{4}. \end{aligned}$$

Thus, the matrix transformation $T : \mathbb{R}^4 \to \mathbb{R}^4$ defined by

$$T(\mathbf{x}) = (\mathbf{I} - \mathbf{A})\mathbf{x} + \mathbf{b}$$

is a $\frac{3}{4}$-contraction under the metric d_∞. Therefore, the system of linear equations given by (6.4) has a unique solution.

Exercise 6.1 Let $X = \mathbb{R}^n$ be a metric space with the metric $d_1(\mathbf{x}, \mathbf{y}) = \sum_{i=1}^{n} |x_i - y_i|$. If

$$\max_{1 \le j \le n} \sum_{i=1}^{n} |\alpha_{ij}| \le \alpha < 1, \quad \text{where } \alpha_{ij} = -a_{ij} + \delta_{ij}, \tag{6.6}$$

then prove that the linear system (6.1) of n linear equations in n unknowns has a unique solution.

Hint:

$$\begin{aligned}
d_1(T(\mathbf{x}), T(\mathbf{y})) &= \sum_{i=1}^{n} \left| \sum_{j=1}^{n} \alpha_{ij} \left(x_j - y_j \right) \right| \\
&\leq \sum_{i=1}^{n} \sum_{j=1}^{n} \left| \alpha_{ij} \right| \, \left| x_j - y_j \right| \\
&= \sum_{j=1}^{n} \sum_{i=1}^{n} \left| \alpha_{ij} \right| \, \left| x_j - y_j \right| \\
&\leq \max_{1 \leq j \leq n} \sum_{i=1}^{n} \left| \alpha_{ij} \right| d_1(\mathbf{x}, \mathbf{y}) \\
&\leq \alpha d_1(\mathbf{x}, \mathbf{y}).
\end{aligned}$$

Exercise 6.2 Let $X = \mathbb{R}^n$ be a metric space with the metric $d_2(\mathbf{x}, \mathbf{y}) = \left(\sum_{i=1}^{n} |x_i - y_i|^2 \right)^{1/2}$. If

$$\sum_{i=1}^{n} \sum_{j=1}^{n} \left| \alpha_{ij} \right|^2 \leq \alpha^2 < 1, \quad \text{where } \alpha_{ij} = -a_{ij} + \delta_{ij}, \tag{6.7}$$

then prove that the linear system (6.1) of n linear equations in n unknowns has a unique solution.
Hint:

$$\begin{aligned}
\left[d_2(T(\mathbf{x}), T(\mathbf{y})) \right]^2 &= \sum_{i=1}^{n} \left| \sum_{j=1}^{n} \alpha_{ij} \left(x_j - y_j \right) \right|^2 \\
&\leq \sum_{i=1}^{n} \left(\sum_{j=1}^{n} \left| \alpha_{ij} \right| \, \left| x_j - y_j \right| \right)^2 \\
&\leq \sum_{i=1}^{n} \left(\sum_{j=1}^{n} \left| \alpha_{ij} \right|^2 \sum_{j=1}^{n} \left| x_j - y_j \right|^2 \right).
\end{aligned}$$

So,

$$\left[d_2(T(\mathbf{x}), T(\mathbf{y})) \right]^2 \leq \sum_{i=1}^{n} \sum_{j=1}^{n} \left| \alpha_{ij} \right|^2 \left(d_2(\mathbf{x}, \mathbf{y}) \right)^2 \leq \alpha^2 \left(d_2(\mathbf{x}, \mathbf{y}) \right)^2.$$

6.2 Applications to Differential Equations

We give an application of Banach contraction principle (Theorem 2.1) to prove the existence of a unique solution of the following ordinary differential equation with an initial condition:

$$y' = f(x, y), \quad y(x_0) = y_0. \tag{6.8}$$

Here we use $y' = \frac{dy}{dx}$.

Theorem 6.2 (Picard Theorem) *Let $f(x,y)$ be a real-valued continuous function of two variables defined on a rectangle $A = \{(x,y) : a \le x \le b, c \le y \le d\}$ such that the following Lipschitz condition in the second variable holds: there exists a constant $q > 0$ such that*

$$|f(x,y) - f(x,\hat{y})| \le q\,|y - \hat{y}|, \quad \text{for all } y, \hat{y} \in [c,d]. \tag{6.9}$$

Further, let (x_0, y_0) be an interior point of A. Then, the differential equation (6.8) with the given initial condition has a unique solution.

Proof First of all, we show that the problem of determining the solution of differential equation (6.8) is equivalent to find a solution of an integral equation. If $y = g(x)$ satisfies the differential equation (6.8) and has the property that $g(x_0) = y_0$, then by integrating differential equation (6.8) from x_0 to x, we obtain

$$\left.\begin{aligned} g(x) - g(x_0) &= \int_{x_0}^{x} f(t, g(t))dt \\ g(x) &= y_0 + \int_{x_0}^{x} f(t, g(t))dt. \end{aligned}\right\} \tag{6.10}$$

Thus, a unique solution of the differential equation (6.8) with the given initial condition is equivalent to a unique solution of (6.10). We apply Banach contraction principle (Theorem 2.1) to determine the solution of (6.10).

Since $f(x,y)$ is continuous on a compact subset A of $\mathbb{R}^2$, it is bounded and so there exists a positive constant m such that $|f(x,y)| \le m$ for all $(x,y) \in A$.

Choose a positive constant p such that $pq < 1$ and the rectangle $B = \{(x,y) : -p + x_0 \le x \le p + x_0, -pm + y_0 \le y \le pm + y_0\}$ is contained in A.

Let X be the set of all real-valued continuous functions $y = g(x)$ defined on $[-p + x_0, p + x_0]$ such that $d(g(x), y_0) \le mp$. The set X is a closed subset of the metric space $C[x_0 - p, x_0 + p]$ with the sup metric. Since $C[x_0 - p, x_0 + p]$ is complete, X is complete.

Consider the mapping $T : X \to X$ defined by

$$T(g) = h, \quad \text{where } h(x) = y_0 + \int_{x_0}^{x} f(t, g(t))dt.$$

Since

$$d(h(x), y_0) = \sup\left|\int_{x_0}^{x} f(t, g(t))dt\right| \le m\,(x - x_0) \le m\,p,$$

we have $h(x) \in X$, and so, T is well defined. For all $g, g_1 \in X$, we have

$$\begin{aligned} d(T(g), T(g_1)) = d(h, h_1) &= \sup\left|\int_{x_0}^{x} \left[f(t, g(t)) - f(t, g_1(t))\right] dt\right| \\ &\le \int_{x_0}^{x} \sup\left|f(t, g(t)) - f(t, g_1(t))\right| dt \\ &\le q \int_{x_0}^{x} |g(t) - g_1(t)|\, dt \\ &\le q\,p\,d(g, g_1), \end{aligned}$$

that is,

$$d(T(g), T(g_1)) \leq \alpha d(g, g_1),$$

where $0 \leq \alpha = qp < 1$.

Hence T is a contraction mapping on X into itself. By Banach contraction principle (Theorem 2.1), T has a unique fixed point $g^* \in X$. This unique fixed point g^* is the unique solution of the differential equation (6.8) and satisfies the given initial condition. ∎

6.2.1 An Application to Delay Differential Equations

The ordinary differential equations in which the derivative of the unknown function at a certain time is given in terms of the values of the function at previous times are called delay differential equations, also known as time-delay systems. Many practical problems concerning heat flow, species interaction microbiology, neural networks, etc., can be formulated in the form of the delay differential equations.

In this subsection, we study the following *delay differential equation*:

$$x'(t) = f(t, x(t), x(t-\tau)), \quad t \in [t_0, b], \tag{6.11}$$

with the initial condition:

$$x(t) = \phi(t), \quad t \in [t_0 - \tau, t_0], \tag{6.12}$$

where

(C1) $t_0, b \in \mathbb{R}$ and $\tau \in (0, \infty)$;
(C2) $f \in C([t_0, b] \times \mathbb{R}^2)$, that is, f is a real-valued continuous function defined on $[t_0, b] \times \mathbb{R}^2$;
(C3) $\phi \in C[t_0 - \tau, b]$, that is, ϕ is a real-valued continuous function defined on $[t_0 - \tau, b]$.

Assume that the following conditions hold:

(C4) There exists $L_f > 0$ such that

$$|f(t, u_1, v_1) - f(t, u_2, v_2)| \leq L_f(|u_1 - u_2| + |v_1 - v_2|),$$

for all $t \in \mathbb{R}$ and all $u_1, v_1, u_2, v_2 \in \mathbb{R}$;

(C5) $2L_f(b - t_0) < 1$.

By a solution of the problem (6.11)–(6.12), we mean a function $\bar{x} \in C[t_0 - \tau, b] \cap C^1[t_0, b]$ that satisfies (6.11)–(6.12), where $C^1[t_0, b] = \{f : [t_0, b] \to \mathbb{R} : f' \text{ is continuous}\}$.

The problem (6.11)–(6.12) can be reformulated as follows:

$$x(t) = \begin{cases} \phi(t), & \text{if } t \in [t_0 - \tau, t_0], \\ \phi(t_0) + \int_{t_0}^{t} f(t, x(s), x(s-\tau))ds, & \text{if } t \in [t_0, b]. \end{cases}$$

Define a mapping $T : C[t_0 - \tau, b] \cap C^1[t_0, b] \to C[t_0 - \tau, b] \cap C^1[t_0, b]$ by

$$T(x(t)) = \begin{cases} \phi(t), & \text{if } t \in [t_0 - \tau, t_0], \\ \phi(t_0) + \int_{t_0}^{t} f(t, x(s), x(s-\tau))ds, & \text{if } t \in [t_0, b]. \end{cases} \tag{6.13}$$

Theorem 6.3 *Assume that the conditions (C1)–(C5) are satisfied. Then, the following assertions hold:*

(a) *The problem* (6.11)–(6.12) *has a unique solution* $\bar{x}$ *in* $C[t_0 - \tau, b] \cap C^1[t_0, b]$.

(b) *For* $x_0 \in C[t_0 - \tau, b] \cap C^1[t_0, b]$, *the sequence* $\{x_n\}$ *in* $C[t_0 - \tau, b] \cap C^1[t_0, b]$ *defined by*

$$x_{n+1}(s) = T(x_n(s)), \quad \textit{for all } n = 0, 1, 2, \ldots \tag{6.14}$$

converges to $\bar{x}$, *where* T *is defined by* (6.13).

Proof (a) Let $x, y \in C[t_0 - \tau, b] \cap C^1[t_0, b]$. Then, we have

$$\begin{aligned}
& d_\infty(T(x), T(y)) \\
&= \sup_{t\in[t_0-\tau,b]} |T(x(t)) - T(y(t))| \\
&= \sup_{t\in[t_0-\tau,b]} \left| \phi(t_0) + \int_{t_0}^{t} f(t, x(s), x(s-\tau))ds - \left(\phi(t_0) + \int_{t_0}^{t} f(t, y(s), y(s-\tau))ds\right)\right| \\
&= \sup_{t\in[t_0-\tau,b]} \left| \int_{t_0}^{t} [f(t, x(s), x(s-\tau)) - f(t, y(s), y(s-\tau))]ds \right| \\
&\le \sup_{t\in[t_0-\tau,b]} \int_{t_0}^{t} |f(t, x(s), x(s-\tau)) - f(t, y(s), y(s-\tau))|\, ds \\
&\le L_f \sup_{t\in[t_0-\tau,b]} \int_{t_0}^{t} [|x(s) - y(s)| + |x(s-\tau)) - y(s-\tau))|]ds \\
&\le L_f \sup_{t\in[t_0-\tau,b]} \int_{t_0}^{t} \left[\sup_{s\in[t_0-\tau,b]} |x(s) - y(s)| + \sup_{s\in[t_0-\tau,b]} |x(s-\tau)) - y(s-\tau))| \right] ds \\
&= 2L_f d_\infty(x, y) \sup_{t\in[t_0-\tau,b]} \int_{t_0}^{t} ds \\
&= 2L_f d_\infty(x, y) \sup_{t\in[t_0-\tau,b]} (t - t_0) \\
&\le 2L_f(b - t_0) d_\infty(x, y).
\end{aligned}$$

Thus, T is a contraction mapping. Since $C[t_0 - \tau, b] \cap C^1[t_0, b]$ is complete as a metric space, by Banach contraction principle (Theorem 2.1), T has a unique fixed point $\bar{x}$ in $C[t_0 - \tau, b] \cap C^1[t_0, b]$. Therefore, the problem (6.11)–(6.12) has a unique solution $\bar{x}$ in $C[t_0 - \tau, b] \cap C^1[t_0, b]$.

(b) Since the mapping T defined by (6.13) is a contraction, as in the proof of Banach Contraction Principle 2.1, the sequence $\{x_n\}$ defined by (6.14) converges to $\bar{x}$. ■

6.3 Applications to Second-order Two-point Boundary-value Problems

Second-order linear and nonlinear differential equations with various types of boundary conditions occur in many problems from science, engineering, medicine, social sciences, etc. In particular, the second-order two-point boundary-value problems are encountered in many engineering fields including optimal control, beam deflections, heat flow, and various dynamical systems. These problems can be solved either analytically or numerically.

In this section, we study the existence of a unique solution of the following second-order two-point boundary-value problem:

$$\begin{cases} y''(t) + f(t, y(t), y'(t)) = 0, \\ y(a) = p, y(b) = q, \end{cases} \tag{6.15}$$

where $p, q \in \mathbb{R}$ and $f : [a, b] \times \mathbb{R} \times \mathbb{R} \to \mathbb{R}$ is a continuous function which satisfies the following uniform Lipschitz condition. There exist nonnegative real numbers K and L such that

$$|f(t, \beta, \nu) - f(t, \mu, \zeta)| \leq K|\beta - \mu| + L|\nu - \zeta|, \tag{6.16}$$

for all $t \in [a, b]$ and all $\beta, \mu, \nu, \zeta \in \mathbb{R}$. Here we use $y'(t)$ for $\frac{dy}{dt}$ and $y''(t)$ for $\frac{d^2y}{dt^2}$.

The *Green's function* is defined by

$$G(t, s) = \begin{cases} \dfrac{(b-t)(s-a)}{b-a}, & \text{if } a \leq s \leq t \leq b, \\ \dfrac{(b-s)(t-a)}{b-a}, & \text{if } a \leq t \leq s \leq b. \end{cases} \tag{6.17}$$

The following result provides the estimate of upper bounds for $\int_a^b G(t, s)ds$ and $\int_a^b \frac{\partial G(t, s)}{\partial t} ds$.

Proposition 6.1 [123, Lemma 2.1] *For the green's function G defined by* (6.17), *we have*

$$\int_a^b G(t, s)ds \leq \frac{1}{8}(b-a)^2 \quad \text{and} \quad \int_a^b \left|\frac{\partial G(t, s)}{\partial t}\right| ds \leq \frac{1}{2}(b-a).$$

Proof We have

$$\begin{aligned} \int_a^b G(t, s)ds &= \frac{b-t}{b-a}\int_a^t (s-a)ds + \frac{t-a}{b-a}\int_t^b (b-s)ds \\ &= \frac{b-t}{b-a}\frac{1}{2}(t-a)^2 + \frac{t-a}{b-a}\frac{1}{2}(b-t)^2 \\ &= \frac{1}{2}(t-a)(b-t) \\ &\leq \frac{1}{8}(b-a)^2. \end{aligned}$$

Further,

$$\begin{aligned} \int_a^b \left|\frac{\partial G(t, s)}{\partial t}\right| ds &= \frac{1}{b-a}\left[\int_a^t (s-a)ds + \int_t^b (b-s)ds\right] \\ &= \frac{1}{b-a}\left[\frac{1}{2}(t-a)^2 + \frac{1}{2}(b-t)^2\right] \\ &\leq \frac{1}{2}(b-a). \end{aligned}$$

■

Let $X = C^1[a,b]$ be the space of real-valued continuously differentiable functions defined on $[a,b]$ with the metric

$$d_\infty(x,y) = \max\left\{ \sup_{t\in[a,b]} |x(t)-y(t)|,\ \frac{b-a}{4} \sup_{t\in[a,b]} |x'(t)-y'(t)| \right\}, \tag{6.18}$$

for all $x, y \in X$. Then, (X, d_∞) is a complete metric space. Define T on $C^1[a,b]$ by

$$T(y(t)) = \int_a^b G(t,s)f(s,y(s),y'(s))ds, \quad \text{for all } y \in X. \tag{6.19}$$

In order to see that T maps X into itself, we need to verify that $T(y(t))$ is continuously differentiable whenever $y(t)$ is so. For this, let $y \in X$. Then,

$$\frac{T(y(t+h)) - T(y(t))}{h} = \int_a^b \frac{G(t+h,s)-G(t,s)}{h} f(s,y(s),y'(s))ds.$$

Taking limit as $h \to 0$, we get

$$\frac{d}{dt}T(y(t)) = \int_a^b \frac{\partial G(t,s)}{\partial t} f(s,y(s),y'(s))ds,$$

which is continuous in t. Therefore, T maps X into itself.

Proposition 6.2 *Let $X = C^1[a,b]$ be a metric space with the metric d_∞ defined by* (6.18). *Then, the mapping $T : X \to X$ defined by* (6.19) *is Lipschitz continuous with Lipschitz constant* $\left(\frac{(b-a)^2}{8}K + \frac{b-a}{2}L\right)$.

Proof Let $y, z \in X$. Then for all $t \in [a,b]$, we have

$$\begin{aligned}
|T(y(t)) - T(z(t))| &= \left| \int_a^b G(t,s)\left[f(s,y(s),y'(s)) - f(s,z(s),z'(s))\right] ds \right| \\
&\le \int_a^b G(t,s)\left|f(s,y(s),y'(s)) - f(s,z(s),z'(s))\right| ds \\
&\le \int_a^b G(t,s)\left(K|y(s)-z(s)| + L|y'(s)-z'(s)|\right) ds \\
&\le \int_a^b G(t,s)\left(Kd_\infty(y,z) + \frac{4}{b-a}L d_\infty(y,z)\right) ds \\
&= \left(K + \frac{4}{b-a}L\right) d_\infty(y,z) \int_a^b G(t,s)ds \\
&\le \frac{1}{8}(b-a)^2\left(K + \frac{4}{b-a}L\right) d_\infty(y,z) \\
&= \left(\frac{(b-a)^2}{8}K + \frac{b-a}{2}L\right) d_\infty(y,z).
\end{aligned}$$

Note that

$$\begin{aligned}\left|\frac{d}{dt}T(y(t))-\frac{d}{dt}T(z(t))\right| &= \left|\int_a^b \frac{\partial G(t,s)}{\partial t}\left[f(s,y(s),y'(s))-f(s,z(s),z'(s))\right]ds\right| \\ &\leq \int_a^b \frac{\partial G(t,s)}{\partial t}\left|f(s,y(s),y'(s))-f(s,z(s),z'(s))\right| ds \\ &\leq \int_a^b \frac{\partial G(t,s)}{\partial t}\left(K|y(s)-z(s)|+L|y'(s)-z'(s)|\right) ds \\ &\leq \int_a^b \frac{\partial G(t,s)}{\partial t}\left(Kd_\infty(y,z)+\frac{4}{b-a}Ld_\infty(y,z)\right) ds \\ &\leq \frac{1}{2}(b-a)\left(K+\frac{4}{b-a}L\right)d_\infty(y,z) \\ &= \left(\frac{b-a}{2}K+2L\right)d_\infty(y,z).\end{aligned}$$

Hence, we have

$$\begin{aligned}& d_\infty(T(y),T(z)) \\ &= \max\left\{\sup_{t\in[a,b]}|T(y(t))-T(z(t))|,\ \frac{b-a}{4}\sup_{t\in[a,b]}|T(y'(t))-T(z'(t))|\right\} \\ &\leq \max\left\{\frac{(b-a)^2}{8}K+\frac{b-a}{2}L,\ \frac{b-a}{4}\left(\frac{b-a}{2}K+2L\right)\right\}d_\infty(y,z) \\ &= \left(\frac{(b-a)^2}{8}K+\frac{b-a}{2}L\right)d_\infty(y,z).\end{aligned}$$

Therefore, the mapping T defined by (6.19) is Lipschitz continuous with Lipschitz constant $\left(\frac{(b-a)^2}{8}K+\frac{b-a}{2}L\right)$. ■

By using Proposition 6.2, we establish the following existence and uniqueness result for a solution of the problem (6.15).

Theorem 6.4 *Let $X = C^1[a,b]$ be a metric space with the metric d_∞ defined by* (6.18). *Let $f: [a,b]\times\mathbb{R}\times\mathbb{R}\to\mathbb{R}$ be a real-valued continuous function such that the Lipschitz condition* (6.16) *is satisfied. Suppose that*

$$\alpha = \frac{(b-a)^2}{8}K+\frac{b-a}{2}L < 1. \tag{6.20}$$

Then, the problem (6.15) *has a unique solution in X.*

Proof From (6.20) and Proposition 6.2, we see that the mapping T defined by (6.19) is a contraction. By Banach contraction principle (Theorem 2.1), T has a unique solution in X. Therefore, the problem (6.15) has a unique fixed point in X. ■

Exercise 6.3 Consider the following second-order two-point boundary-value problem:

$$\begin{cases} y''(t) + \cos y(t) = 0, \\ y(0) = 0,\ y(1) = 0. \end{cases} \tag{6.21}$$

Show that the boundary-value problem (6.21) has a unique solution.

Proof Note that

$$|f(t, y(t)) - f(t, z(t))| = |\cos y(t) - \cos z(t)| \leq |y(t) - z(t)|.$$

Here $K = 1$ and $b - a = 1$. Hence, from (6.20), we get

$$\alpha = \frac{(b-a)^2}{8} K = \frac{1}{8} < 1.$$

The boundary-value problem (6.21) has a unique solution by Theorem 6.4. ∎

Remark 6.1 If ϕ is a unique solution of the boundary-value problem (6.21), then for $y_0 \in C^1[a,b]$ with $y_1 = T(y_0)$, we have

$$\begin{aligned} d_\infty(y_1, \phi) &= d_\infty(T(y_0), \phi) \\ &\leq d_\infty(T(y_0), T(y_1)) + d_\infty(T(y_1), \phi) \\ &\leq \alpha d_\infty(y, y_0) + \alpha d_\infty(y_1, \phi), \end{aligned}$$

which implies that

$$d_\infty(y_1, \phi) \leq \frac{\alpha}{1-\alpha} d_\infty(y_0, y_1).$$

Consider $y_0 \equiv 0$. Then, $y_1(t)$ is the solution of the following problem:

$$\begin{cases} y_1''(t) + 1 = 0, \\ y_1(0) = 0,\ y_1(1) = 0. \end{cases}$$

Hence $y_1(t) = \frac{t(1-t)}{2}$, which is a single iterate approximation to the solution.

We now establish existence and uniqueness theorem for a solution of the problem (6.15), where the solution lies in a certain domain.

Theorem 6.5 *Let $X = C^1[a,b]$ be a metric space with the metric d_∞ defined by (6.18). Let $r > 0$ and let $f(t, y(t), y'(t))$ be a real-valued continuous function defined on $D = [a,b] \times [-r, r] \times \left[-\frac{4r}{b-a}, \frac{4r}{b-a}\right]$ such that the following Lipschitz condition holds. There exist $K \geq 0$ and $L \geq 0$ such that*

$$|f(t, \beta, \gamma) - f(t, \mu, \zeta)| \leq K|\beta - \mu| + L|\gamma - \zeta|,$$

for all $t \in [a,b]$ *and all* $\beta, \gamma, \mu, \zeta \in D$. *Assume that* $m = \sup_{t\in[a,b]} |f(t,0,0)|$ *and*

$$M = \sup\left\{|f(t,\beta,\mu)| \,:\, |\beta| \le r,\ |\mu| \le \frac{4r}{b-a} \text{ and } t \in [a,b]\right\}.$$

Suppose that

$$\alpha = \frac{(b-a)^2}{8}K + \frac{b-a}{2}L < 1. \tag{6.22}$$

If either

$$\frac{(b-a)^2}{8}m \le (1-\alpha)r, \tag{6.23}$$

or

$$\frac{(b-a)^2}{8}M \le r, \tag{6.24}$$

then the problem (6.15) *has one and only one solution* $y(t)$ *such that*

$$|y(t)| \le r \quad \text{and} \quad |y'(t)| \le \frac{4}{b-a}r, \quad \text{for all } t \in [a,b].$$

Proof Let $y_0 \equiv 0$ and let $S_r[y_0] = \{y \in X : d(y,y_0) \le r\}$ be the closed ball. It is easy to see, from Proposition 6.2, that the mapping T defined by (6.19) is α-Lipschitz continuous. Since $\alpha < 1$, it infers that $T : S_r[y_0] \to X$ is a contraction mapping with constant $\alpha < 1$.

First assume that the condition (6.23) holds. In order to apply Exercise 2.2, we need to show that

$$d_\infty(T(y_0), y_0) \le (1-\alpha)r. \tag{6.25}$$

Note that

$$\begin{aligned}
|T(y_0(t))| &= \left|\int_a^b G(t,s)f(s,y_0(s),y_0'(s))ds\right| \\
&\le \int_a^b G(t,s)|f(s,0,0)|ds \\
&\le \frac{(b-a)^2}{8}m,
\end{aligned}$$

and

$$\begin{aligned}
\left|\frac{d}{dt}T(y_0(t))\right| &= \left|\int_a^b \frac{\partial G(t,s)}{\partial t}f(s,0,0)ds\right| \\
&\le \int_a^b \frac{\partial G(t,s)}{\partial t}|f(s,0,0)|\,ds \\
&\le m\int_a^b \frac{\partial G(t,s)}{\partial t}ds \\
&\le \frac{b-a}{2}m.
\end{aligned}$$

Hence, from (6.22) and (6.23), we get

$$d_\infty(T(y_0), y_0) = \max\left\{\frac{1}{8}(b-a)^2 m, \frac{b-a}{4} \times \frac{b-a}{2} m\right\}$$
$$= \frac{(b-a)^2}{8} m \leq (1-\alpha)r.$$

This shows that the inequality (6.25) holds. By Exercise 2.2, we conclude that the problem (6.15) has a unique solution in $S_r[y_0]$.

Now we assume that the condition (6.24) holds. Let $u \in S_r[y_0]$. We next show that $T(u) \in S_r[y_0]$. Note that

$$\max\left\{\sup_{t\in[a,b]} |u(t)|, \frac{b-a}{4} \sup_{t\in[a,b]} |u'(t)|\right\} \leq r.$$

It follows that

$$|u(t)| \leq r \quad \text{and} \quad |u'(t)| \leq \frac{4}{b-a} r, \qquad \text{for all } t \in [a,b].$$

By the assumption, we see that $|f(t, u(t), u'(t))| \leq M$ for all $t \in [a, b]$. Thus,

$$|T(u(t))| = \left|\int_a^b G(t,s) f(s, u(s), u'(s)) ds\right|$$
$$\leq \int_a^b G(t,s)|f(s, u(s), u'(s))| ds$$
$$\leq \frac{(b-a)^2}{8} M \leq r,$$

and

$$\left|\frac{d}{dt} T(u(t))\right| = \left|\int_a^b \frac{\partial G(t,s)}{\partial t} f(s, u(s), u'(s)) ds\right|$$
$$\leq \int_a^b \frac{\partial G(t,s)}{\partial t} |f(s, u(s), u'(s))|\, ds$$
$$\leq M \int_a^b \frac{\partial G(t,s)}{\partial t} ds$$
$$\leq \frac{b-a}{2} M$$
$$= \frac{(b-a)^2}{8} M \times \frac{4}{b-a}$$
$$\leq \frac{4}{b-a} r.$$

Hence,

$$d_\infty(T(u), y_0) = \max\left\{\sup_{t\in[a,b]} |Tu(t)|, \frac{b-a}{4} \sup_{t\in[a,b]} |T(u'(t))|\right\} \leq r.$$

This shows that T maps from $S_r[y_0]$ into itself. Since $S_r[y_0]$ is closed, it follows that $S_r[y_0]$ is a complete metric space. Therefore, from Banach contraction principle (Theorem 2.1), T has a unique fixed point in $S_r[y_0]$. ■

Exercise 6.4 Consider the following second-order two-point boundary-value problem:

$$\begin{cases} y''(t) + e^{y(t)} = 0, \\ y(0) = 0, \; y(1) = 0. \end{cases} \tag{6.26}$$

Show that the boundary-value problem (6.26) has a unique solution $y(t)$ for which $|y(t)| \leq 1$.

Proof Here $f(t, y(t)) = e^{y(t)}$, which is not Lipschitz continuous for all $y(t)$. Thus, Theorem 6.4 is not applicable in this case. But by using Theorem 6.5, we can obtain a solution in the restricted domain. Assume that $r > 0$, $|y(t)| \leq r$ and $|z(t)| \leq r$. Note that

$$|f(t, y(t)) - f(t, z(t))| = \left| e^{y(t)} - e^{z(t)} \right| \leq e^r |y(t) - z(t)|.$$

Here $K = e^r$ and $b - a = 1$. Hence,

$$\alpha = \frac{(b-a)^2}{8} e^r = \frac{1}{8} e^r < 1,$$

which gives that $r < \log 8 = 2.0794$. Note $f(t, 0) = 1 = m$ and $M = e^r$. Observe that

$$\frac{(b-a)^2}{8} m = \frac{1}{8} = 0.125 \quad \text{and} \quad (1-\alpha)r = \left(1 - \frac{1}{8} e^r\right) r.$$

In particular, choose $r = 1$. Then, $(1-\alpha)r = \left(1 - \frac{1}{8} e^1\right) 1 = 0.660\,21$, which implies that

$$\frac{(b-a)^2}{8} m < (1-\alpha)r.$$

Thus, all the assumptions of Theorem 6.5 are satisfied. Therefore, we conclude that the boundary-value problem (6.26) has a unique solution $y(t)$ for which $|y(t)| \leq 1$. ■

6.4 Applications to Integral Equations

It is well known that many real life problems can be modelled in the form of integral equations. In this section, we study the existence of a unique solution of some known integral equations, namely, Volterra integral equations and Fredholm integral equations.

6.4.1 An Application to Volterra Integral Equations

Vito Volterra (1860–1940) presented his celebrated theory of integral equations in his four famous papers in 1896. The Volterra integral equations are a special type of integral equations. In this subsection, we discuss the existence and uniqueness of the Volterra integral equations by using Proposition 2.1 and Banach contraction principle (Theorem 2.1).

Let $f : [a,b] \to \mathbb{R}$ and $k : [a,b] \times [a,b] \to \mathbb{R}$ be continuous functions. A *linear Volterra integral equation of first kind* is defined by

$$f(s) = \int_a^s k(s,t)x(t)dt, \quad \text{for all } s \in [a,b],$$

where $x : [a,b] \to \mathbb{R}$ is an unknown continuous function.

A *linear Volterra integral equation of second kind* is defined by

$$x(s) = f(s) + \lambda \int_a^s k(s,t)x(t)dt, \quad \text{for all } s \in [a,b], \tag{6.27}$$

where λ is a nonzero parameter in $\mathbb{R}$.

The following result provides the existence of a unique solution of the linear Volterra integral equation (6.27).

Theorem 6.6 *Let $X = C[a,b]$ be the set of all continuous real-valued functions defined on $[a,b]$ with the uniform metric d_∞ defined by*

$$d_\infty(u,v) = \sup_{t\in[a,b]} |u(t) - v(t)|, \quad \text{for all } u,v \in X. \tag{6.28}$$

Then, for each $\lambda \in \mathbb{R}$ with $\lambda \neq 0$, the linear Volterra equation (6.27) has a unique solution $\bar{x} \in C[a,b]$.

Proof Since k is continuous, there exists a constant $M > 0$ such that $|k(s,t)| \leq M$ for all $s,t \in [a,b]$. Define the mapping $T : X \to X$ by

$$T(x(s)) = f(s) + \lambda \int_a^s k(s,t)x(t)dt, \quad \text{for all } x \in X.$$

For all $x, y \in X$, we have

$$\begin{aligned} |T(x(s)) - T(y(s))| &= \left| \lambda \int_a^s k(s,t)(x(t) - x(t))dt \right| \\ &\leq |\lambda|\, M\, (s-a)\, d_\infty(x,y), \quad \text{for all } s \in [a,b]. \end{aligned}$$

Hence,

$$\begin{aligned} d_\infty(T(x),T(y)) &= \sup_{s\in[a,b]} |T(x(s)) - T(y(s))| \\ &\leq \sup_{s\in[a,b]} |\lambda| M\, d_\infty(x,y)(s-a) \\ &\leq |\lambda| M(b-a) d_\infty(x,y). \end{aligned}$$

Note that

$$\begin{aligned}\left|T^2(x(s)) - T^2(y(s))\right| &= \left|\lambda \int_a^s k(s,t)(T(x(t)) - T(y(t)))dt\right| \\ &\leq |\lambda| \int_a^s |k(s,t)|\,|T(x(t)) - T(y(t))|dt \\ &\leq |\lambda| \int_a^s |k(s,t)| \left(|\lambda|\, M\, (t-a)\, d_\infty(x,y)\right) dt \\ &\leq |\lambda|^2 M \int_a^s |k(s,t)|\, (t-a)\, d_\infty(x,y) dt \\ &\leq |\lambda|^2 M^2\, d_\infty(x,y) \int_a^s (t-a)dt \\ &\leq \frac{|\lambda|^2\, M^2\, (s-a)^2}{2} d_\infty(x,y).\end{aligned}$$

Continuing in this way, we obtain

$$\left|T^n(x(s)) - T^n(y(s))\right| \leq \frac{|\lambda|^n\, M^n\, (s-a)^n}{n!} d_\infty(x,y), \quad \text{for all } s \in [a,b].$$

Hence,

$$d_\infty\left(T^n(x), T^n(y)\right) \leq \frac{\left[|\lambda|\, M\, (b-a)\right]^n}{n!} d_\infty(x,y).$$

Recalling that $\frac{r^n}{n!} \to 0$ as $n \to \infty$ for any $r \in \mathbb{R}$, we conclude that there exists n such that T^n is a contraction mapping. Taking n sufficiently large to have $\frac{[|\lambda|\, k\, (b-a)]^n}{n!} < 1$. Then, by Banach contraction principle (Theorem 2.1), T^n has a unique fixed point $\bar{x} \in X$, and thus by Proposition 2.1, $T(\bar{x}) = \bar{x}$. Obviously, $\bar{x}$ solves (6.27). ∎

Now we consider the following *nonlinear Volterra integral equation:*

$$x(s) = f(s) + \lambda \int_a^s k(s,t,x(t))dt, \quad \text{for all } s \in [a,b], \tag{6.29}$$

where $f : [a,b] \to \mathbb{R}$, $k : [a,b] \times [a,b] \times \mathbb{R} \to \mathbb{R}$ are continuous functions and $x : [a,b] \to \mathbb{R}$ is an unknown continuous function. If $k(s,t,x(t))$ is linear with respect to the third argument, then it reduces to the linear Volterra integral equation (6.27).

We give the existence result for a uniqueness solution of the nonlinear Volterra integral equation (6.29).

Theorem 6.7 *Let $X = C[a,b]$ be a metric space with uniform metric d_∞ defined by* (6.28). *Let $f : [a,b] \to \mathbb{R}$ and $k : [a,b] \times [a,b] \times \mathbb{R} \to \mathbb{R}$ be continuous real-valued function. Suppose that there exists a constant $M \geq 0$ such that*

$$\left|k(s,t,\beta) - k(s,t,\mu)\right| \leq M|\beta - \mu|, \quad \textit{for all } a \leq t \leq s \leq b \textit{ and all } \beta, \mu \in \mathbb{R}.$$

Define the operator $T : C[a,b] \to C[a,b]$ *by*

$$T(x(s)) = f(s) + \lambda \int_a^s k(s,t,x(t))dt, \quad \text{for all } x \in X \text{ and all } s \in [a,b]. \tag{6.30}$$

Then, the operator T *has a unique fixed point* $\bar{x} \in C[a,b]$, *that is, the nonlinear Volterra integral equation* (6.29) *has a unique solution* $\bar{x} \in C[a,b]$.

Proof Let $x, y \in C[a,b]$. Then, we have

$$\begin{aligned}
|T(x(s)) - T(y(s))| &= \left|\lambda \int_a^s \big(k(s,t,x(t)) - k(s,t,y(t))\big)\, dt\right| \\
&\leq |\lambda| \int_a^s |k(s,t,x(t)) - k((s,t,y(t))|\, dt \\
&\leq |\lambda| \int_a^s M\, |x(t) - y(t)| dt \\
&\leq |\lambda|\, M\, d_\infty(x,y)\,(s-a),
\end{aligned}$$

and therefore,

$$\begin{aligned}
d_\infty(T(x), T(y)) &= \sup_{s\in[a,b]} |T(x(s)) - T(y(s))| \\
&\leq \sup_{s\in[a,b]} |\lambda|\, M\, d_\infty(x,y)\,(s-a) \\
&\leq |\lambda|\, M\, d_\infty(x,y)\,(b-a).
\end{aligned}$$

Now,

$$\begin{aligned}
|T^2(x(s)) - T^2(y(s))| &\leq |\lambda| \int_a^s |k(s,t,T(x(t))) - k(s,t,T(y(t)))|\, dt \\
&\leq |\lambda| M \int_a^s |T(x(t)) - T(y(t))|\, dt \\
&\leq |\lambda| M \int_a^s (|\lambda| M\, d_\infty(x,y)\,(t-a))dt \\
&= |\lambda|^2 M^2\, d_\infty(x,y) \int_a^s (t-a)dt \\
&= |\lambda|^2 M^2\, d_\infty(x,y) \frac{(s-a)^2}{2!},
\end{aligned}$$

and hence,

$$\begin{aligned}
d_\infty\left(T^2(x), T^2(y)\right) &\leq \sup_{s\in[a,b]} |\lambda|^2 M^2\, d_\infty(x,y) \frac{(s-a)^2}{2!} \\
&\leq |\lambda|^2 M^2\, d_\infty(x,y) \frac{(b-a)^2}{2!}.
\end{aligned}$$

Similarly, we have

$$d_\infty\left(T^m(x), T^m(y)\right) \le |\lambda|^m M^m \frac{(b-a)^m}{m!} d_\infty(x, y).$$

For any fixed λ and sufficiently large m, we have $\alpha := |\lambda|^m M^m \frac{(b-a)^m}{m!} < 1$. Hence, the corresponding mapping T^m is contraction on $C[a, b]$. Therefore, by Banach contraction principle (Theorem 2.1), T^m has a unique fixed point, and thus by Proposition 2.1, T has a unique fixed point $\bar{x}$ in $C[a, b]$. ■

6.4.2 An Application to Fredholm Integral Equations

A *Fredholm equation* is an integral equation in which integration limits are constant. Fredholm was a Swedish mathematician who developed the theory of Fredholm integral equations published in Acta Mathematica in 1903.

Let $f : [a, b] \to \mathbb{R}$ and $k : [a, b] \times [a, b] \to \mathbb{R}$ be continuous functions. Consider the following *Fredholm integral equation*:

$$x(s) = f(s) + \int_a^b k(s, t)x(t)dt, \quad \text{for all } s \in [a, b]. \tag{6.31}$$

To find a solution of the equation (6.31) is equivalent to find a fixed point $x \in C[a, b]$ of the mapping $T : C[a, b] \to C[a, b]$ defined by

$$T(x(s)) = f(s) + \int_a^b k(s, t)x(t)dt, \quad \text{for all } x \in C[a, b] \text{ and all } s \in [a, b]. \tag{6.32}$$

We show that the operator $T : C[a, b] \to C[a, b]$ defined by (6.32) is Lipschitz continuous under a suitable condition.

Theorem 6.8 *Let $f : [a, b] \to \mathbb{R}$ and $k : [a, b] \times [a, b] \to \mathbb{R}$ be continuous functions such that*

$$\kappa := \sup_{s \in [a,b]} \int_a^b |k(s, t)|dt < \infty.$$

Let $X = C[a, b]$ be a metric space with uniform metric d_∞ defined by (6.28) and $T : C[a, b] \to C[a, b]$ be an operator defined by (6.32). Then, T is Lipschitz continuous with Lipschitz constant κ.

Proof For $x, y \in C[a, b]$, we have

$$\begin{aligned} d_\infty(T(x), T(y)) &= \sup_{s \in [a,b]} |T(x(s)) - T(y(s))| \\ &= \sup_{s \in [a,b]} \left| \int_a^b k(s, t)x(t)dt - \int_a^b k(s, t)y(t)dt \right| \\ &= \sup_{s \in [a,b]} \left| \int_a^b k(s, t)(x(t) - y(t))dt \right| \end{aligned}$$

$$\leq \sup_{s\in[a,b]} \int_a^b |k(s,t)|\,|x(t)-y(t)|dt$$

$$\leq d_\infty(x,y) \sup_{s\in[a,b]} \int_a^b |k(s,t)|dt$$

$$\leq \kappa d_\infty(x,y).$$

Therefore, T is κ-Lipschitz continuous with Lipschitz constant κ. ■

We now establish the existence and uniqueness theorem for a solution of the Fredholm integral equation (6.31).

Theorem 6.9 *Let $f : [a,b] \to \mathbb{R}$ and $k : [a,b] \times [a,b] \to \mathbb{R}$ be continuous functions such that*

$$\kappa := \sup_{s\in[a,b]} \int_a^b |k(s,t)|dt < 1. \tag{6.33}$$

Let $X = C[a,b]$ be a metric space with uniform metric d_∞ defined by (6.28). Then, the following assertions hold:

(a) *There exists a unique solution $\bar{x} \in C[a,b]$ of the Fredholm integral equation* (6.31).

(b) *For any $x_0 \in X$, the sequence $\{x_n\}$ defined by*

$$x_{n+1}(s) = f(s) + \int_a^b k(s,t)x_n(t)dt, \quad \textit{for all } n = 0,1,2,... \tag{6.34}$$

converges to $\bar{x}$.

Proof (a) Since $\kappa < 1$, it follows from Theorem 6.8 that the operator $T : C[a,b] \to C[a,b]$ defined by (6.32) is a contraction. Therefore, from Banach contraction principle (Theorem 2.1), we conclude that there exists a unique function $\bar{x} \in C[a,b]$ which is a unique solution of the Fredholm integral equation (6.31).

(b) Since the operator T defined by (6.32) is a contraction, from the proof of Banach contraction principle (Theorem 2.1), the sequence $\{x_n\}$ defined by (6.34) converges to $\bar{x}$. ■

Exercise 6.5 Solve the following Fredholm integral equation

$$x(s) = 1 + \int_0^\alpha s\,x(t)dt, \quad \text{for all } s \in [0,\alpha], \tag{6.35}$$

where $\alpha \in (0,1)$.

Solution Note that

$$\kappa := \sup_{s\in[0,\alpha]} \int_0^\alpha |k(s,t)|dt = \sup_{s\in[0,\alpha]} \int_0^\alpha s\,dt = \alpha^2.$$

Hence, from Theorem 6.9, the Fredholm integral equation (6.35) has a unique solution $\bar{x} \in C[0,\alpha]$.

In order to compute $\bar{x}$, we may choose initial point $x_0 \in C[a,b]$. Take $x_0(s) = 1$. Then,

$$
\begin{aligned}
T(x_0(s)) &= 1 + \int_0^{\alpha} s\, x_0(t)dt = 1 + \alpha s, \\
T^2(x_0) &= T(T(x_0)) = 1 + \int_0^{\alpha} s(1 + \alpha t)dt \\
&= 1 + \left(\alpha + \frac{\alpha^3}{2}\right)s = 1 + \alpha s\left(1 + \frac{\alpha^2}{2}\right), \\
T^3(x_0) &= T(T^2(x_0)) = 1 + \int_0^{\alpha} s\left[1 + \left(\alpha + \frac{\alpha^3}{2}\right)t\right]dt \\
&= 1 + s\left[\alpha + \left(\alpha + \frac{\alpha^3}{2}\right)\frac{\alpha^2}{2}\right] \\
&= 1 + s\left(\alpha + \frac{\alpha^3}{2} + \frac{\alpha^5}{4}\right) = 1 + \alpha s\left(1 + \frac{\alpha^2}{2} + \frac{\alpha^4}{4}\right), \\
T^4(x_0) &= T(T^3(x_0)) = 1 + \int_0^{\alpha} s\left[1 + \left(\alpha + \frac{\alpha^3}{2} + \frac{\alpha^5}{4}\right)t\right]dt \\
&= 1 + s\left[t + \left(\alpha + \frac{\alpha^3}{2} + \frac{\alpha^5}{4}\right)\frac{t^2}{2}\right]_{t=0}^{\alpha} \\
&= 1 + s\left[\alpha + \frac{\alpha^3}{2} + \frac{\alpha^5}{2^2} + \frac{\alpha^7}{2^3}\right] = 1 + \alpha s\left[1 + \frac{\alpha^2}{2} + \frac{\alpha^4}{2^2} + \frac{\alpha^6}{2^3}\right], \\
&\vdots \\
T^n(x_0) &= T(T^{n-1}(x_0)) \\
&= 1 + \alpha s\left[1 + \frac{\alpha^2}{2} + \frac{\alpha^4}{2^2} + \frac{\alpha^6}{2^3} + \cdots + \frac{\alpha^{2n}}{2^n}\right], \quad \text{for all } n \in \mathbb{N}.
\end{aligned}
$$

Therefore,

$$
\begin{aligned}
\bar{x}(s) &= \lim_{n\to\infty} T^n(x_0(s)) \\
&= 1 + \alpha s \sum_{n=0}^{\infty} \left(\frac{\alpha^2}{2}\right)^n \\
&= 1 + \frac{\alpha s}{1 - \frac{\alpha^2}{2}}.
\end{aligned}
$$

■

Remark 6.2 If the upper limit α is equal to 1, then

$$
\kappa := \sup_{s\in[0,1]} \int_0^1 |k(s,t)|dt = \sup_{s\in[0,1]} \int_0^1 s\,dt = 1,
$$

and therefore, Theorem 6.9 is not applicable in this case.

Exercise 6.6 Consider the integral equation

$$x(s) = \sin s + \int_0^{\pi/2} \sin s \cos t\, x(t)dt. \tag{6.36}$$

Show that $\sup_{s\in[0,\pi/2]} \int_0^{\pi/2} |k(s,t)|dt = 1.$

Proof Here $k(s,t) = \sin s \cos t$. Note that

$$\begin{aligned}
\sup_{s\in[0,\pi/2]} \int_0^{\pi/2} |k(s,t)|dt &= \sup_{s\in[0,\pi/2]} \int_0^{\pi/2} |\sin s|\,|\cos t|dt \\
&= \sup_{s\in[0,\pi/2]} \int_0^{\pi/2} \sin s\ \cos t dt \\
&= \sup_{s\in[0,\pi/2]} \sin s \int_0^{\pi/2} \cos t dt \\
&= \sup_{s\in[0,\pi/2]} \sin s = 1.
\end{aligned}$$

■

Remark 6.3 In Exercise 6.6, $\varkappa = \sup_{s\in[0,\pi/2]} \int_0^{\pi/2} |k(s,t)|dt = 1$, and therefore, Theorem 6.9 is not applicable. However, if we consider the different range of integration in the Fredholm integral equation (6.35), then Theorem 6.9 is applicable as shown in the following exercise.

Exercise 6.7 Consider the following Fredholm integral equation

$$x(s) = \sin s + \int_0^{\pi/4} \sin s \cos t\, x(t)dt. \tag{6.37}$$

Find the fourth successive approximation of solution of integral equation (6.37).

Solution Note that

$$\begin{aligned}
\sup_{s\in[0,\pi/4]} \int_0^{\pi/4} |k(s,t)|dt &= \sup_{s\in[0,\pi/4]} \int_0^{\pi/4} |\sin s|\,|\cos t|dt \\
&= \sup_{s\in[0,\pi/4]} \int_0^{\pi/4} \sin s\ \cos t dt \\
&= \sup_{s\in[0,\pi/4]} \sin s \int_0^{\pi/4} \cos t dt \\
&= \frac{1}{\sqrt{2}} \sup_{s\in[0,\pi/4]} \sin s \\
&= \frac{1}{\sqrt{2}}\frac{1}{\sqrt{2}} = \frac{1}{2} < 1.
\end{aligned}$$

Hence, from Theorem 6.9, the Fredholm integral equation (6.35) has a unique solution $\bar{x} \in C[0, \alpha]$. Indeed, $\bar{x}(s) = \frac{4}{3} \sin s$ is the unique solution of

$$x(s) = \sin s + \int_0^{\pi/4} \sin s \cos t\, x(t) dt.$$

Consider $x_0 = 1$. Then, we obtain the approximations of $\bar{x}$ as follows:

$$\begin{aligned}
x_1(s) &= \sin s + \int_0^{\pi/4} \sin s \cos t\, x_0(t) dt \\
&= \sin s + \sin s \int_0^{\pi/4} \cos t dt = \left(1 + \frac{1}{\sqrt{2}}\right) \sin s, \\
x_2(s) &= \sin s + \left(1 + \frac{1}{\sqrt{2}}\right) \int_0^{\pi/4} \sin s \cos t \, \sin t dt \\
&= \sin s + \frac{1}{4}\left(1 + \frac{1}{\sqrt{2}}\right) \sin s = \frac{1}{4}\left(\frac{1}{\sqrt{2}} + 5\right) \sin s, \\
x_3(s) &= \sin s + \left(\frac{1}{4\sqrt{2}} + \frac{5}{4}\right) \int_0^{\pi/4} \sin s \cos t \, \sin t dt \\
&= \sin s + \frac{1}{4}\left(\frac{1}{4\sqrt{2}} + \frac{5}{4}\right) \sin s \\
&= \frac{1}{16}\left(\frac{1}{\sqrt{2}} + 21\right) \sin s, \\
x_4(s) &= \sin s + \frac{1}{16}\left(\frac{1}{\sqrt{2}} + 21\right) \int_0^{\pi/4} \sin s \cos t \, \sin t dt \\
&= \sin s + \frac{1}{64}\left(\frac{1}{\sqrt{2}} + 21\right) \sin s \\
&= \frac{1}{64}\left(\frac{1}{\sqrt{2}} + 85\right) \sin s.
\end{aligned}$$

■

Let $f : [a, b] \to \mathbb{R}$ and $k : [a, b] \times [a, b] \times \mathbb{R} \to \mathbb{R}$ be continuous real-valued functions. An integral equation of the form

$$x(s) = f(s) + \lambda \int_a^b k(s, t, x(t)) dt, \quad \text{for all } x \in C[a, b] \text{ and all } s \in [a, b], \tag{6.38}$$

is called a *nonlinear Fredholm integral equation.* If $k(s, t, x(t))$ is linear with respect to the third argument, then the Fredholm integral equation (6.38) reduces to the linear Fredholm integral equation (6.31).

Theorem 6.10 *Let $f : [a,b] \to \mathbb{R}$ and $k : [a,b] \times [a,b] \times \mathbb{R} \to \mathbb{R}$ be continuous real-valued functions. Suppose that there exists a constant $M \geq 0$ such that*

$$|k(s,t,\beta) - k(s,t,\mu)| \leq M|\beta - \mu|,$$

for all $s,t \in [a,b]$ and all $\beta, \mu \in \mathbb{R}$. Let $X = C[a,b]$ be a metric space with uniform metric d_∞ defined by (6.28). Define an operator $T : C[a,b] \to C[a,b]$ by

$$T(x(s)) = f(s) + \lambda \int_a^b k(s,t,x(t))dt, \quad \textit{for all } s \in [a,b].$$

Then, the following assertions hold:

(a) *T is $|\lambda|M(b-a)$-Lipschitz continuous.*
(b) *If $|\lambda|M(b-a) < 1$, then T has a unique fixed point $\bar{x} \in C[a,b]$.*
(c) *For $x_0 \in C[a,b]$, the sequence $\{x_n\}$ in $C[a,b]$ defined by*

$$x_{n+1}(s) = f(s) + \lambda \int_a^b k(s,t,x_n(t))dt, \quad \textit{for all } n = 0,1,2,...$$

converges to $\bar{x}$.

Proof (a) For $x, y \in X$, we have

$$\begin{aligned} |T(x(s)) - T(y(s))| &= \left| \lambda \int_a^b \left(K(s,t,x(t)) - K(s,t,y(t)) \right) dt \right| \\ &\leq |\lambda| \int_a^b |K(s,t,x(t)) - K(s,t,y(t))| \, dt \\ &\leq M|\lambda| \int_a^b |x(t) - y(t)| dt \\ &\leq M|\lambda|(b-a) d_\infty(x,y), \qquad \text{for all } s \in [a,b], \end{aligned}$$

and hence,

$$\begin{aligned} d_\infty(T(x), T(y)) &= \sup_{t \in [a,b]} |T(x(t)) - T(y(t))| \\ &\leq M|\lambda|(b-a) d_\infty(x,y). \end{aligned}$$

Thus, T is $|\lambda|M(b-a)$-Lipschitz continuous.

(b) We see that T is a contraction if

$$|\lambda| < \frac{1}{M(b-a)}.$$

Suppose that $|\lambda| < \frac{1}{M(b-a)}$. Then, by Banach contraction principle (Theorem 2.1), T has a unique fixed point, says $\bar{x} \in C[a,b]$. Since T is a contraction, from the proof of Banach contraction principle (Theorem 2.1), the function $\bar{x}$ is the limit point of the iterative sequence generated by

$$x_{n+1}(s) = T(x_n)(s) = f(s) + \lambda \int_a^b k(s,t,x_n(t))dt.$$

■

Example 6.2 Consider the following integral equation

$$x(s) = 2(1-2s^2) + \lambda \int_0^1 s\,t\,x(t)dt, \quad \text{for all } s \in [0,1], \tag{6.39}$$

where λ is a nonzero real number.

Here

$$f(s) = 2(1-2s^2) \text{ and } k(s,t,u) = s\,t\,u, \quad \text{for all } s,t \in [0,1] \text{ and all } u \in \mathbb{R}.$$

Note that

$$|k(s,t,\beta) - k(s,t,\mu)| = s\,t|\beta - \mu| \le |\beta - \mu|, \quad \text{for all } s,t \in [0,1] \text{ and all } \beta,\mu \in \mathbb{R}.$$

Here $M = 1$. By Theorem 6.10, we see that if $|\lambda| = |\lambda|M(b-a) < 1$, then the integral equation (6.39) has a unique solution $\bar{x} \in C[a,b]$. For finding this solution, the sequence $\{x_n\}$ in $C[a,b]$ is defined recursively as follows:

$$\begin{cases} x_0(s) = 1, \\ x_{n+1}(s) = 2(1-2s^2) + \lambda \int_0^1 s\,t\,x_n(t)dt, \quad \text{for all } s \in [0,1] \text{ and ll } n = 0,1,2,\dots. \end{cases} \tag{6.40}$$

Now,

$$\begin{aligned} x_1(s) &= f(s) + \lambda \int_0^1 s\,t\,dt = f(s) + \frac{1}{2}\lambda s \\ x_2(s) &= f(s) + \lambda \int_0^1 s\,t\left(2 - 4t^2 + \frac{1}{2}\lambda t\right)dt = f(s) + \frac{1}{2\cdot 3}s\lambda^2 \\ x_3(s) &= f(s) + \lambda \int_0^1 s\,t\left(2(1-2t^2) + \frac{1}{2}t\lambda^2\right)dt = f(s) + \frac{1}{2.3^2}s\lambda^3 \\ x_4(s) &= f(s) + \lambda \int_0^1 s\,t\left(2(1-2t^2) + \frac{1}{18}t\lambda^3\right)dt = f(s) + \frac{1}{2.3^3}s\lambda^4. \end{aligned}$$

Assume that

$$x_n(s) = f(s) + \frac{\lambda s}{2}\left(\frac{\lambda}{3}\right)^{n-1}, \tag{6.41}$$

for fixed $n \in \mathbb{N}$. Now

$$\begin{aligned} x_{n+1}(s) &= f(s) + \lambda \int_0^1 s\,t\,x_n(t)dt \\ &= f(s) + \lambda \int_0^1 s\,t\left(2(1-2t^2) + \frac{\lambda t}{2}\left(\frac{\lambda}{3}\right)^{n-1}\right)dt \\ &= f(s) + \frac{\lambda s}{2}\left(\frac{\lambda}{3}\right)^n. \end{aligned}$$

Thus, by the mathematical induction principle, the inequality (6.41) is true for all $n \in \mathbb{N}$. Note that

$$d(x_n, f)_\infty = \sup_{s\in[0,1]} \left| \frac{\lambda s}{2} \left(\frac{\lambda}{3}\right)^n \right| = \sup_{s\in[0,1]} \frac{|\lambda|}{2} \left|\frac{\lambda}{3}\right|^n |s| \le \frac{|\lambda|}{2} \left|\frac{\lambda}{3}\right|^n.$$

Thus, for an arbitrary nonzero parameter λ with $|\lambda| < 3$, the sequence $\{x_n\}$ in $C[a,b]$ defined by (6.40) converges uniformly to $\bar{x}$, where

$$\bar{x}(s) = f(s) = 2(1 - 2s^2).$$

Exercise 6.8 Consider the following nonlinear integral equation

$$x(s) = s\left(\pi s - \frac{\pi}{5}\right) + \frac{\pi}{5} \int_0^1 st \sin(x(t))dt, \quad \text{for all } s \in [0,1]. \tag{6.42}$$

Show that the nonlinear integral equation (6.42) has a unique solution in $C([0,1], \mathbb{R})$.

Proof Here $\lambda = \frac{\pi}{5}$. Note that

$$f(s) = s\left(\pi s - \frac{\pi}{5}\right) \text{ and } k(s,t,\beta) = s\,t \sin(u), \text{ for all } s,t \in [0,1] \text{ and all } \beta \in \mathbb{R}.$$

For $\beta, \mu \in \mathbb{R}$, we have

$$|k(s,t,\beta) - k(s,t,\mu)| = s\,t|\sin(\beta) - \sin(\mu)| \le |\beta - \mu|,$$

for all $s,t \in [0,1]$ and all $\beta, \mu \in \mathbb{R}$. Clearly, $M = 1$. Note that

$$\kappa = |\lambda| M(b-a) = \frac{\pi}{5} = 0.62832 < 1.$$

Therefore, by Theorem 6.10, the nonlinear integral equation (6.42) has a unique solution $\bar{x} \in C([0,1], \mathbb{R})$. ■

6.4.3 An Application to Nonlinear Volterra–Fredholm Integral Equations

The Volterra–Fredholm integral equations [122, 172] arise in the mathematical modeling of the spatio-temporal development of an epidemic. Various physical and biological models can be written in the form of nonlinear Volterra–Fredholm integral equations; see, for example, [177].

Let $f : [a,b] \to \mathbb{R}$ and $k_1, k_2 : [a,b] \times [a,b] \times \mathbb{R} \to \mathbb{R}$ be continuous functions. The integral equation defined by

$$x(s) = f(s) + \int_a^s k_1(s,t,x(t))dt + \int_a^b k_2(s,t,x(t))dt, \quad \text{for all } s \in [a,b], \tag{6.43}$$

is known as *nonlinear Volterra–Fredholm integral equation*.

To find the solution of the integral equation (6.43) is equivalent to find the fixed point $\bar{x} \in C[a,b]$ of the mapping $T : C[a,b] \to C[a,b]$ defined by

$$T(x(s)) = f(s) + \int_a^s k_1(s,t,x(t))dt + \int_a^b k_2(s,t,x(t))dt, \tag{6.44}$$

for all $s \in [a,b]$.

The following theorem shows that the operator $T : C[a,b] \to C[a,b]$ defined by (6.44) is Lipschitz continuous.

Theorem 6.11 *Let $X = C[a,b]$ be a metric space with uniform metric d_∞ defined by* (6.28). *Let $f : [a,b] \to \mathbb{R}$ and $k_1, k_2 : [a,b] \times [a,b] \times \mathbb{R} \to \mathbb{R}$ be continuous functions such that*

$$|k_i(s,t,\beta) - k_i(s,t,\mu)| \le L_i\,|\beta - \mu|, \text{ for all } s,t \in [a,b],\ \beta,\mu \in \mathbb{R} \text{ and for some } L_i > 0, \tag{6.45}$$

and $T : C[a,b] \to C[a,b]$ be an operator defined by (6.44). *Then, T is Lipschitz continuous with Lipschitz constant $\kappa = (L_1 + L_2)(b-a)$.*

Proof Let $x, y \in X$. Then, for $s \in [a,b]$, we have

$$\begin{aligned}
&|T(x(s)) - T(y(s))| \\
&= \left| \int_a^s (k_1(s,t,x(t)) - k_1(s,t,y(t)))dt + \int_a^b (k_2(s,t,x(t)) - k_2(s,t,y(t)))dt \right| \\
&\le \int_a^s |k_1(s,t,x(t)) - k_1(s,t,y(t))|\,dt + \int_a^b |k_2(s,t,x(t)) - k_2(s,t,y(t))|\,dt \\
&\le L_1 \int_a^s |x(t) - y(t)|\,dt + L_2 \int_a^b |x(t) - y(t)|\,dt \\
&\le L_1 \int_a^s d_\infty(x,y)dt + L_2 \int_a^b d_\infty(x,y)dt \\
&\le (L_1 + L_2)(b-a)d_\infty(x,y),
\end{aligned}$$

and therefore,

$$\begin{aligned}
d_\infty(T(x), T(y)) &= \sup_{s \in [a,b]} |T(x(s)) - T(y(s))| \\
&\le (L_1 + L_2)(b-a)d_\infty(x,y).
\end{aligned}$$

Hence, T is Lipschitz continuous with Lipschitz constant $\kappa = (L_1 + L_2)(b-a)$. ∎

We now establish existence and uniqueness theorem for a solution of Volterra–Fredholm integral equation (6.43).

Theorem 6.12 *Let $X = C[a,b]$ be a metric space with uniform metric d_∞ defined by* (6.28). *Let $f : [a,b] \to \mathbb{R}$ and $k_1, k_2 : [a,b] \times [a,b] \times \mathbb{R} \to \mathbb{R}$ be continuous functions such that the inequality* (6.45) *holds. Assume that $\kappa := (L_1 + L_2)(b-a) < 1$. Then, the following assertions hold:*

(a) *There exists a unique solution $\bar{x} \in C[a,b]$ of the Volterra–Fredholm integral equation* (6.43).

(b) *For $x_0 \in C[a,b]$, the sequence $\{x_n\}$ in $C[a,b]$ defined by*

$$x_{n+1}(s) = T(x_n(s)), \quad \text{for all } n = 0, 1, 2, ...$$

converges to $\bar{x}$, where $T : C[a,b] \to C[a,b]$ is defined by (6.44).

Proof (a) Since $\kappa = (L_1 + L_2)(b - a) < 1$, it follows from Theorem 6.11 that the operator $T : C[a,b] \to C[a,b]$ defined by (6.44) is contraction. Therefore, from Banach contraction principle (Theorem 2.1), there exists a unique function $\bar{x} \in C[a,b]$ such that $\bar{x}$ is the solution of Volterra–Fredholm integral equation (6.43).

Part (b) follows from the proof of Banach Contraction Principle 2.1 as the operator $T : C[a,b] \to C[a,b]$ defined by (6.44) is a contraction. ■

We now establish existence and uniqueness theorem for a solution of the Volterra–Fredholm integral equation (6.43), where the solution is determined in some closed sphere.

Theorem 6.13 *Let $X = C[a,b]$ be a metric space with uniform metric d_∞ defined by (6.28). Let $f : [a,b] \to \mathbb{R}$ and $k_1, k_2 : [a,b] \times [a,b] \times \mathbb{R} \to \mathbb{R}$ be continuous functions such that the condition* (6.45) *holds. Define $r_1 = \inf_{x \in [a,b]} f(x)$, $r_2 = \sup_{x \in [a,b]} f(x)$ and the closed sphere*

$$S_r[f] := \{u \in C[a,b] : d_\infty(u,f) \leq r\}, \quad \text{for } r > 0.$$

Suppose that there exist constants $M_1, M_2 > 0$ such that

$$|k_i(s,t,\lambda)| \leq M_i, \quad \text{for all } s,t \in [a,b] \text{ and all } \lambda \in \mathbb{R}.$$

Suppose that $(M_1 + M_2)(b - a) \leq r$ and $\kappa =: (L_1 + L_2)(b - a) < 1$. Then, the following assertions hold:

(a) *There exists a unique solution $\bar{x} \in S_r[f]$ of the Volterra–Fredholm integral equation* (6.43).

(b) *For $x_0 \in S_r[f]$, the sequence $\{x_n\}$ in $S_r[f]$ defined by*

$$x_{n+1}(s) = T(x_n(s)), \quad \text{for all } n = 0, 1, 2, ...$$

converges to $\bar{x} \in S_r[f]$, where $T : C[a,b] \to C[a,b]$ is defined by (6.44).

Proof We first show that the operator T defined by (6.44) maps from $S_r[f]$ into itself. Let $u \in S_r[f]$ be arbitrary. Then, $d_\infty(u,f) \leq r$, and therefore,

$$u(s) \in [r_1 - r, r_2 + r], \quad \text{for all } s \in [a,b].$$

Now, let $s \in [a,b]$. Then, we have

$$\begin{aligned} |T(u(s)) - f(s)| &= \left| \int_a^s k_1(s,t,u(t))dt + \int_a^b k_2(s,t,u(t))dt \right| \\ &\leq \int_a^s |k_1(s,t,u(t))|\, dt + \int_a^b |k_2(s,t,u(t))|\, dt \end{aligned}$$

$$\begin{aligned}
&\leq \int_a^s M_1 dt + \int_a^b M_2 dt \\
&\leq (M_1 + M_2)(b-a) \\
&\leq r.
\end{aligned}$$

Thus, $d_\infty(T(u), f) \leq r$, that is, $T(u) \in S_r[f]$.

(a) Since $\kappa =: (L_1 + L_2)(b-a) < 1$, it follows from Theorem 6.12 that the operator $T : S_r[f] \to S_r[f]$ defined by (6.44) is a contraction. Therefore, from Banach Contraction Principle 2.1, we conclude that there exists a unique function $\bar{x} \in S_r[f]$ which is a unique solution of the Volterra–Fredholm integral equation (6.43).

Part (b) follows from the proof of Banach Contraction Principle 2.1 as the operator $T : S_r[f] \to S_r[f]$ defined by (6.44) is a contraction. ■

6.4.4 An Application to Mixed Volterra–Fredholm Integral Equations

The mixed nonlinear Volterra–Fredholm integral equation occurs in the study of population dynamics, parabolic boundary value problems, etc. A study of the formulation of such models is given in [172]. In this subsection, we study existence of solutions of the linear and nonlinear mixed Volterra–Fredholm integral equations by using fixed point results.

Let $f : [a,b] \to \mathbb{R}$ and $k : [a,b] \times [a,b] \to \mathbb{R}$ be continuous functions and let λ be a nonzero real number. The integral equation

$$u(s) = f(x) + \lambda \int_a^x \int_a^b k(s,t)u(t)dt\,ds, \quad \text{for all } x \in [a,b], \tag{6.46}$$

is known as *linear mixed Volterra–Fredholm integral equation.*

Finding the solution of the integral equation (6.46) is equivalent to find the fixed point $u \in C[a,b]$ of the mapping $T : C[a,b] \to C[a,b]$ defined by

$$T(u(x)) = f(x) + \lambda \int_a^x \int_a^b k(s,t)u(t)dt\,ds, \tag{6.47}$$

for all $u \in C[a,b]$ and all $x \in [a,b]$.

We first show that the operator $T : C[a,b] \to C[a,b]$ defined by (6.47) is Lipschitz continuous.

Theorem 6.14 *Let λ be a nonzero real number and let $f : [a,b] \to \mathbb{R}$ and $k : [a,b] \times [a,b] \to \mathbb{R}$ be continuous functions such that*

$$|k(s,t)| \leq M, \quad \textit{for all } s,t \in [a,b] \textit{ and for some } M > 0. \tag{6.48}$$

Let $X = C[a,b]$ be a metric space with uniform metric d_∞ defined by (6.28) and $T : C[a,b] \to C[a,b]$ be an operator defined by (6.47). Then, T is Lipschitz continuous with Lipschitz constant $\kappa = M|\lambda|(b-a)^2$.

Proof Let $u, v \in C[a,b]$. Then,

$$\begin{aligned} d_\infty(T(u), T(v)) &= \sup_{x\in[a,b]} |T(u(x)) - T(v(x))| \\ &= \sup_{x\in[a,b]} |\lambda| \left| \int_a^x \int_a^b k(s,t)u(t)dt\,ds - \int_a^x \int_a^b k(s,t)v(t)dt\,ds \right| \\ &= \sup_{x\in[a,b]} |\lambda| \int_a^x \int_a^b |k(s,t)|\,|u(t) - v(t)|dt\,ds \\ &\leq M|\lambda| d_\infty(u,v) \sup_{x\in[a,b]} \int_a^x \int_a^b dt\,ds \\ &\leq M|\lambda|(b-a)^2 d_\infty(u,v). \end{aligned}$$

Therefore, T is Lipschitz continuous with Lipschitz constant $\kappa = M|\lambda|(b-a)^2$. ∎

We now establish the existence and uniqueness theorem for a solution of the linear mixed Volterra–Fredholm integral equation (6.46).

Theorem 6.15 *Let $X = C[a,b]$ be a metric space with uniform metric d_∞ defined by* (6.28). *Let λ be nonzero real number and let $f : [a,b] \to \mathbb{R}$ and $k : [a,b] \times [a,b] \to \mathbb{R}$ be continuous functions such that the inequality* (6.48) *holds. Assume that $\kappa := M|\lambda|(b-a)^2 < 1$. Then, the following assertions hold:*

(a) *There exists a unique solution $\bar{x} \in C[a,b]$ of the linear mixed Volterra–Fredholm integral equation* (6.46).

(b) *For any $x \in X$,*

$$\bar{x} = \lim_{n\to\infty} T^n(x),$$

where $T : C[a,b] \to C[a,b]$ is defined by (6.47).

Proof (a) Since $\kappa = M|\lambda|(b-a)^2 < 1$, it follows from Theorem 6.14 that the operator $T : C[a,b] \to C[a,b]$ defined by (6.47) is a contraction. Therefore, from Banach contraction principle (Theorem 2.1), there exists a unique function $\bar{u} \in C[a,b]$ which is a unique solution of the linear mixed Volterra–Fredholm integral equation (6.46).

Part (b) follows from the proof of Banach contraction principle (Theorem 2.1) since the operator $T : C[a,b] \to C[a,b]$ defined by (6.47) is a contraction. ∎

Let $f : [a,b] \to \mathbb{R}$ and $k : [a,b] \times [a,b] \times [a,b] \times \mathbb{R} \to \mathbb{R}$ be continuous functions. The integral equation

$$u(s) = f(x) + \int_a^x \int_a^b k(x,s,t,u(t))dt\,ds, \quad \text{for all } x \in [a,b]. \tag{6.49}$$

is called *nonlinear mixed Volterra–Fredholm integral equation*.

To find the solution of the integral equation (6.49) is equivalent to find the fixed point $u \in C[a,b]$ of the mapping $T : C[a,b] \to C[a,b]$ defined by

$$T(u(x)) = f(x) + \int_a^x \int_a^b k(x,s,t,u(t))dt\,ds, \tag{6.50}$$

for all $u \in C[a,b]$ and all $x \in [a,b]$.

We first show that the operator $T : C[a,b] \to C[a,b]$ defined by (6.50) is Lipschitz continuous under a suitable condition.

Theorem 6.16 *Let $f : [a,b] \to \mathbb{R}$ and $k : [a,b]\times[a,b]\times[a,b]\times\mathbb{R} \to \mathbb{R}$ be continuous functions. Suppose that there exists a constant $L \geq 0$ such that*

$$|k(x,s,t,\lambda) - k(x,s,t,\mu)| \leq L|\lambda - \mu|, \quad \text{for all } \lambda, \mu \in \mathbb{R} \text{ and } x,s,t \in [a,b]. \tag{6.51}$$

Let $X = C[a,b]$ be a metric space with uniform metric d_∞ defined by (6.28) and $T : C[a,b] \to C[a,b]$ be an operator defined by (6.50). Then, T is Lipschitz continuous with Lipschitz constant $\kappa = L(b-a)^2$.

Proof Let $u, v \in C[a,b]$. Then, we have

$$\begin{aligned}
d_\infty(T(u),T(v)) &= \sup_{x\in[a,b]} |T(u(x)) - T(v(x))| \\
&= \sup_{x\in[a,b]} \left| \int_a^x \int_a^b k(x,s,t,u(t))dt\,ds - \int_a^x \int_a^b k(x,s,t,v(t))dt\,ds \right| \\
&\leq \sup_{x\in[a,b]} \int_a^x \int_a^b |k(x,s,t,u(t))dt\,ds - k(x,s,t,v(t))|\,dt\,ds \\
&\leq L \sup_{x\in[a,b]} \int_a^x \int_a^b |u(s) - v(s)|dt\,ds \\
&\leq L d_\infty(u,v) \sup_{x\in[a,b]} \int_a^x \int_a^b dt\,ds \\
&\leq L(b-a)^2 d_\infty(u,v).
\end{aligned}$$

Therefore, T is Lipschitz continuous with Lipschitz constant $\kappa = L(b-a)^2$. ∎

We now establish the existence of a unique solution of the nonlinear mixed Volterra–Fredholm integral equation (6.49).

Theorem 6.17 *Let $X = C[a,b]$ be a metric space with uniform metric d_∞ defined by (6.28). Let $f : [a,b] \to \mathbb{R}$ and $k : [a,b]\times[a,b]\times[a,b]\times\mathbb{R} \to \mathbb{R}$ be continuous functions such that the condition (6.51) holds. Define $r_1 = \inf_{x\in[a,b]} f(x)$, $r_2 = \sup_{x\in[a,b]} f(x)$ and the closed sphere*

$$S_r[f] := \{u \in C[a,b] : d_\infty(u,f) \leq r\}, \quad \text{for } r > 0.$$

Suppose that there exists a constant $M > 0$ such that

$$|k(x,s,t,\lambda)| \leq M, \quad \text{for all } \lambda \in \mathbb{R} \text{ and } x,s,t \in [a,b].$$

Assume that $M(b-a)^2 \leq r$ and $L(b-a)^2 < 1$. Then, the following assertions hold:

(a) *There exists a unique solution $\bar{u} \in S_r[f]$ of the mixed Volterra–Fredholm integral equation (6.49).*

(b) *For any* $u \in S_r[f]$,

$$\bar{u} = \lim_{n\to\infty} T^n(u),$$

where T *is defined by* (6.50).

Proof We show that the operator T defined by (6.50) maps from $S_r[f]$ to itself. Let u be an arbitrary element in $S_r[f]$. Then $d_\infty(u,f) \le r$, and therefore,

$$u(s) \in [r_1 - r, r_2 + r], \quad \text{for all } s \in [a,b].$$

For $x \in [a,b]$, we have

$$\begin{aligned} |Tu(x) - f(x)| &= \left| \int_a^x \int_a^b k(x,s,t,u(t))dt\, ds \right| \\ &\le \int_a^x \int_a^b |k(x,s,t,u(t))|\, ds\, dt \\ &\le M(b-a)^2 \\ &\le r, \end{aligned}$$

it follows that $d_\infty(T(u),f) \le r$. Hence, $T(u) \in S_r[f]$.

(a) Since $\varkappa := L(b-a)^2 < 1$, it follows from Theorem 6.8 that the operator $T : S_r[f] \to S_r[f]$ defined by (6.50) is a contraction. Therefore, from Banach contraction principle (Theorem 2.1), there exists a unique function $\bar{u} \in S_r[f]$ which is a unique solution of the mixed Volterra–Fredholm integral equation (6.49).

(b) It follows from the proof of Banach contraction principle (Theorem 2.1) since the operator $T : S_r[f] \to S_r[f]$ defined by (6.50) is a contraction.

■

Appendix A

Some Basic Concepts and Inequalities

Let $x = (x_1, x_2, \ldots, x_n)$ and $y = (y_1, y_2, \ldots, y_n)$ be the vectors in an n-dimensional Euclidean space $\mathbb{R}^n$. Then, the inner product of x and y is defined as

$$\langle x, y\rangle = \sum_{i=1}^{n} x_i y_i.$$

and norm on $\mathbb{R}^n$ is defined by

$$\|x\| = \sqrt{\sum_{i=1}^{n} x_i^2}.$$

The Cauchy–Schwarz inequality in $\mathbb{R}^n$ is given by

$$|\langle x, y\rangle| \leq \|x\|\,\|y\|, \quad \text{for all } x, y \in \mathbb{R}^n.$$

Proposition A.1 *For any $x \geq 0$, the function $f(x) = \frac{x}{1+x}$ is monotonically increasing.*

Proof Let $y > x \geq 0$. Then,

$$\frac{1}{1+y} < \frac{1}{1+x} \text{ and so } 1 - \frac{1}{1+y} > 1 - \frac{1}{1+x},$$

that is,

$$\frac{y}{1+y} > \frac{x}{1+x}.$$

∎

Theorem A.1 *For any two real numbers x and y, the following inequality holds:*

$$\frac{|x+y|}{1+|x+y|} \leq \frac{|x|}{1+|x|} + \frac{|y|}{1+|y|}. \tag{A.1}$$

Proof Let x and y have the same sign. Without loss of generality, we may assume that $x \geq 0$ and $y \geq 0$. Then,

$$\begin{aligned}\frac{|x+y|}{1+|x+y|} &= \frac{x+y}{1+x+y} = \frac{x}{1+x+y} + \frac{y}{1+x+y} \\ &\leq \frac{x}{1+x} + \frac{y}{1+y} = \frac{|x|}{1+|x|} + \frac{|y|}{1+|y|}.\end{aligned}$$

If x and y have different signs, then we may assume that $|x| > |y|$. Then, $|x+y| \leq |x|$. From Proposition A.1, we have

$$\frac{|x+y|}{1+|x+y|} \leq \frac{|x|}{1+|x|} \leq \frac{|x|}{1+|x|} + \frac{|y|}{1+|y|}.$$

■

Theorem A.2 (Hölder's Inequality) *Let $x_i \geq 0$ and $y_i \geq 0$ for all $i = 1, 2, \ldots, n$ and let $p > 1$ and $q > 1$ such that $\frac{1}{p} + \frac{1}{q} = 1$. Then,*

$$\sum_{i=1}^{n} x_i y_i \leq \left(\sum_{i=1}^{n} x_i^p\right)^{1/p} \left(\sum_{i=1}^{n} y_i^q\right)^{1/q}. \tag{A.2}$$

When $p = q = 2$, then the above inequality becomes

$$\sum_{i=1}^{n} x_i y_i \leq \left(\sum_{i=1}^{n} x_i^2\right)^{1/2} \left(\sum_{i=1}^{n} y_i^2\right)^{1/2}. \tag{A.3}$$

This inequality is known as Cauchy–Schwarz inequality.

Theorem A.3 (Minkowski's Inequality) *Let $x_i \geq 0$ and $y_i \geq 0$ for all $i = 1, 2, \ldots, n$ and let $p > 1$. Then,*

$$\left(\sum_{i=1}^{n} (x_i + y_i)^p\right)^{1/p} \leq \left(\sum_{i=1}^{n} x_i^p\right)^{1/p} + \left(\sum_{i=1}^{n} y_i^p\right)^{1/p}. \tag{A.4}$$

Proof If $p = 1$, then there is nothing to prove. So, we assume that $p > 1$. Note that

$$\sum_{i=1}^{n} (x_i + y_i)^p = \sum_{i=1}^{n} x_i (x_i + y_i)^{p-1} + \sum_{i=1}^{n} y_i (x_i + y_i)^{p-1}. \tag{A.5}$$

Let $q > 1$ be such that $\frac{1}{p} + \frac{1}{q} = 1$. From equation (A.5) and Hölder's inequality, we obtain

$$\begin{aligned}\sum_{i=1}^{n} (x_i + y_i)^p &\leq \left(\sum_{i=1}^{n} x_i^p\right)^{1/p} \left(\sum_{i=1}^{n} (x_i + y_i)^{(p-1)q}\right)^{1/q} \\ &\quad + \left(\sum_{i=1}^{n} y_i^p\right)^{1/p} \left(\sum_{i=1}^{n} (x_i + y_i)^{(p-1)q}\right)^{1/q} \\ &= \left\{\left(\sum_{i=1}^{n} x_i^p\right)^{1/p} + \left(\sum_{i=1}^{n} y_i^p\right)^{1/p}\right\} \left(\sum_{i=1}^{n} (x_i + y_i)^p\right)^{1/q}.\end{aligned}$$

If $\sum_{i=1}^{n}(x_i+y_i)^p = 0$, then obviously (A.4) holds. If $\sum_{i=1}^{n}(x_i+y_i)^p \neq 0$, then by dividing the above inequality by $\left(\sum_{i=1}^{n}(x_i+y_i)^p\right)^{1/q}$, we obtain (A.4). ■

Theorem A.4 (Minkowski's Inequality for Infinite Sums) *Let $p > 1$ and let $\{x_n\}$ and $\{y_n\}$ be sequences of nonnegative terms such that $\sum_{n=1}^{\infty} x_n^p$ and $\sum_{n=1}^{\infty} y_n^p$ are convergent. Then, $\sum_{n=1}^{\infty}(x_n+y_n)^p$ is convergent. Moreover,*

$$\left(\sum_{n=1}^{\infty}(x_n+y_n)^p\right)^{1/p} \leq \left(\sum_{n=1}^{\infty} x_n^p\right)^{1/p} + \left(\sum_{n=1}^{\infty} y_n^p\right)^{1/p}. \tag{A.6}$$

Proof For any positive integer m, from Theorem A.3, we have

$$\begin{aligned}\left(\sum_{n=1}^{m}(x_n+y_n)^p\right)^{1/p} &\leq \left(\sum_{n=1}^{m} x_n^p\right)^{1/p} + \left(\sum_{n=1}^{m} y_n^p\right)^{1/p}\\ &\leq \left(\sum_{n=1}^{\infty} x_n^p\right)^{1/p} + \left(\sum_{n=1}^{\infty} y_n^p\right)^{1/p}.\end{aligned}$$

Since $\left\{\left(\sum_{n=1}^{m}(x_n+y_n)^p\right)^{1/p}\right\}$ is an increasing sequence of nonnegative real numbers, it is bounded above by the sum

$$\left(\sum_{n=1}^{\infty} x_n^p\right)^{1/p} + \left(\sum_{n=1}^{\infty} y_n^p\right)^{1/p}.$$

It follows that $\sum_{n=1}^{\infty}(x_n+y_n)^p$ is convergent and that the inequality (A.6) holds. ■

Theorem A.5 *Let $p > 1$. For any $a \geq 0$ and $b \geq 0$, we have*

$$(a+b)^p \leq 2^{p-1}(a^p+b^p). \tag{A.7}$$

Proof If either a or b is 0, then nothing to prove. Suppose that $a > 0$ and $b > 0$. Since the function $x \mapsto x^p$ defined on the set of all positive numbers is convex when $p > 1$, we have

$$\left(\frac{a+b}{2}\right)^p \leq \frac{a^p+b^p}{2}, \quad \text{equivalently,} \quad (a+b)^p \leq 2^{p-1}(a^p+b^p).$$ ■

Theorem A.6 (Weierstrass's Theorem) *Every continuous real-valued function defined on a compact set has a maximum and a minimum.*

In particular, we have the following result.

Theorem A.7 (Weierstrass's Extreme Value Theorem) *If a real-valued function* $f : [a, b] \to \mathbb{R}$ *is continuous on the closed interval* $[a, b]$, *then* f *must attain a maximum and a minimum, each at least once. That is, there exist numbers* c *and* d *in* $[a, b]$ *such that* $f(c) \leq f(x) \leq f(d)$ *for all* $x \in [a, b]$.

Theorem A.8 (Weierstrass's Theorem) *Let* $f : [a, b] \to \mathbb{R}$ *be a continuous function. Then, there exists a sequence* $\{p_n\}$ *of polynomials with real coefficients that converges uniformly to* f *on* $[a, b]$, *that is, for all* $\varepsilon > 0$, *there exists a positive integer* N *such that for all* $t \in [a, b]$,

$$|p_n(t) - f(t)| < \varepsilon, \quad \textit{whenever } n > N.$$

Appendix B

Partial Ordering

Definition B.1 (Partial Ordering) A binary relation $\preccurlyeq$ on a nonempty set X is called *partial ordering* if the following conditions hold:

(i) For all $a \in X$, $a \preccurlyeq a$; (Reflexivity)

(ii) If $a \preccurlyeq b$ and $b \preccurlyeq a$, then $a = b$; (Antisymmetry)

(iii) If $a \preccurlyeq b$ and $b \preccurlyeq c$, then $a \preccurlyeq c$. (Transitivity)

Definition B.2 (Partially Ordered Set) A nonempty set X is called *partially ordered* if there is a partial ordering on X.

Remark B.1 The word 'partially' emphasizes that X may contain elements a and b for which neither $a \preccurlyeq b$ nor $b \preccurlyeq a$ holds. In this case, a and b are called *incomparable elements*. If $a \preccurlyeq b$ or $b \preccurlyeq a$ (or both), then a and b are called *comparable elements*.

Definition B.3 (Totally Ordered Set) A partially ordered set X is said to be *totally ordered* if every two elements of X are comparable.

In other word, partially ordered set X is totally ordered if it has no incomparable elements.

Definition B.4 Let M be a nonempty subset of a partially ordered set X.

(a) The element $x \in X$ is called an *upper bound* of M if $a \preccurlyeq x$ for all $a \in M$.

(b) The element $x \in X$ is called a *lower bound* of M if $x \preccurlyeq a$ for all $a \in M$.

(c) The element $x \in M$ is called a *maximal element* of M if $x \preccurlyeq a$ implies $x = a$.

Remark B.2 A subset of a partially ordered set M may or may not have an upper or lower bound. Also, M may or may not have maximal elements. Note that a maximal element need not be an upper bound.

Example B.1 **(a)** Let X be the set of all real numbers and let $a \leq b$ have its usual meaning. Then, X is totally ordered and it has no maximal element.

(b) Let $X = \mathbb{N}$, the set of all natural numbers. Let $m \preccurlyeq n$ mean that m divides n. Then $\preccurlyeq$ is a partial ordering on X.

(c) Let $\mathscr{P}(X)$ be the power set, that is, set of all subsets of a set X. Let $A \preccurlyeq B$ mean $A \subset B$. Then, $\mathscr{P}(X)$ is a partially ordered set and the only maximal element of $\mathscr{P}(X)$ is X.

(d) Let X be the set of all ordered n-tuples $x = (x_1, x_2, \dots x_n)$, $y = (y_1, y_2, \dots y_n)$, ... of real numbers and let $x \preccurlyeq y$ mean $x_i \leq y_i$ for all $i = 1, 2, \dots, n$. Then, $\preccurlyeq$ is a partial ordering on X.

Lemma B.1 (Zorn's Lemma) *Let X be a nonempty partially ordered set in which every totally ordered set has an upper bound. Then, X has at least one maximal element.*

References

1. R.P. Agarwal, D. O'Regan, and D.R. Sahu, *Fixed Point Theory for Lipschitzian-type Mappings with Applications*, Springer, Dordrecht, Heidelberg, London, New York (2009).
2. Ya. I. Alber and S. Guerre-Delabriere, Principle of weakly contractive maps in Hilbert spaces. In *Operator Theory: Advanced and Applications*, edited by I. Gohberg and Yu Lyubich, Vol. 98, Birkhäuser Verlag, Basel, Switzerland, pp. 7–22 (1997).
3. S. Al-Homidan, M. AlShahrani, and Q.H. Ansari, Weak sharp solutions for equilibrium problems in metric spaces, *J. Nonlinear Convex Anal.*, **16**(7) (2015), 1185–1193.
4. S. Al-Homidan, Q.H. Ansari, and J.-C. Yao, Some generalizations of Ekeland-type variational principle with applications to equilibrium problems and fixed point theory, *Nonlinear Anal.*, **69** (2008), 126–139.
5. B. Alleche and V.D. Rădulescu, The Ekeland variatonal principle for equilibrium problems revsited and applications, *Nonlinear Anal. RWA*, **23** (2015), 17–25.
6. M. Altman, An integral test for series and generalized contractions, *Amer. Math. Monthly*, **82**(8) (1975), 827–829.
7. A. Amini-Harandi, Q.H. Ansari, and A.P. Farajzadeh, Existence of equilibria in complete metric spaces, *Taiwanese J. Math.*, **16** (2012), 777–785.
8. Q.H. Ansari, *Metric Spaces: Including Fixed Point Theory and Set-valued Maps*, Narosa Publishing House, New Delhi (2010).
9. Q.H. Ansari, Ekeland's variational principle and its extensions with applications. In *Fixed Point Theory and Applications*, edited by S. Almezel, Q.H. Ansari, and M.A. Khamsi, Springer International Publishing Switzerland, pp. 65–99 (2014).
10. Q.H. Ansari, E. Köbis, and J.-C. Yao, *Vector Variational Inequalities and Vector Optimization - Theory and Applications*, Springer (2018).
11. Q.H. Ansari, C.S. Lalitha, and M. Mehta, *Generalized Convexity, Nonsmooth Variational Inequalities, and Nonsmooth Optimization*, CRC Press, Taylor and Francis, London, New York (2013).
12. Q.H. Ansari and L.-J. Lin, Ekeland-type variational principle and equilibrium problems. In *Topics in Nonconvex Optimization*, edited by S.K. Mishra, Springer Optimization and Its Applications, No. 59, Springer, New York, pp. 147–174 (2011).
13. Q.H. Ansari, N.C. Wong, and J.-C. Yao, The existence of nonlinear inequalities, *Appl. Math. Lett.*, **12**(5) (1999), 89–92.
14. N.A. Assad and W.A. Kirk, Fixed point theorems for set-valued mappings of contractive type, *Pacific J. Math.*, **43** (1972), 553–562.
15. J.-P. Aubin, *Mathematical Methods of Game and Economic Theory*, North-Holland Publishing Company, Amsterdam, New York, Oxford (1979).
16. J.-P. Aubin and A. Cellina, *Differential Inclusions, Set-Valued Maps and Viability Theory*, Springer-Verlag, New York, Heidelberg, Berlin (1984).

17. J.-P. Aubin and I. Ekeland, *Applied Nonlinear Analysis*, John Wiley & Sons, Inc., New York, Chichester, Brisbane, Toronto, Singapore (1984).

18. J.-P. Aubin and H. Frankowska, *Set-Valued Analysis*, Birkhäuser, Boston, Basel, Berlin (1990).

19. J.-P. Aubin and J. Siegel, Fixed points and stationary points of dissipative multivalued maps, *Proc. Amer. Math. Soc.*, **78**(3) (1980), 391–398.

20. J.S. Bae, Fixed point theorems for weakly contractive multivalued maps, *J. Math. Anal. Appl.*, **284** (2003), 690–697.

21. J.S. Bae and S. Park, Remarks on the Caristi-Kirk fixed point theorem, *Bull. Korean Math. Soc.*, **19**, 57–60 (1983).

22. J. Banaś and A. Martión, Some properties of the Hausdorff distance in metric spaces, *Bull. Austral. Math. Soc.*, **42** (1990), 511–516.

23. T.Q. Bao and P.Q. Khanh, Are several recent generalizations of Ekeland's variational principle more general than the original principle? *Acta Math Vietnam.*, **28** (2003), 345–350.

24. K. Barich, *Proving completeness of the Hausdorff induced metric space*, Whitman College, Washington (2011).

25. M. Barnsley, *Fractals Everywhere*, Academic Press, Inc., Harcourt Brace Jovanovich, Publishers, Boston, New York, London, Tokyo, Toronto (1988).

26. C. Berge, *Topological Spaces: Including a Treatment of Multi-valued Functions, Vector Spaces and Convexity*, Oliver & Boyd Ltd., Edinburgh and London (1963).

27. V. Berinde, On the approximation of fixed points of weakly φ-contractive operators, *Fixed Point Theory*, **4**(2) (2003), 131–142.

28. V. Berinde, On the approximation of fixed points of weakly contractive mappings, *Carpathian J. Math.*, **19**(1) (2003), 7–22.

29. V. Berinde, Approximation fixed points of weakly contractions using the Picard iteration, *Nonlinear Anal. Forum*, **9**(1) (2004), 43–53.

30. V. Berinde, *Iterative Approximation of Fixed Points*, Springer-Verlag, Berlin, Heidelberg (2007).

31. C. Bessaga, On the converse of Banach "fixed-point principle", *Coll. Math.*, **VII** (1959), 41–43.

32. M. Bianchi, G. Kassay, and R. Pini, Existence of equilibria via Ekeland's principle, *J. Math. Anal. Appl.*, **305** (2005), 502–512.

33. M. Bianchi and R. Pini, A note on equilibrium problems with properly quasimonotone functions, *J. Global Optim.*, **20** (2001), 67–76.

34. M. Bianchi and R. Pini, Coercivity conditions for equilibrium problems, *J. Optim. Theory Appl.*, **124** (2005), 79–92.

35. M. Bianchi and S. Schaible, Generalized monotone bifunctions and equilibrium problems, *J. Optim. Theory Appl.*, **90** (1996), 31–43.

36. M. Bianchi and S. Schaible, Equilibrium problems under generalized convexity and generalized monotonicity, *J. Global Optim.*, **30** (2004), 121–134.

37. E. Bishop and R.R. Phelps, The support function of a convex set. In *Convexity*, edited by Klee, Proc. Sympo. Pure Math. Vol. VII, Amer. Math. Soc. Providence, RI, pp. 27–35, (1963).

38. E. Blum and W. Oettli, From optimization and variational inequalities to equilibrium problems, *Math. Student*, **63** (1994), 123–145.

39. J.M. Borwein and D. Preiss, A smooth variational principle with applications to subdiffferentiability and to differentiability of convex functions, *Trans. Amer. Math. Soc.*, **303** (1987), 517–527.

40. J.M. Borwein and Q.J. Zhu, *Techniques of Variational Analysis*, Springer-Verlag, Berlin, Heidelberg, New York (2005).

41. C. Bosch, A. Garcia, and C.L. Garcia, An extension of Ekeland's variational principle to locally complete spaces, *J. Math. Anal. Appl.*, **328** (2007), 106–108.

42. G. Bouligand, Sur la semi-continuité d'inclusions et quelques sujets connexes, *Enseignement Math.*, **31** (1932), 14–22.

43. D. Boyd and J.S.W. Wong, On nonlinear contractions, *Proc. Amer. Math. Soc.*, **20** (1969), 458–464.

44. H. Brézis and F. Browder, A general principle on ordered sets in nonlinear functional analysis, *Adv. Math.*, **21** (1976), 355–364.

45. H. Brézis, L. Nirenberg, and G. Stampacchia, A remark on Ky Fan's minimax principle, *Boll. Un. Mat. Ital.*, **6** (1972), 293–300.

46. V. Bryant, *Metric Spaces, Iteration and Applications*, Cambridge University Press, Cambridge, New York, Sydney (1985).

47. R.S. Burachik and A.N. Iusem, *Set-Valued Mappings and Enlargments of Monotone Operators*, Springer-Verlag, New York (2008).

48. J.V. Burke and S. Deng, Weak sharp minima revisited, part I: Basic theory, *Control Cybern.*, **31** (2002), 439–469.

49. J.V. Burke and S. Deng, Weak sharp minima revisited, part II: Application to linear regularity and error bounds, *Math. Program. Ser. B*, **104** (2005), 235–261.

50. J.V. Burke and S. Deng, Weak sharp minima revisited, part III: Error bounds for differentiable convex inclusions, *Math. Program. Ser. B*, **116** (2009), 37–56.

51. J.V. Burke and M.C. Ferris, Weak sharp minima in mathematical programming, *SIAM J. Control Optim.*, **31** (1993), 1340–1359.

52. J. Caristi, Fixed point theorems for mappings satisfying inwardness conditions, *Trans. Amer. Math. Soc.*, **215** (1976), 241–251.

53. J. Caristi and W.A. Kirk, Geometric fixed point theory and inwardness conditions. In *The Geometry of Metric and Linear Spaces*, Lecture Notes in Mathematics, **490**, Springer-Verlag, Berlin, Heidelberg, New York, pp. 74–83 (1975).

54. M. Castellani and M. Giuli, Ekeland's principle for cyclically antimonotone equilibrium problems, *Nonlinear Anal. RWA*, **32** (2016), 213–228.

55. M. Castellani, M. Pappalardo, and M. Passacantando, Existence results for nonconvex equilibrium problems, *Optim. Meth. Software*, **25**(1) (2010), 49–58.

56. O. Chadli, Z. Chbani, and H. Riahi, Equilibrium problems with generalized monotone bifunctions and applications to variational inequalities, *J. Optim. Theory Appl.*, **105** (2000), 299–323.

57. O. Chadli, N.C. Wong, and J.-C. Yao, Equilibrium problems with applications to eigenvalue problems, *J. Optim. Theory Appl.*, **117** (2003), 245–266.

58. Y. Chen, Y.J. Cho, and L. Yang, Note on the results with lower semicontinuity, *Bull. Korean Math. Soc.*, **39** (2002), 535–541.

59. E. Cinlar and R. J. Vanderbei, *Mathematical Methods of Engineering Analysis*, Lecture Notes (2000).

60. F. Clarke, Pointwise contraction criteria for the existence of fixed points, *Canad. Math. Bull.*, **21** (1978), 7–11.

61. E.T. Copson, *Metric Spaces*, Cambridge University Press, Cambridge, U.K. (1968).

62. J. Cotrina and Y. García, Equilibrium problems: Exsitence results and applications, *Set-Valued Var. Anal.*, **26** (2018), 159–177.

63. J. Cotrina and A. Svensson, The finite intersection property for equilibrium problems, *J. Global Optim.*, **79** (2021), 941–957.

64. J. Cotrina, M. Théra, and J. Zúñiga, An existence result for quasi-equilibrium problems via Ekelnad's variational principle, *J. Optim. Theory Appl.*, **187** (2020), 336–355.

65. P.Z. Daffer and H. Kaneko, Fixed points of generalized contractive multi-valued mappings, *J. Math. Anal. Appl.*, **192** (1995), 655–666.

66. P.Z. Daffer, H. Kaneko, and W. Li, Variational principle and fixed points, *SIMAA*, **4** (2002), 129–136.

67. S. Dancs, M. Hegedűs, and P. Medvegyev, A general ordering and fixed-point principle in complete metric space, *Acta Sci. Math.*, **46** (1983), 381–388.

68. J. Daneš, A geometric theorem useful in nonlinear analysis, *Boll. Un. Mat. Ital.*, **6** (1972), 369–372.

69. J. Daneš, Equivalence of some geometric and related results of nonlinear functional analysis, *Comment. Math. Univ. Carolinae*, **26** (1985), 443–454.

70. D.G. De Figueiredo, *The Ekeland Variational Principle with Applications and Detours*, Tata Institute of Fundamental Research, Bombay (1989).

71. K. Deimling, *Nonlinear Functional Analysis*, Springer-Verlag, Berlin (1985).

72. M. Edelstein, An extension of Banach contraction principle, *Proc. Amer. Math. Soc.*, **1** (1961), 7–10.

73. M. Edelstein, On fixed points and period points under contractive mappings, *J. London Math. Soc.*, **37** (1962), 74–79.

74. I. Ekeland, Sur les prolèms variationnels. *C. R. Acad. des Sci. Paris*, **275**, 1057–1059 (1972).

75. I. Ekeland, On the variational principle. *J. Math. Anal. Appl.*, **47**, 324–353 (1974).

76. I. Ekeland, On convex minimization problems. *Bull. Amer. Math. Soc.*, **1**, 445–474 (1979).

77. F. Facchinei and J.-S. Pang, *Finite Dimensional Variational Inequalities and Complementarity Problems*, I and II, Springer-Verlag, New York, Berlin, Heidelberg (2003).

78. K. Fan, Extensions of two point theorems of F.E. Browder, *Math. Z.*, **112** (1969), 234–240.

79. K. Fan, A minimax inequlity and applications. In *Inequality III*, edited by O. Shisha, Academic Press, New York, pp. 103–113 (1972).

80. C. Farkas, A generalizated form of Ekeland's variational principle, *An. Şt. Univ. Ovidius Constanţa*, **20** (2012), 101–112.

81. Y.Q. Feng and S.Y. Liu, Fixed point theorems for multi-valued contractive mappings and multi-valued Caristi type mappings, *J. Math. Anal. Appl.*, **317** (2006), 103–112.

82. M.C. Ferris, *Weak Sharp Minima and Penalty Functions in Mathematical Programming*, Ph.D. Thesis, University of Cambridge (1988).

83. F. Flores-Bazán, Existence theorems for generalized noncoercive equilibrium problems: The quasi-convex case, *SIAM J. Optim.*, **11** (2000), 675–690.

84. F. Flores-Bazán, Existence theory for finite-dimensional pseudomonotone equilibrium problems, *Acta Appl. Math.*, **77** (2003), 249–297.

85. P.G. Georgiev, The strong Ekeland variational principle, the strong drop theorem and applications, *J. Math. Anal. Appl.*, **131** (1988), 1–21.

86. M.A. Geraghty, On contractive mappings, *Proc. Amer. Math. Soc.*, **40**(2) (1973), 604–608.

87. J.R. Giles, *Convex Analysis with Application in the Differention of Convex Functions*, Research Notes in Mathematics, **58**, Pitman Advanced Publishing Program, Boston, London, Melbourne (1982).

88. L. Górniewicz, *Topological Fixed Point Theory of Multivalued Mappings*, Kluwer Academic Publishers, Dordrecht, Boston, London (1999).

89. C.-L. Guo, Fixed point theorems for singlevalued mappings and multivalued mappings on complete metric spaces, *Chinese J. Math.*, **19**(1) (1991), 31–53.

90. T.X.D. Ha, A remark on the lower semicontinuity assumption in the Ekeland variational principle, *Optimization*, **65**(10) (2016), 1781–1789.

91. A. Hamel, Remarks to an equivalent formulation of Ekeland's variational principle, *Optimization*, **31** (1994), 233–238.

92. A. Hamel, Phelps' lemma, Daneš' drop theorem and Ekeland principle in locally convex spaces, *Proc. Amer. Math. Soc.*, **131** (2003), 3025–3038.

93. A. Hamel, Equivalents to Ekeland's variational principle in uniform spaces, *Nonlinear Anal.*, **62** (2005), 913–924.

94. Y.H. Hu and W. Song, Weak sharp solutions for variational inequalities in Banach spaces, *J. Math. Anal. Appl.*, **374** (2011), 118–132.

95. V.I. Istrătescu, *Fixed Point Theory: An Introduction*, D. Reidel Publicating Company, Dordrecht, Boston, London (1981).

96. A. Iusem, G. Kassay, and W. Sosa, On certain conditions for the existence of solutions of equilibrium problems, *Math. Program.*, **116** (2009), 259–273.

97. A. Iusem, G. Kassay, and W. Sosa, An existence result for equilibrium with surjectivity consequences, *J. Convex Anal.*, **16** (2009), 807–826.

98. J. Jachymski, Caristi's fixed point theorem and selections of set-valued contractions, *J. Math. Anal. Appl.*, **227**(1) (1998), 55–67.

99. J. Jachymski, A short proof of the converse to the contraction principle and some related results, *Top. Meth. Nonlinear Anal.*, **15**(1) (2000), 179–186.

100. J. Jachymski, A stationary point theorem characterizing metric completeness, *Appl. Math. Lett.*, **24**(2) (2011), 169–171.

101. J. Jachymski and I. Jóźwik, Nonlinear contractive conditions: A comparison and related problems, *Fixed Point Theory Appl.*, Banach Center Publication, Vol. **77**, Institute of Mathematics, Polish Academy of Sciences, Warszawa, pp. 123–146 (2007).

102. L. Janos, A converse of Banach's contraction principle, *Proc. Amer. Math. Soc.*, **18** (1967), 287–289.

103. L. Janos, An application of combinatorial techniques to a topological problem, *Bull. Austral. Math. Soc.*, **9** (1973), 287–289.

104. M. C. Joshi and R. K. Bose, *Some Topics in Nonlinear Functional Analysis*, Wiley Eastern Limited, New Delhi, Bangalore, Bombay, Calcutta, Madras, Hyderabad (1985).

105. O. Kada, T. Suzuki, and W. Takahashi, Nonconvex minimization theorems and fixed point theorems in complete metric spaces, *Math. Japonica*, **44** (1996), 381–391.

106. P. Kas, G. Kassay, and Z.B. Sensoy, On generalized equilibrium points, *J. Math. Anal. Appl.*, **296** (2004), 619–633.

107. G. Kassay and V. Rădulescu, *Equilibrium Problems and Applications*, Elsevier, Academic Press (2019).

108. J. L. Kelley, *General Topology*, VanNostrand-Reinholt, Princeton, NJ (1955).

109. S. Kakutani, A generalization of Brouwer's fixed point theorem, *Duke Math. J.*, **8** (1941), 457–459.

110. M.A. Khamsi and W.A. Kirk, *An Introdunction to Metric Spaces and Fixed Point Theory*, John Wiley & Sons, Inc., New York, Chichester, Brisbane, Singapore, Toronto (2001).

111. P.Q. Khanh and N.H. Quan, Versions of the Weierstrass theorem for bifunctions and solution existence in optimization, *SIAM J. Optim.*, **29**(2) (2019), 1502–1523.

112. E. Klein and A. C. Thompson, *Theory of Correspondences, Including Applications to Mathematical Economics*, John Wiley & Sons, Inc., New York, Chichester, Brisbane, Toronto, Singapore (1984).

113. M.A. Krasnosel'skii, G.M. Vainikko, P.P. Zabreiko, Ya. B. Ruttski, and V. Ya. Stetsenko, *Approximate Solution of Operator Equations*, Wolters-Nordhoff Publishing, Groningen (1972).

114. K. Kuratowski, Les fonctions semi-continues dans l'espace des ensembles fermés, *Fund. Math.*, **18** (1932), 148–159.

115. S. Lipschutz, *Theory and Problems of General Topology*, Schaum's Outlines Series, McGraw-Hill Book Company, New York, Toronto, Sydney (1965).

116. Y.-X. Li and S.-Z. Shi, A generalization of Ekeland's ϵ-variational principle and its Borwein-Preiss smooth variant, *J. Math. Anal. Appl.*, **246** (2000), 308–319.

117. L.-J. Lin and W.S. Du, Ekeland's variational principle, minimax theorems and existence of nonconvex equilibria in complete metric spaces, *J. Math. Anal. Appl.*, **323** (2006), 360–370.

118. L.-J. Lin and W.S. Du, Some equivalent formulations of generalized Ekeland's variational principle and their applications, *Nonlinear Anal.*, **67** (2007), 187–199.

119. L.-J. Lin and W.S. Du, On maximal element theorems, varinats of Ekeland's variational principle and their applications, *Nonlinear Anal.*, **68** (2008), 1246–1262.

120. D.T. Luc, E. Sarabi, and A. Soubeyran, Existence of solutions in variational relation problems without convexity, *J. Math. Anal. Appl.*, **364** (2010), 544–555.

121. V. L. Makarov and A. M. Rubinov, *Mathematical Theory of Economic, Dynamics and Equilibria*, Springer-Verlag, New York, Heidelberg, Berlin (1977).

122. K. Maleknejad and M. Hadizadeh, A new computational method for Volterra-Fredholm integral equations, *Computers Math. Appl.*, **37**(9) (1999), 1–8.

123. R.L. Manicke, Unique solutions of systems of second-order nonlinear two-point boundary-value problems, *J. Computat. Appl. Math.*, **10**(2) (1984), 175–178.

124. P. Marcotte and D.L. Zhu, Weak sharp solutions of variational inequalities, *SIAM J. Optim.*, **9**(1) (1998), 179–189.

125. G. Mastroeni, Gap functions for equilibrium problems, *J. Global Optim.*, **27** (2003), 411–426.

126. J. Matkowski, Nonlinear contractions in metrically convex spaces, *Publ. Math. Debrecen*, **45** (1994), 103–114.

127. L. McLinden, An application of Ekeland's theorem to minimax problems, *Nonlinear Anal. Theory, Meth. Appl.*, **6** (1982), 189–196.

128. I. Meghea, *Ekeland Variational Principle with Generalizations and Variants*, Old City Publishing, Inc., Philadelphia (2009).

129. N. Mizoguchi and W. Takahashi, Fixed point theorems for mutlivalued mappings on complete metric spaces, *J. Math. Anal. Appl.*, **141** (1989), 177–188.

130. L.D. Muu and W. Oettli, Convergence of an adaptive penalty scheme for finding constrained equilibria, *Nonlinear Anal.*, **18** (1992), 1159–1166.

131. S.B. Jr. Nadler, Multivalued contraction mappings, *Pacific J. Math.*, **30** (1969), 475–488.

132. J.F. Nash, Equilibrium points in n-person games, *Proc. Nat. Acad. Sci. USA*, **36** (1950), 48–49.

133. J.F. Nash, Noncooperative games. *Ann. Math.*, **54** (1951), 286–295.

134. H. Nikaido and K. Isoda, Note on noncooperative convex games, *Pacific J. Math.*, **5** (1955), 807–815.

135. W. Oettli and M. Théra, Equivalents of Ekeland's principle, *Bull. Austral. Math. Soc.*, **48** (1993), 385–392.

136. S. Park, On fixed points of set-valued directional contractions. *Internat. J. Math. Math. Sci.*, **8** (1985), 663–667.

137. S. Park, Some applications of Ekeland's variational principle to fixed point theory. In *Approximation Theory and Applications*, edited by S.P. Singh, Research Notes in Mathematics, No. 133, Pitman, Boston, pp. 159–172 (1985).

138. S. Park, Equivalent formulations of Ekeland's variational principle for approximate solutions of minimization problems and their applications. In *Operator Equations and Fixed Point Theorems*, edited by S.P. Singh, V.M. Sehgal, and J.H.W. Burry, MSRI-Korea Publishers 1, Soul, pp. 55–68 (1986).

139. S. Park, On generalizations of the Ekeland-type variational principles, *Nonlinear Anal.*, **39** (2000), 881–889.

140. H. K. Pathak, *An Introduction to Nonlinear Analysis and Fixed Point Theory*, Springer Nature, Singapore (2018).

141. Penot, The drop theorem, the petal theorem and Ekeland's variational principle, *Nonlinear Anal. Theory, Meth. Appl.*, **10** (1986), 813–822.

142. J.-P. Penot and M. Théra, Semi-continuous mappings in general topology, *Arch. Math.*, **38** (1982), 158–166.

143. B.T. Polyak, *Sharp minima*, Institute of Control Sciences Lecture Notes, Moscow, USSR (1979). Presented at the IIASA workshop on generalized Lagrangians and their applications, IIASA, Laxenburg, Austria (1979).

144. B.T. Polyak, *Introduction to Optimization*, Optimization Software, Inc., Publications Division, New York (1987).

145. S. Radenović, Z. Kadelburg, D. Jandrlić, and A. Jandrlić, Some results on weakly contractive maps, *Bull. Iranian Math. Soc.*, **38**(3) (2012), 625–645.

146. E. Rakotch, A note on contractive mappings, *Proc. Amer. Math. Soc.*, **13** (1962), 459–465.

147. J.S. Raymond, Multivalued contractions, *Set-Valued Anal.*, **2** (1994), 559–571.

148. B.E. Rhoades, Some theorems on weakly contractive maps, *Nonlinear Anal.*, **47** (2001), 2683–2693.

149. R.T. Rockafellor, Directionally Lipschitz functions and subdifferential calculus, *Proc. London Math. Soc.*, **39** (1979), 331–355.

150. M.Ó. Searcóid, *Metric Spaces*, Springer-Verlag, London (2007).

151. V.M. Sehgal, Some fixed point theorems for set-valued directional contraction mappings, *Internat. J. Math. Math. Sci.*, **3** (1980), 455–460.

152. P.V. Semenov, Fixed points of multivlaued contractions, *Funct. Anal. Appl.*, **36**(2) (2002), 159–161.

153. N. Shioji, T. Suzuki, and W. Takahashi, Contractive mappings, Kannan mappings and metric completeness, *Proc. Amer. Math. Soc.*, **126** (1998), 3117–3124.

154. S. Shirali and H. Vasudeva, *Metric Spaces*, Springer-Verlag, London (2006).

155. S. Singh, B. Watson, and P. Srivastava, *Fixed Point Theory and Best Approximation: The KKM-Map Principle*, Kluwer Academic Publishers, Dordrecht, Boston, London (1997).

156. R.E. Smithson, Fixed points for contractive multifunctions, *Proc. Amer. Math. Soc.*, **27**(1) (1971), 192–194.

157. R.E. Smithson, Multifunctions, *Niew Archief Voor-Wiskunde*, **(3)XX** (1972), 31–53.

158. W. Song, A generalization of Clarke's fixed point theorem, *Appl. Math. J. Chinese Univ. Ser. B*, **10**(4) (1995), 463–466.

159. M. Squassina, On Ekeland's variational principle, *J. Fixed Point Theory Appl.*, **10** (2011), 191–195.

160. G. Stampacchia, Formes bilinéaires coercitives sur les ensembles convexes, *C. R. Acad. Sci. Paris*, **258** (1964), 4413–4416.

161. F. Sullivan, A characterization of complete metric spaces, *Proc. Amer. Math. Soc.*, **83** (1981), 345–346.

162. T. Suzuki, Generalized distance and existence theorems in complete metric spaces, *J. Math. Anal. Appl.*, **253** (2001), 440–458.

163. T. Suzuki, On Downing-Kirk's theorem, *J. Math. Anal. Appl.*, **286** (2003), 453–458.

164. T. Suzuki, Generalized Caristi's fixed point theorems by Bae and others, *J. Math. Anal. Appl.*, **302** (2005), 502–508.

165. T. Suzuki, The strong Ekeland variational principle, *J. Math. Anal. Appl.*, **320** (2006), 787–794.

166. T. Suzuki, Mizoguchi-Takahashi's fixed point theorem is a real generalization of Nadler's, *J. Math. Anal. Appl.*, **340** (2008), 752–755.

167. T. Suzuki and W. Takahashi, Fixed point theorems and characterizations of metric completeness, *Top. Meth. Nonlinear Anal.*, **8** (1996), 371–382.

168. W. Takahashi, Existence theorems generalizing fixed point theorems for multivalued mappings. In *Fixed Point Theory and Applications*, edited by J.-B. Baillon and M. Théra, Pitman Research Notes in Mathematics, **252**, Longman, Harlow, pp. 397–406 (1991).

169. Chr. Tammer, A generalization of Ekeland's variational principle, *Optimization*, **25** (1992), 129–141.

170. D. Tataru, Viscosity solutions of Hamilton-Jacobi equations with unbounded nonlinear terms, *J. Math. Anal. Appl.*, **163** (1992), 345–392.

171. A. Uderzo, Fixed points for directional multi-valued $k(\cdot)$-contractions, *J. Global Optim.*, **31** (2005), 455–469.

172. A.-M. Wazwaz, A reliable treatment for mixed Volterra-Fredholm integral equations, *Appl. Math. Computat.*, **127** (2-3) (2002), 405–414.

173. J.D. Weston, A characterization of metric completeness, *Proc. Amer. Math. Soc.*, **64**(1) (1977), 186–188.

174. C.S. Wong, On a fixed point theorem of contractive type, *Proc. Amer. Math. Soc.*, **57**(2) (1976), 283–284.

175. J.S.W. Wong, Generalizations of the converse of the contraction mapping principle, *Canad. J. Math.*, **18** (1969), 1095–1104.

176. Z.L. Wu and S.Y. Wu, Weak sharp solutions of variational inequalities in Hilbert spaces, *SIAM J. Optim.*, **14**(4) (2004), 1011–1027.

177. S. Yalçinbaş, Taylor polynomial solutions of nonlinear Volterra-Fredholm integral equations, *Appl. Math. Computat.*, **127**(2-3) (2002), 195–206.

178. L. Yongsin and S. Shuzhong, A generalization of Ekeland's ε-variational principle and its Borwein-Preiss variant, *J. Math. Anal. Appl.*, **246** (2000), 308–319.

179. P.P. Zabreiko and M.A. Krasnoselskii, On the sovability of nonlinear operator equations, *Functional Anal. Prilož*, **5** (1971), 42–44. (In Russian)

180. J. Zhang, C. Wan, and N. Xiu, The dual gap function for variational inequalities, *Appl. Math. Optim.*, **48** (2003), 129–148.

181. C.-K. Zhong, On Ekeland's variational principle and a minimax theorem, *J. Math. Anal. Appl.*, **205** (1997), 239–250.

182. C.-K. Zhong, A generalization of Ekeland's variational principle and application to the study of the relation between the P.S. condition and coercivity, *Nonlinear Anal., Theory Meth. Appl.*, **29** (1997), 1421–1431.

183. C.-K. Zhong, J. Zhu, and P.-H. Zhao, An extension of multi-valued contraction mappings and fixed points, *Proc. Amer. Math. Soc.*, **128**(8) (2000), 2439–2444.

184. J. Zhou and C. Wang, A note on finite termination of iterative algorithms in mathematical programming, *Oper. Res. Lett.*, **36** (2008), 715–717.

Index